Animal Behavior

Animal Behavior

Edited by
Barrett Adkins

Larsen & Keller
www.larsen-keller.com

Animal Behavior
Edited by Barrett Adkins
ISBN: 978-1-63549-025-1 (Hardback)

Larsen & Keller
Published by Larsen and Keller Education,
5 Penn Plaza,
19th Floor,
New York, NY 10001, USA

Cataloging-in-Publication Data

Animal behavior / edited by Barrett Adkins.
p. cm.
Includes bibliographical references and index.
ISBN 978-1-63549-025-1
1. Animal behavior. 2. Animal psychology.
I. Adkins, Barrett.
QL751 .A55 2017
591.5--dc23

Printed and bound in the United States of America.

For more information regarding Larsen and Keller Education and its products, please visit the publisher's website www.larsen-keller.com

Table of Contents

Preface

Animal behavior, ethology or behaviorism is a scientific field that studies the behavior of animals under normal circumstances. It includes topics like observational learning, habituation, imprinting, and associative learning, etc. This book aims to shed light on some of the unexplored aspects of animal behavior. Such selected concepts that redefine the field have been presented in it. This book is a valuable compilation of topics, ranging from the basic to the most complex theories and principles in the field of animal behavior. For someone with an interest and eye for detail, this text covers the most significant topics in this field. It will provide comprehensive knowledge to the students.

To facilitate a deeper understanding of the contents of this book a short introduction of every chapter is written below:

Chapter 1- The study of animal behavior is termed as ethology. It studies the behavior of animals under natural conditions. Cognitive ethology is a branch of ethology that is mainly concerned with the influence of conscious awareness of an animal. This chapter will provide an integrated understanding of animal behavior.

Chapter 2- The key concepts of animal behavior that have been elucidated in this section are behaviorism, animal cognition and adaption. Behaviorism is the theory that studies and understands the behavior of animals and humans. Behaviorism as a subject includes elements of psychological theory, philosophy and methodology. The major components of animal behavior are discussed in the following section.

Chapter 3- Behavioral ecology is the study of the development in the change of animal behavior due to the ecological pressures. The features explained in this section are habituation, Lamarckism, natural selection and neuroethology. This chapter is an overview of the subject matter incorporating all the major aspects of the evolutionary basis for animal behavior.

Chapter 4- Animal communication is the communication between one or a group of animals. This subject is rapidly growing as a field of study. Bee learning and communication, deception in animals, ultrasound avoidance, waggle dance and bird migration are some of the topics discussed in this section. The chapter strategically encompasses and incorporates the major components and key concepts of animal communication and animal migration, providing a complete understanding.

Chapter 5- Animal sexual behavior is the reproductive process that takes place in animals. The systems involved in animal sexual behavior include monogamy, polyandry, polygamy and promiscuity. The topics discussed in the section are of great importance to broaden the existing knowledge on animal sexual behavior.

Chapter 6- Predation is the process of a predator hunting and eating its prey. They may or may not kill their prey before eating them but the act of predation usually does result in the death of the prey. The alternative eating behavior of animals are hoarding, regurgitation and cannibalism. This chapter helps the readers in developing an in depth understanding the eating behavior of animals.

Chapter 7- Agonistic behavior is the behavior that animals represent during a fight. It also includes threats, displays, placation and conciliation. Animal psychopathology, broodiness, licking, feather pecking and vent pecking are some of the topics explained in the chapter. This section on abnormal behavior in animals offers an insightful focus, keeping in mind the subject matter.

I would like to share the credit of this book with my editorial team who worked tirelessly on this book. I owe the completion of this book to the never-ending support of my family, who supported me throughout the project.

Editor

1

Understanding Animal Behavior

The study of animal behavior is termed as ethology. It studies the behavior of animals under natural conditions. Cognitive ethology is a branch of ethology that is mainly concerned with the influence of conscious awareness of an animal. This chapter will provide an integrated understanding of animal behavior.

Ethology

A range of animal behaviours

Ethology is the scientific and objective study of animal behaviour, usually with a focus on behaviour under natural conditions, and viewing behaviour as an evolutionarily adaptive trait. Behaviourism is a term that also describes the scientific and objective study of animal behaviour, usually referring to measured responses to stimuli or trained behavioural responses in a laboratory context, without a particular emphasis on evolutionary adaptivity. Many naturalists have studied aspects of animal behaviour throughout history. Ethology has its scientific roots in the work of Charles Darwin and of American and German ornithologists of the late 19th and early 20th century, including Charles O. Whitman, Oskar Heinroth, and Wallace Craig. The modern discipline of ethology is generally considered to have begun during the 1930s with the work of Dutch biologist Nikolaas Tinbergen and by Austrian biologists Konrad Lorenz and Karl von Frisch, joint awardees of the 1973 Nobel Prize in Physiology or Medicine. Ethology is a combination of laboratory and field science, with a strong relation to some other disciplines such as neuroanatomy, ecology, and evolutionary biology. Ethologists are typically interested in a behavioural process rather than in a particular animal group, and often study one type of behaviour, such as aggression, in a number of unrelated animals.

Ethology is a rapidly growing field. Since the dawn of the 21st century, many aspects of animal communication, emotions, culture, learning and sexuality that the scientific community long thought it understood have been re-examined, and new conclusions reached. New fields, such as neuroethology, have developed.

Understanding ethology or animal behaviour can be important in animal training. Considering the natural behaviours of different species or breeds enables the trainer to select the individuals best suited to perform the required task. It also enables the trainer to encourage the performance of naturally occurring behaviours and also the discontinuance of undesirable behaviours.

Etymology

The term was first popularized by American myrmecologist (a person who studies ants) William Morton Wheeler in 1902. An earlier, slightly different sense of the term was proposed by John Stuart Mill in his 1843 *System of Logic*. He recommended the development of a new science, "ethology", the purpose of which would be explanation of individual and national differences in character, on the basis of associationistic psychology. This use of the word was never adopted.

Relationship with Comparative Psychology

Comparative psychology also studies animal behaviour, but, as opposed to ethology, is construed as a sub-topic of psychology rather than as one of biology. Historically, where comparative psychology has included research on animal behaviour in the context of what is known about human psychology, ethology involves research on animal behaviour in the context of what is known about animal anatomy, physiology, neurobiology, and phylogenetic history. Furthermore, early comparative psychologists concentrated on the study of learning and tended to research behaviour in artificial situations, whereas early ethologists concentrated on behaviour in natural situations, tending to describe it as instinctive.

The two approaches are complementary rather than competitive, but they do result in different perspectives, and occasionally conflicts of opinion about matters of substance. In addition, for most of the twentieth century, comparative psychology developed most strongly in North America, while ethology was stronger in Europe. From a practical standpoint, early comparative psychologists concentrated on gaining extensive knowledge of the behaviour of very few species. Ethologists were more interested in understanding behaviour across a wide range of species to facilitate principled comparisons across taxonomic groups. Ethologists have made much more use of such cross-species comparisons than comparative psychologists have.

History

Scala Naturae And Lamarck's Theories

Until the 19th century, the most common theory among scientists was still the concept of *scala naturae*, proposed by Aristotle. According to this theory, living beings were classified on an ideal pyramid that represented non-living things (such as minerals and sediment) and the simplest animals on the lower levels, with complexity increasing progressively towards the top, occupied by human beings. In the Western world of the time, people believed animal species were eternal and immutable, created with a specific purpose, as this seemed the only possible explanation for the incredible variety of living beings and their surprising adaptation to their habitats.

Jean-Baptiste Lamarck (1744–1829)

Jean-Baptiste Lamarck (1744 - 1829) was the first biologist to describe a complex theory of evolution. His theory substantially comprised two statements: first, that animal organs and behaviour can change according to the way they are used; and second, that those characteristics can transmit from one generation to the next (the example of the giraffe whose neck becomes longer while trying to reach the upper leaves of a tree is well-known). The second statement is that every living organism, humans included, tends to reach a greater level of perfection. When Charles Darwin went to the Galapagos Islands, he was well aware of Lamarck's theories and was influenced by them.

Theory of Evolution by Natural Selection and the Beginnings of Ethology

Because ethology is considered a topic of biology, ethologists have been concerned particularly with the evolution of behaviour and the understanding of behaviour in terms of the theory of natural selection. In one sense, the first modern ethologist was Charles Darwin, whose book *The Expression of the Emotions in Man and Animals* influenced many ethologists. He pursued his interest in behaviour by encouraging his protégé George Romanes, who investigated animal learning and intelligence using an anthropomorphic method, anecdotal cognitivism, that did not gain scientific support.

Charles Darwin (1809–1882)

Other early ethologists, such as Charles O. Whitman, Oskar Heinroth, Wallace Craig and Julian Huxley, instead concentrated on behaviours that can be called instinctive, or natural, in that they occur in all members of a species under specified circumstances. Their beginning for studying the behaviour of a new species was to construct an ethogram (a description of the main types of behaviour with their frequencies of occurrence). This provided an objective, cumulative data-base of behaviour, which subsequent researchers could check and supplement.

Social Ethology and Recent Developments

In 1970, the English ethologist John H. Crook published an important paper in which he distinguished comparative ethology from social ethology, and argued that much of the ethology that had existed so far was really comparative ethology—examining animals as individuals—whereas, in the future, ethologists would need to concentrate on the behaviour of social groups of animals and the social structure within them.

Also in 1970, Robert Ardrey's book *The Social Contract: A Personal Inquiry into the Evolutionary Sources of Order and Disorder* was published. The book and study investigated animal behaviour and then compared human behaviour to it as a similar phenomenon.

E. O. Wilson's book *Sociobiology: The New Synthesis* appeared in 1975, and since that time, the study of behaviour has been much more concerned with social aspects. It has also been driven by the stronger, but more sophisticated, Darwinism associated with Wilson, Robert Trivers, and William Hamilton. The related development of behavioural ecology has also helped transform ethology. Furthermore, a substantial reapprochement with comparative psychology has occurred, so the modern scientific study of behaviour offers a more or less seamless spectrum of approaches: from animal cognition to more traditional comparative psychology, ethology, sociobiology, and behavioural ecology.

Growth of the Field

Due to the work of Lorenz and Tinbergen, ethology developed strongly in continental Europe during the years prior to World War II. After the war, Tinbergen moved to the University of Oxford, and ethology became stronger in the UK, with the additional influence of William Thorpe, Robert Hinde, and Patrick Bateson at the Sub-department of Animal Behaviour of the University of Cambridge, located in the village of Madingley. In this period, too, ethology began to develop strongly in North America.

Lorenz, Tinbergen, and von Frisch were jointly awarded the Nobel Prize in Physiology or Medicine in 1973 for their work of developing ethology.

Ethology is now a well-recognized scientific discipline, and has a number of journals covering developments in the subject, such as *Animal Behaviour*, *Animal Welfare*, *Applied Animal Behaviour Science*, *Behaviour*, *Behavioral Ecology* and *Journal of Ethology*. In 1972, the International Society for Human Ethology was founded to promote exchange of knowledge and opinions concerning human behaviour gained by applying ethological principles and methods and published their journal, *The Human Ethology Bulletin*. In 2008, in a paper published in the journal *Behaviour*, ethologist Peter Verbeek introduced the term "Peace Ethology" as a sub-discipline of Human Ethology that is concerned with issues of human conflict, conflict resolution, reconciliation, war, peacemaking, and peacekeeping behaviour.

Today, along with ethologists, many biologists, zoologists, primatologists, anthropologists, veterinarians, and physicians study ethology and other related fields such as animal psychology, the study of animal social groups, animal cognition and animal welfare science.

Instinct

Kelp gull chicks peck at red spot on mother's beak to stimulate regurgitating reflex

The Merriam-Webster dictionary defines instinct as "A largely inheritable and unalterable tendency of an organism to make a complex and specific response to environmental stimuli without involving reason".

Fixed Action Patterns

An important development, associated with the name of Konrad Lorenz though probably due more to his teacher, Oskar Heinroth, was the identification of fixed action patterns (FAPs). Lorenz popularized FAPs as instinctive responses that would occur reliably in the presence of identifiable stimuli called sign stimuli or "releasing stimuli". FAPs are now considered to be instinctive behavioural sequences that are relatively invariant within the species and almost inevitably run to completion.

One example of a releaser is the beak movements of many bird species performed by newly hatched chicks, which stimulates the mother to regurgitate food for her offspring. Other examples are the classic studies by Tinbergen on the egg-retrieval behaviour and the effects of a "supernormal stimulus" on the behaviour of graylag geese.

One investigation of this kind was the study of the waggle dance ("dance language") in bee communication by Karl von Frisch. Lorenz subsequently developed a theory of the evolution of animal communication based on his observations of fixed action patterns and the circumstances in which they are expressed.

Learning

Habituation

Habituation is a simple form of learning and occurs in many animal taxa. It is the process whereby an animal ceases responding to a stimulus. Often, the response is an innate behaviour. Essentially, the animal learns not to respond to irrelevant stimuli. For example, prairie dogs (*Cynomys ludovicianus*) give alarm calls when predators approach, causing all individuals in the group to quickly scramble down burrows. When prairie dog towns are located near trails used by humans, giving alarm calls every time a person walks by is expensive in terms of time and energy. Habituation to humans is therefore an important adaptation in this context.

Associative Learning

Associative learning in animal behaviour is any learning process in which a new response becomes associated with a particular stimulus. The first studies of associative learning were made by Russian physiologist Ivan Pavlov. Examples of associative learning include when a goldfish swims to the water surface when a human is going to feed it, or the excitement of a dog whenever it sees a leash as a prelude for a walk.

Imprinting

Being able to discriminate the members of one's own species is also of fundamental importance for reproductive success. Such discrimination can be based on a number of factors. However, this important type of learning only takes place in a very limited period of time. This kind of learning is called imprinting, and was a second important finding of Lorenz. Lorenz observed that the young of birds such as geese and chickens followed their mothers spontaneously from almost the first day after they were hatched, and he discovered that this response could be imitated by an arbitrary stimulus if the eggs were incubated artificially and the stimulus were presented during a critical period that continued for a few days after hatching.

Example of imprinting in a moose

Cultural Learning

Imitation

Imitation is an advanced behaviour whereby an animal observes and exactly replicates the behaviour of another. The National Institutes of Health reported that capuchin monkeys preferred the company of researchers who imitated them to that of researchers who did not. The monkeys not only spent more time with their imitators but also preferred to engage in a simple task with them even when provided with the option of performing the same task with a non-imitator. Imitation has been observed in recent research on chimpanzees; not only did these chimps copy the actions of another individual, when given a choice, the chimps preferred to imitate the actions of the higher-ranking elder chimpanzee as opposed to the lower-ranking young chimpanzee.

Stimulus and Local Enhancement

There are various ways animals can learn using observational learning but without the process of imitation. One of these is *stimulus enhancement* in which individuals become interested in an object as the result of observing others interacting with the object. Increased interest in an object can result in object manipulation which allows for new object-related behaviours by trial-and-error learning. Haggerty (1909) devised an experiment in which a monkey climbed up the side of a cage, placed its arm into a wooden chute, and pulled a rope in the chute to release food. Another monkey was provided an opportunity to obtain the food after watching a monkey go through this process on four separate occasions. The monkey performed a different method and finally succeeded after trial-and-error. Another example familiar to some cat and dog owners is the ability of their animals to open doors. The action of humans operating the handle to open the door results in the animals becoming interested in the handle and then by trial-and-error, they learn to operate the handle and open the door.

In local enhancement, a demonstrator attracts an observer's attention to a particular location. Local enhancement has been observed to transmit foraging information among birds, rats and pigs. The stingless bee (*Trigona corvina*) uses local enhancement to locate other members of their colony and food resources.

Social Transmission

A well-documented example of social transmission of a behaviour occurred in a group of macaques on Hachijojima Island, Japan. The macaques lived in the inland forest until the 1960s, when a group of researchers started giving them potatoes on the beach: soon, they started venturing onto the beach, picking the potatoes from the sand, and cleaning and eating them. About one year later, an individual was observed bringing a potato to the sea, putting it into the water with one hand, and cleaning it with the other. This behaviour was soon expressed by the individuals living in contact with her; when they gave birth, this behaviour was also expressed by their young - a form of social transmission.

Teaching

Teaching is a highly specialized aspect of learning in which the "teacher" (demonstrator) adjusts their behaviour to increase the probability of the "pupil" (observer) achieving the desired end-result of the behaviour. For example, killer whales are known to intentionally beach themselves to catch pinniped prey. Mother killer whales teach their young to catch pinnipeds by pushing them onto the shore and encouraging them to attack the prey. Because the mother killer whale is altering her behaviour to help her offspring learn to catch prey, this is evidence of teaching. Teaching is not limited to mammals. Many insects, for example, have been observed demonstrating various forms of teaching to obtain food. Ants, for example, will guide each other to food sources through a process called "tandem running," in which an ant will guide a companion ant to a source of food. It has been suggested that the pupil ant is able to learn this route to obtain food in the future or teach the route to other ants.This behaviour of teaching is also exemplified by crows. Specifically New Caledonian crows. The adults (whether individual or in families) teach their young adolescent offspring how to construct and utilize tools. For example; *Pandanus* branches are used to extract insects and other larvae from holes within trees.

Mating and the Fight for Supremacy

Individual reproduction is the most important phase in the proliferation of individuals or genes within a species: for this reason, there exist complex mating rituals, which can be very complex even if they are often regarded as FAPs. The stickleback's complex mating ritual, studied by Tinbergen, is regarded as a notable example of a FAP.

Often in social life, animals fight for the right to reproduce, as well as social supremacy. A common example of fighting for social and sexual supremacy is the so-called pecking order among poultry. Every time a group of poultry cohabitate for a certain time length, they establish a pecking order. In these groups, one chicken dominates the others and can peck without being pecked. A second chicken can peck all the others except the first, and so on. Higher level chickens are easily distinguished by their well-cured aspect, as opposed to lower level chickens. While the pecking order is establishing, frequent and violent fights can happen, but once established, it is broken only when other individuals enter the group, in which case the pecking order re-establishes from scratch.

Living in Groups

Several animal species, including humans, tend to live in groups. Group size is a major aspect of their social environment. Social life is probably a complex and effective survival strategy. It may be regarded as a sort of symbiosis among individuals of the same species: a society is composed of a group of individuals belonging to the same species living within well-defined rules on food management, role assignments and reciprocal dependence.

When biologists interested in evolution theory first started examining social behaviour, some apparently unanswerable questions arose, such as how the birth of sterile castes, like in bees, could be explained through an evolving mechanism that emphasizes the reproductive success of as many individuals as possible, or why, amongst animals living in small groups like squirrels, an individual would risk its own life to save the rest of the group. These behaviours may be examples of altruism. Of course, not all behaviours are altruistic, as indicated by the table below. For example, revengeful behaviour was at one point claimed to have been observed exclusively in *Homo sapiens*. However, other species have been reported to be vengeful, including reports of vengeful camels and chimpanzees.

Classification of social behaviours		
Type of behaviour	Effect on the donor	Effect on the receiver
Egoistic	Increases fitness	Decreases fitness
Cooperative	Increases fitness	Increases fitness
Altruistic	Decreases fitness	Increases fitness
Revengeful	Decreases fitness	Decreases fitness

Altruistic behaviour has been explained by the gene-centred view of evolution.

Benefits and Costs of Group Living

One advantage of group living can be decreased predation. If the number of predator attacks stays the same despite increasing prey group size, each prey may have a reduced risk of predator attacks through the dilution effect. Additionally, a predator that is confused by a mass of individuals can find it more difficult to single out one target. For this reason, the zebra's stripes offer not only camouflage in a habitat of tall grasses, but also the advantage of blending into a herd of other zebras. In groups, prey can also actively reduce their predation risk through more effective defense tactics, or through earlier detection of predators through increased vigilance.

Another advantage of group living can be an increased ability to forage for food. Group members may exchange information about food sources between one another, facilitating the process of resource location. Honeybees are a notable example of this, using the waggle dance to communicate the location of flowers to the rest of their hive. Predators also receive benefits from hunting in groups, through using better strategies and being able to take down larger prey.

Some disadvantages accompany living in groups. Living in close proximity to other animals can facilitate the transmission of parasites and disease, and groups that are too large may also experience greater competition for resources and mates.

Group Size

Theoretically, social animals should have optimal group sizes that maximize the benefits and minimize the costs of group living. However, in nature, most groups are stable at slightly larger than optimal sizes. Because it generally benefits an individual to join an optimally-sized group, despite slightly decreasing the advantage for all members, groups may continue to increase in size until it is more advantageous to remain alone than to join an overly full group.

Tinbergen's Four Questions for Ethologists

Niko Tinbergen argued that ethology always needed to include four kinds of explanation in any instance of behaviour:

- Function – How does the behaviour affect the animal's chances of survival and reproduction? Why does the animal respond that way instead of some other way?
- Causation – What are the stimuli that elicit the response, and how has it been modified by recent learning?
- Development – How does the behaviour change with age, and what early experiences are necessary for the animal to display the behaviour?
- Evolutionary history – How does the behaviour compare with similar behaviour in related species, and how might it have begun through the process of phylogeny?

These explanations are complementary rather than mutually exclusive—all instances of behaviour require an explanation at each of these four levels. For example, the function of eating is to acquire nutrients (which ultimately aids survival and reproduction), but the immediate cause of eating is hunger (causation). Hunger and eating are evolutionarily ancient and are found in many species (evolutionary history), and develop early within an organism's lifespan (development). It is easy to confuse such questions—for example, to argue that people eat because they're hungry and not to acquire nutrients—without realizing that the reason people experience hunger is because it causes them to acquire nutrients.

Cognitive Ethology

Cognitive ethology is a branch of ethology concerned with the influence of conscious awareness and intention on the behaviour of an animal. Donald Griffin, a zoology professor in the United States, set up the foundations for researches in the cognitive awareness of animals within their habitats.

The fusion of cognitive science and classical ethology into cognitive ethology "emphasizes observing animals under more-or-less natural conditions, with the objective of understanding the evolution, adaptation (function), causation, and development of the species-specific behavioral repertoire" - (Niko Tinbergen 1963).

According to Jamieson & Bekoff (1993), "Tinbergen's four questions about the evolution, adaptation, causation and development of behavior can be applied to the cognitive and mental abilities of animals." Allen & Bekoff (1997, chapter 5) attempt to show how cognitive ethology can take on the central questions of cognitive science, taking as their starting point the four questions described by Barbara Von Eckardt in her 1993 book *What is Cognitive Science?*, generalizing the four questions and adding a fifth. Kingstone, Smilek & Eastwood (2008) suggested that cognitive ethology should include human behavior. They proposed that researchers should firstly study how people behave in their natural, real world environments and then move to the lab. Anthropocentric claims for the ways non-human animals interact in their social and non-social worlds are often used to influence decisions on how the non-human animals can or should be used by humans.

Relation to Laboratory Experimental Psychology

Traditionally, cognitive ethologists have questioned research methods that isolate animals in unnatural surroundings and present them with a limited set of artificial stimuli, arguing that such techniques favor the study of artificial issues that are not relevant to an understanding of the natural behavior of animals. However, many modern researchers favor a judicious combination of field and laboratory methods.

Relation to Ethics

Bekoff, M and Allen, C (1997) "identify three major groups of people (among some of whose members there are blurred distinctions) with different views on cognitive ethology, namely, slayers, skeptics, and proponents." The latter seemingly convergent with animal rights thinking in seeing animal experience as worthy in itself.

Ethicist Peter Singer is an example of a "proponent" in this sense, as is biologist E. O. Wilson who coined the term biophilia to describe the basis of a direct moral cognition, that 'higher' animals would use to perceive moral implications in the environment directly.

Three Views

According to Marc Bekoff, there are three different views towards whether a science of cognitive ethology is even possible. Slayers deny any possibility of success in cognitive ethology, proponents keep an open mind about animal cognition and the utility of cognitive ethological investigation, and skeptics stand somewhere in between.

References

- McGreevy, Paul; Robert Boakes (2011). Carrots and Sticks: Principles of Animal Training. Darlington Press. pp. xi–23. ISBN 978-1-921364-15-0. Retrieved 9 September 2016.
- Bourg, Julian (2007). From Revolution to Ethics: May 1968 and Contemporary French Thought. McGill-Queen's Press - MQUP. p. 155. ISBN 978-0-7735-7621-6.
- Bateson, P. (1991). The Development and Integration of Behaviour: Essays in Honour of Robert Hinde. Cambridge University Press. p. 479. ISBN 978-0-521-40709-0.
- Buchmann, Stephen (2006). Letters from the Hive: An Intimate History of Bees, Honey, and Humankind. Random House of Canada. p. 105. ISBN 978-0-553-38266-2.
- Keil, Frank C.; Robert Andrew Wilson (2001). The MIT encyclopedia of the cognitive sciences. MIT Press. p. 184. ISBN 978-0-262-73144-7.

- Mercer, Jean (2006). Understanding attachment: parenting, child care, and emotional development. Greenwood Publishing Group. p. 19. ISBN 978-0-275-98217-1.
- Hoppitt, W.; Laland, K.N. (2013). Social Learning: An Introduction to Mechanisms, Methods, and Models. Princeton University Press. ISBN 978-1-4008-4650-4.
- Wilson, Edward O. (2000). Sociobiology: the new synthesis. Harvard University Press. p. 170. ISBN 978-0-674-00089-6.
- Cummings, Mark; Carolyn Zahn-Waxler; Ronald Iannotti (1991). Altruism and aggression: biological and social origins. Cambridge University Press. p. 7. ISBN 978-0-521-42367-0.
- Collins, Harber (2012). Collins COBUILD Advanced Dictionary of English. Higher Education Press. ISBN 9787040327878.

2

Key Concepts of Animal Behavior

The key concepts of animal behavior that have been elucidated in this section are behaviorism, animal cognition and adaption. Behaviorism is the theory that studies and understands the behavior of animals and humans. Behaviorism as a subject includes elements of psychological theory, philosophy and methodology. The major components of animal behavior are discussed in the following section.

Behaviorism

Behaviorism (or behaviourism) is a systematic approach to the understanding of human and animal behavior. It assumes that all behavior are either reflexes produced by a response to certain stimuli in the environment, or a consequence of that individual's history, including especially reinforcement and punishment, together with the individual's current motivational state and controlling stimuli. Thus, although behaviorists generally accept the important role of inheritance in determining behavior, they focus primarily on environmental factors.

Behaviorism combines elements of philosophy, methodology, and psychological theory. It emerged in the late nineteenth century as a reaction to depth psychology and other traditional forms of psychology, which often had difficulty making predictions that could be tested experimentally. The earliest derivatives of Behaviorism can be traced back to the late 1800s where Edward Thorndike pioneered the law of effect (a process that involved strengthening behavior through the use of reinforcement).

During the first half of the twentieth century, John B. Watson devised methodological behaviorism, which rejected introspective methods and sought to understand behavior by only measuring observable behaviors and events. It was not until the 1930s that B. F. Skinner suggested that private events—including thoughts and feelings—should be subjected to the same controlling variables as observable behavior which became the basis for his philosophy called radical behaviorism. While Watson and Ivan Pavlov investigated the stimulus-response procedures of classical conditioning, Skinner assessed the controlling nature of consequences and also the antecedents (or discriminative stimuli) that signal the behavior; the technique became known as operant conditioning.

The application of radical behaviorism—known as applied behavior analysis—is used in a variety of settings, including, for example, organizational behavior management, to the treatment of mental disorders, such as autism and substance abuse. In addition, while behaviorism and cognitive schools of psychological thought may not agree theoretically, they have complemented each other in cognitive behavior therapies, which have demonstrated utility in treating certain pathologies, including simple phobias, PTSD, and mood disorders.

Versions

There is no universally agreed-upon classification, but some titles given to the various branches of behaviorism include:

- Methodological behaviorism: Watson's behaviorism states that only public events (behaviors of an individual) can be objectively observed, and that therefore private events (thoughts and feelings) should be ignored. It also became the basis for the early approach behavior modification in the late 1970s and early 1980s.
- Radical behaviorism: Skinner's behaviorism theorizes that processes within the organism should be acknowledged, particularly the presence of private events (such as thoughts and feelings), and suggests that environmental variables also control these internal events just as they control observable behaviors. Radical behaviorism forms the core philosophy behind behavior analysis. Willard Van Orman Quine used many of radical behaviorism's ideas in his study of knowledge and language.
- Teleological behaviorism: Post-Skinnerian, purposive, close to microeconomics. Focuses on objective observation as opposed to cognitive processes.
- Theoretical behaviorism: Post-Skinnerian, accepts observable internal states ("within the skin" once meant "unobservable", but with modern technology we are not so constrained); dynamic, but eclectic in choice of theoretical structures, emphasizes parsimony.
- Biological behaviorism: Post-Skinnerian, centered on perceptual and motor modules of behavior, theory of behavior systems.
- Psychological behaviorism: As proposed by Arthur W. Staats, this version of behaviorism centers on the practical control of human behavior. It is noted for its use of time-outs, token-reinforcement and other methods, which importantly influenced modern approaches to child development, education, and abnormal psychology.

Two subtypes are:

- Hullian and post-Hullian: theoretical, group data, not dynamic, physiological;
- Purposive: Tolman's behavioristic anticipation of cognitive psychology

Radical Behaviorism

B. F. Skinner proposed radical behaviorism as the conceptual underpinning of the experimental analysis of behavior. This view differs from other approaches to behavioral research in various ways but, most notably here, it contrasts with methodological behaviorism in accepting feelings, states of mind and introspection as behaviors subject to scientific investigation. Like methodological behaviorism it rejects the reflex as a model of all behavior, and it defends the science of behavior as complementary to but independent of physiology. Radical behaviorism overlaps considerably with other western philosophical positions such as American pragmatism.

Experimental and Conceptual Innovations

This essentially philosophical position gained strength from the success of Skinner's early experimental work with rats and pigeons, summarized in his books *The Behavior of Organisms* and *Schedules of Reinforcement*. Of particular importance was his concept of the operant response, of which the canonical example was the rat's lever-press. In contrast with the idea of a physiological or reflex response, an operant is a class of structurally distinct but functionally equivalent responses. For example, while a rat might press a lever with its left paw or its right paw or its tail, all of these responses operate on the world in the same way and have a common consequence. Operants are often thought of as species of responses, where the individuals differ but the class coheres in its function-shared consequences with operants and reproductive success with species. This is a clear distinction between Skinner's theory and S–R theory.

Skinner's empirical work expanded on earlier research on trial-and-error learning by researchers such as Thorndike and Guthrie with both conceptual reformulations—Thorndike's notion of a stimulus–response "association" or "connection" was abandoned; and methodological ones—the use of the "free operant", so called because the animal was now permitted to respond at its own rate rather than in a series of trials determined by the experimenter procedures. With this method, Skinner carried out substantial experimental work on the effects of different schedules and rates of reinforcement on the rates of operant responses made by rats and pigeons. He achieved remarkable success in training animals to perform unexpected responses, to emit large numbers of responses, and to demonstrate many empirical regularities at the purely behavioral level. This lent some credibility to his conceptual analysis. It is largely his conceptual analysis that made his work much more rigorous than his peers', a point which can be seen clearly in his seminal work *Are Theories of Learning Necessary?* in which he criticizes what he viewed to be theoretical weaknesses then common in the study of psychology. An important descendant of the experimental analysis of behavior is the Society for Quantitative Analysis of Behavior.

Relation to Language

As Skinner turned from experimental work to concentrate on the philosophical underpinnings of a science of behavior, his attention turned to human language with his 1957 book *Verbal Behavior* and other language-related publications; *Verbal Behavior* laid out a vocabulary and theory for functional analysis of verbal behavior, and was strongly criticized in a review by Noam Chomsky.

Skinner did not respond in detail but claimed that Chomsky failed to understand his ideas, and the disagreements between the two and the theories involved have been further discussed. Innateness theory is opposed to behaviorist theory which claims that language is a set of habits that can be acquired by means of conditioning. According to some, this process that the behaviorists define is a very slow and gentle process to explain a phenomenon as complicated as language learning. What was important for a behaviorist's analysis of human behavior was not language acquisition so much as the interaction between language and overt behavior. In an essay republished in his 1969 book *Contingencies of Reinforcement*, Skinner took the view that humans could construct linguistic stimuli that would then acquire control over their behavior in the same way that external stimuli could. The possibility of such "instructional control" over behavior meant that contingencies of reinforcement would not always produce the same effects on human behavior as they reliably do in other animals. The focus of a radical behaviorist analysis of human behavior therefore shifted to an attempt to understand the interaction between instructional control and contingency control, and also to understand the behavioral processes that determine what instructions are constructed and what control they acquire over behavior. Recently, a new line of behavioral research on language was started under the name of relational frame theory.

Education

Behaviourism focuses on one particular view of learning: a change in external behaviour achieved through using reinforcement and repetition (Rote learning) to shape behavior. Skinner found that behaviors could be shaped when the use of reinforcement was implemented. Desired behavior is rewarded, while the undesired behavior is punished. Incorporating behaviorism into the classroom allowed educators to assist their students in excelling both academically and personally. In the field of language learning, this type of teaching was called the audio-lingual method, characterised by the whole class using choral chanting of key phrases, dialogues and immediate correction.

Within the behaviourist view of learning, the "teacher" is the dominant person in the classroom and takes complete control, evaluation of learning comes from the teacher who decides what is right or wrong. The learner does not have any opportunity for evaluation or reflection within the learning process, they are simply told what is right or wrong. The conceptualization of learning using this approach could be considered "superficial" as the focus is on external changes in behaviour i.e. not interested in the internal processes of learning leading to behaviour change and has no place for the emotions involved the process.

Whether this approach is right or wrong, it cannot be denied that an aspect of memorization is regarded by key scholars as critical in any language learning.

Operant Conditioning

Operant conditioning was developed by B.F. Skinner in 1937 and deals with the modification of "voluntary behaviour" or operant behaviour. Operant behavior operates on the environment and is maintained by its consequences. Reinforcement and punishment, the core tools of operant conditioning, are either positive (delivered following a response), or negative (withdrawn following a response). Skinner created the *Skinner Box* or operant conditioning chamber to test the effects of operant conditioning principles on rats. From this study, he discovered that the rats learned very effectively if they were rewarded frequently. Skinner also found that he could shape the rats' behavior through the use of rewards, which could, in turn, be applied to human learning as well.

Classical Conditioning

Although operant conditioning plays the largest role in discussions of behavioral mechanisms, classical conditioning (or Pavlovian conditioning or respondent conditioning) is also an important behavior-analytic process that need not refer to mental or other internal processes. Pavlov's experiments with dogs provide the most familiar example of the classical conditioning procedure. In simple conditioning, the dog was presented with a stimulus such as a light or a sound, and then food was placed in the dog's mouth. After a few repetitions of this sequence, the light or sound by itself caused the dog to salivate. Although Pavlov proposed some tentative physiological processes that might be involved in classical conditioning, these have not been confirmed. The idea of classical conditioning helped behaviorist John Watson discover the key mechanism behind how humans acquire the behaviors that they do, which was to find a natural reflex that produces the response being considered.

Watson's "Behaviourist Manifesto" has three aspects that deserve special recognition: one is that psychology should be purely objective, with any interpretation of conscious experience being removed, thus leading to psychology as the "science of behaviour"; the second one is that the goals of psychology should be to predict and control behaviour (as opposed to describe and explain conscious mental states; the third one is that there is no notable distinction between human and non-human behaviour. Following Darwin's theory of evolution, this would simply mean that human behaviour is just a more complex version in respect to behaviour displayed by other species.

Molecular Versus Molar Behaviorism

Skinner's view of behavior is most often characterized as a "molecular" view of behavior; that is, behavior can be decomposed into atomistic parts or molecules. This view is inconsistent with Skinner's complete description of behavior as delineated in other works, including his 1981 article "Selection by Consequences". Skinner proposed that a complete account of behavior requires understanding of selection history at three levels: biology (the natural selection or phylogeny of the animal); behavior (the reinforcement history or ontogeny of the behavioral repertoire of the animal); and for some species, culture (the cultural practices of the social group to which the animal belongs). This whole organism then interacts with its environment. Molecular behaviorists use notions from melioration theory, negative power function discounting or additive versions of negative power function discounting.

Molar behaviorists, such as Howard Rachlin, Richard Herrnstein, and William Baum, argue that behavior cannot be understood by focusing on events in the moment. That is, they argue that behavior is best understood as the ultimate product of an organism's history and that molecular behaviorists are committing a fallacy by inventing fictitious proximal causes for behavior. Molar behaviorists argue that standard molecular constructs, such as "associative strength", are better replaced by molar variables such as rate of reinforcement. Thus, a molar behaviorist would describe "loving someone" as a pattern of loving behavior over time; there is no isolated, proximal cause of loving behavior, only a history of behaviors (of which the current behavior might be an example) that can be summarized as "love".

In Philosophy

Behaviorism is a psychological movement that can be contrasted with philosophy of mind. The basic premise of *radical behaviorism* is that the study of behavior should be a natural science, such as chemistry or physics, without any reference to hypothetical inner states of organisms as causes for their behavior. Less radical varieties are unconcerned with philosophical positions on internal, mental and subjective experience. Behaviorism takes a functional view of behavior. According to Edmund Fantino and colleagues: "Behavior analysis has much to offer the study of phenomena normally dominated by cognitive and social psychologists. We hope that successful application of behavioral theory and methodology will not only shed light on central problems in judgment and choice but will also generate greater appreciation of the behavioral approach."

Behaviorist sentiments are not uncommon within philosophy of language and analytic philosophy. It is sometimes argued that Ludwig Wittgenstein defended a behaviorist position (e.g., the *beetle in a box* argument)—but while there are important relations between his thought and behaviorism, the claim that he was a behaviorist is quite controversial. Mathematician Alan Turing is also sometimes considered a behaviorist, but he himself did not make this identification. In *logical and empirical positivism* (as held, e.g., by Rudolf Carnap and Carl Hempel), the meaning of psychological statements are their verification conditions, which consist of performed overt behavior. W.V. Quine made use of a type of behaviorism, influenced by some of Skinner's ideas, in his own work on language. Gilbert Ryle defended a distinct strain of philosophical behaviorism, sketched in his book *The Concept of Mind*. Ryle's central claim was that instances of dualism frequently represented "category mistakes", and hence that they were really misunderstandings of the use of ordinary language. Daniel Dennett likewise acknowledges himself to be a type of behaviorist, though he offers extensive criticism of radical behaviorism and refutes Skinner's rejection of the value of intentional idioms and the possibility of free will.

This is Dennett's main point in "Skinner Skinned." Dennett argues that there is a crucial difference between explaining and explaining away... If our explanation of apparently rational behavior turns out to be extremely simple, we may want to say that the behavior was not really rational after all. But if the explanation is very complex and intricate, we may want to say not that the behavior is not rational, but that we now have a better understanding of what rationality consists in. (Compare: if we find out how a computer program solves problems in linear algebra, we don't say it's not really solving them, we just say we know how it does it. On the other hand, in cases like Weizenbaum's ELIZA program, the explanation of how the computer carries on a conversation is so simple that the right thing to say seems to be that the machine isn't really carrying on a conversation, it's just a trick.)

—*Curtis Brown, Philosophy of Mind, "Behaviorism: Skinner and Dennett"*

21st-century Behavior Analysis

The early term behavior modification has been obsolete since the 1990s as it currently refers to the brief revival of methodological behaviorism in the late 1950s and again from the late 1970s to early 1980s. Applied behavior analysis—the term that replaced behavior modification—has emerged into a thriving field.

The Association for Behavior Analysis: International (ABAI) currently has 32 state and regional chapters within the United States. Approximately 30 additional chapters have also developed throughout Europe, Asia, South America, and the South Pacific. In addition to 34 annual conferences held by ABAI in the United States and Canada, ABAI held the 5th annual International conference in Norway in 2009. The independent development of behaviour analysis outside the US also continues to develop. For example, the UK Society for Behaviour Analysis was founded in 2013 to further the advancement of the science and practice of behaviour analysis across the UK. And in terms of motivation, there remains strong interest in the variety of human motivational behaviour factors, e.g., indeed one could argue that the entire career counselling and advisory industry has at least partly been predicated on analysing individual behaviours. Some, may go as far as suggesting that the current rapid change in organisational behaviour could partly be attributed to some of these theories and the theories that are related to it.

The interests among behavior analysts today are wide ranging, as a review of the 30 Special Interest Groups (SIGs) within ABAI indicates. Such interests include everything from developmental disabilities and autism, to cultural psychology, clinical psychology, verbal behavior, Organizational

Behavior Management (OBM; behavior analytic I–O psychology). OBM has developed a particularly strong following within behavior analysis, as evidenced by the formation of the OBM Network and the influential *Journal of Organizational Behavior Management* (JOBM; recently rated the 3rd highest impact journal in applied psychology by ISI JOBM rating).

Applications of behavioral technology, also known as applied behavior analysis or ABA, have been particularly well established in the area of developmental disabilities since the 1960s. Treatment of individuals diagnosed with autism spectrum disorders has grown especially rapidly since the mid-1990s. This demand for services encouraged the formation of a professional credentialing program administered by the Behavior Analyst Certification Board, Inc. (BACB) and accredited by the National Commission for Certifying Agencies. As of early 2012, there are over 300 BACB approved course sequences offered by about 200 colleges and universities worldwide preparing students for this credential and approximately 11,000 BACB certificants, most working in the United States. The Association of Professional Behavior Analysts was formed in 2008 to meet the needs of these ABA professionals.

Modern behavior analysis has also witnessed a massive resurgence in research and applications related to language and cognition, with the development of relational frame theory (RFT; described as a "Post-Skinnerian account of language and cognition"). RFT also forms the empirical basis for the highly successful and data-driven acceptance and commitment therapy (ACT). In fact, researchers and practitioners in RFT/ACT have become sufficiently prominent that they have formed their own specialized organization that is highly behaviorally oriented, known as the Association for Contextual Behavioral Science (ACBS). It has rapidly grown in its few years of existence to reach about 5,000 members worldwide.

Some of the current prominent behavior analytic journals include the *Journal of Applied Behavior Analysis* (JABA), the *Journal of the Experimental Analysis of Behavior* (JEAB) JEAB website, the *Journal of Organizational Behavior Management* (JOBM), *Behavior and Social Issues* (BSI), as well as the *Psychological Record*. Currently, the US has 14 ABAI accredited MA and PhD programs for comprehensive study in behavior analysis.

Behavior Analysis and Culture

Cultural analysis has always been at the philosophical core of radical behaviorism from the early days (as seen in Skinner's *Walden Two*, *Science & Human Behavior*, *Beyond Freedom & Dignity*, and *About Behaviorism*).

During the 1980s, behavior analysts, most notably Sigrid Glenn, had a productive interchange with cultural anthropologist Marvin Harris (the most notable proponent of "Cultural Materialism") regarding interdisciplinary work. Very recently, behavior analysts have produced a set of basic exploratory experiments in an effort toward this end. Behaviorism is also frequently used in game development, although this application is controversial.

Behavior Informatics and Behavior Computing

With the fast growth of big behavioral data and applications, behavior analysis is ubiquitous. Understanding behavior from the informatics and computing perspective becomes increasingly critical for in-depth understanding of what, why and how behaviors are formed, interact, evolve, change and affect business and decision. Behavior informatics and behavior computing deeply explore behavior intelligence and behavior insights from the informatics and computing perspectives.

Criticisms and Limitations of Behaviorism

In the second half of the 20th century, behaviorism was largely eclipsed as a result of the cognitive revolution. This shift was due to methodological behaviorism being highly criticized for not examining mental processes, and this led to the development of the cognitive therapy movement.

In the mid-20th century, three main influences arose that would inspire and shape cognitive psychology as a formal school of thought:

- With the development of new warfare technology during WWII, the need for a greater understanding of human performance came to prominence. Problems such as how to best train soldiers to use new technology and how to deal with matters of attention while under duress became areas of need for military personnel. Behaviorism provided little if any insight into these matters and it was the work of Donald Broadbent, integrating concepts from human performance research and the recently developed information theory, that forged the way in this area.
- Developments in computer science would lead to parallels being drawn between human thought and the computational functionality of computers, opening entirely new areas of psychological thought. Allen Newell and Herbert Simon spent years developing the concept of artificial intelligence (AI) and later worked with cognitive psychologists regarding the implications of AI. The effective result was more of a framework conceptualization of mental functions with their counterparts in computers (memory, storage, retrieval, etc.)
- Noam Chomsky's 1959 critique of behaviorism, and empiricism more generally, initiated what would come to be known as the "cognitive revolution".
- Formal recognition of the field involved the establishment of research institutions such as George Mandler's Center for Human Information Processing in 1964. Mandler described the origins of cognitive psychology in a 2002 article in the Journal of the History of the Behavioral Sciences

In the early years of cognitive psychology, behaviorist critics held that the empiricism it pursued was incompatible with the concept of internal mental states. Cognitive neuroscience, however, continues to gather evidence of direct correlations between physiological brain activity and putative mental states, endorsing the basis for cognitive psychology.

At the start of the 20th century, attitudes in America were characterised by pragmatism, which led to a preference for behaviorism as the primary approach in psychology. J.B. Watson was a key figure with his stimulus-response approach. By conducting experiments on animals he was aiming to be able to predict and control behaviour. Behaviourism eventually failed because it could not provide realistic psychology of human action and thought – it was too based in physical concepts to explain phenomena like memory and thought. This led to what is often termed as the "cognitive revolution".

Animal Cognition

Experiments like the string-pulling task performed here by a Carib grackle provide insights into animal cognition.

A crab-eating macaque using a stone tool to crack open a nut

Animal cognition describes the mental capacities of animals and the study of those capacities. The field developed from comparative psychology, including the study of animal conditioning and learning. It has also been strongly influenced by research in ethology, behavioral ecology, and evolutionary psychology, and hence the alternative name cognitive ethology is sometimes used. Many behaviors associated with the term *animal intelligence* are also subsumed within animal cognition.

Researchers have examined animal cognition in mammals (especially primates, cetaceans, elephants, dogs, cats, horses, livestock, raccoons and rodents), birds (including parrots, corvids and pigeons), reptiles (lizards and snakes), fish and invertebrates (including cephalopods, spiders and insects).

Historical Background

Morgan's Canon

Coined by 19th-century British psychologist C. Lloyd Morgan, Morgan's Canon remains a fundamental precept of comparative (animal) psychology. In its developed form, it states that:

In no case is an animal activity to be interpreted in terms of higher psychological processes if it can be fairly interpreted in terms of processes which stand lower in the scale of psychological evolution and development.

In other words, Morgan believed that anthropomorphic approaches to animal behavior were fallacious, and that people should only consider behaviour as, for example, rational, purposive or affectionate, if there is no other explanation in terms of the behaviours of more primitive life-forms to which we do not attribute those faculties.

From Anecdote to Laboratory

The behavior of non-human animals has captivated human imagination from antiquity, and over the centuries many writers have speculated about the animal mind, or its absence. Speculation about animal intelligence gradually yielded to scientific study after Darwin placed humans and animals on a continuum, although Darwin's largely anecdotal approach to the topic would not pass scientific muster later on. Unsatisfied with the anecdotal method of Darwin and his protégé J. G. Romanes, E. L. Thorndike brought animal behavior into the laboratory for objective scrutiny. Thorndike's careful observations of the escape of cats, dogs, and chicks from puzzle boxes led him to conclude that what appears to the naive human observer to be intelligent behavior may be strictly attributable to simple associations. According to Thorndike, using Morgan's Canon, the inference of animal reason, insight, or consciousness is unnecessary and misleading. At about the same time, I. P. Pavlov began his seminal studies of conditioned reflexes in dogs. Pavlov quickly abandoned attempts to infer canine mental processes; such attempts, he said, led only to disagreement and confusion. He was, however, willing to propose unseen physiological processes that might explain his observations.

The Behavioristic Half-century

The work of Thorndike, Pavlov and a little later of the outspoken behaviorist John B. Watson set the direction of much research on animal behavior for more than half a century. During this time there was considerable progress in understanding simple associations; notably, around 1930 the differences between Thorndike's instrumental (or operant) conditioning and Pavlov's classical (or Pavlovian) conditioning were clarified, first by Miller and Kanorski, and then by B. F. Skinner. Many experiments on conditioning followed; they generated some complex theories, but they made little or no reference to intervening mental processes. Probably the most explicit dismissal of the idea that mental processes control behavior was the radical behaviorism of Skinner. This view seeks to explain behavior, including "private events" like mental images, solely by reference to the environmental contingencies impinging on the human or animal.

Despite the predominantly behaviorist orientation of research before 1960, the rejection of mental processes in animals was not universal during those years. Influential exceptions included, for example, Wolfgang Köhler and his insightful chimpanzees and Edward Tolman whose proposed cognitive map was a significant contribution to subsequent cognitive research in both humans and animals.

The Cognitive Revolution

Beginning around 1960, a "cognitive revolution" in research on humans gradually spurred a similar transformation of research with animals. Inference to processes not directly observable became acceptable and then commonplace. An important proponent of this shift in thinking was Donald O. Hebb, who argued that "mind" is simply a name for processes in the head that control complex behavior, and that it is both necessary and possible to infer those processes from behavior. Animals came to be seen as "goal seeking agents that acquire, store, retrieve, and internally process information at many levels of cognitive complexity.". The remainder of this article touches many areas of research that have appeared or greatly progressed since this seminal change in thinking, and many of the theoretical and empirical findings that have captured wide attention.

Methods

The acceleration of research on animal cognition in the last 50 years or so has led to a rapid expansion in the variety of species studied and methods employed. The remarkable behavior of large-brained animals such as primates and cetacea has claimed special attention, but all sorts of mammals large and small, birds, fish, ants, bees, and others have been brought into the laboratory or observed in carefully controlled field studies. In the laboratory, animals push levers, pull strings, dig for food, swim in water mazes, or respond to images on computer screens in discrimination, attention, memory, and categorization experiments. Careful field studies explore memory for food caches, navigation by stars, communication, tool use, identification of conspecifics, and many other matters. Studies often focus on the behavior of animals in their natural environments and discuss the putative function of the behavior for the propagation and survival of the species. These developments reflect an increased cross-fertilization from related fields such as ethology and behavioral biology. Also, contributions from behavioral neuroscience are beginning to clarify the physiological substrate of some inferred mental process.

Some researchers have made effective use of a Piagetian methodology, taking tasks which human children are known to master at different stages of development, and investigating which of them can be performed by particular species. Others have been inspired by concerns for animal welfare and the management of domestic species: for example Temple Grandin has harnessed her unique expertise in animal welfare and the ethical treatment of farm livestock to highlight underlying similarities between humans and other animals. From a methodological point of view, one of the main risks in this sort of work is anthropomorphism, the tendency to interpret an animal's behavior in terms of human feelings, thoughts, and motivations.

Research Questions

Human and animal cognition have much in common, and this is reflected in the research summarized below; most of the headings found here might also appear in an article on human cognition. Of course, research in the two also differs in important respects. Notably, much research with humans either studies or involves language, and much research with animals is related directly or indirectly to behaviors important to survival in natural settings. Following are summaries of some of the major areas of research in animal cognition.

The common chimpanzee can use tools. This individual is using a stick to get food.

Perception

Animals process information from eyes, ears, and other sensory organs to perceive the environment. Perceptual processes have been studied in many species, with results that are often similar to those in humans. Equally interesting are those perceptual processes that differ from, or go beyond those found in humans, such as echolocation in bats and dolphins, motion detection by skin receptors in fish, and extraordinary visual acuity, motion sensitivity and ability to see ultraviolet light in some birds.

Attention

Much of what is happening in the world at any moment is irrelevant to current behavior. Attention refers to mental processes that select relevant information, inhibit irrelevant information, and switch among these as the situation demands. Often the selective process is tuned before relevant information appears; such expectation makes for rapid selection of key stimuli when they become available. A large body of research has explored the way attention and expectation affect the behavior of non-human animals, and much of this work suggests that attention operates in birds, mammals and reptiles in much the same way that it does in humans.

Selective Learning

Animals trained to discriminate between two stimuli, say black versus white, can be said to attend to the "brightness dimension," but this says little about whether this dimension is selected in preference to others. More enlightenment comes from experiments that allow the animal to choose from several alternatives. For example, several studies have shown that performance is better on, for example, a color discrimination (e.g. blue vs green) after the animal has learned another color discrimination (e.g. red vs orange) than it is after training on a different dimension such as an X shape versus and O shape. The reverse effect happens after training on forms. Thus, the earlier learning appears to affect which dimension, color or form, the animal will attend to.

Other experiments have shown that after animals have learned to respond to one aspect of the environment responsiveness to other aspects is suppressed. In "blocking", for example, an animal is conditioned to respond to one stimulus ("A") by pairing that stimulus with reward or punishment. After the animal responds consistently to A, a second stimulus ("B") accompanies A on additional training trials. Later tests with the B stimulus alone elicit little response, suggesting that learning about B has been blocked by prior learning about A . This result supports the hypothesis that stimuli are neglected if they fail to provide new information. Thus, in the experiment just cited, the animal failed to attend to B because B added no information to that supplied by A. If true, this interpretation is an important insight into attentional processing, but this conclusion remains uncertain because blocking and several related phenomena can be explained by models of conditioning that do not invoke attention.

Divided Attention

Attention is a limited resource and is not an all-or-nothing response: the more attention devoted to one aspect of the environment, the less is available for others. A number of experiments have studied this in animals. In one experiment, a tone and a light are presented simultaneously to pigeons.

The pigeons gain a reward only by choosing the correct combination of the two stimuli (e.g. a high frequency tone together with a yellow light). The birds perform well at this task, presumably by dividing attention between the two stimuli. When only one of the stimuli varies and the other is presented at its rewarded value, discrimination improves on the variable stimulus but discrimination on the alternative stimulus worsens. These outcomes are consistent with the notion that attention is a limited resource that can be more or less focused among incoming stimuli.

Visual Search and Attentional Priming

As noted above, attention functions to select information that is of special use to the animal. Visual search typically calls for this sort of selection, and search tasks have been used extensively in both humans and animals to determine the characteristics of attentional selection and the factors that control it.

Experimental research on visual search in animals was initially prompted by field observations published by Luc Tinbergen (1960). Tinbergen observed that birds are selective when foraging for insects. For example, he found that birds tended to catch the same type of insect repeatedly even though several types were available. Tinbergen suggested that this prey selection was caused by an attentional bias that improved detection of one type of insect while suppressing detection of others. This "attentional priming" is commonly said to result from a pretrial activation of a mental representation of the attended object, which Tinbergen called a "searching image."

Tinbergen's field observations on priming have been supported by a number of experiments. For example, Pietrewicz and Kamil (1977, 1979) presented blue jays with pictures of tree trunks upon which rested either a moth of species A, a moth of species B, or no moth at all. The birds were rewarded for pecks at a picture showing a moth. Crucially, the probability with which a particular species of moth was detected was higher after repeated trials with that species (e.g. A, A, A,...) than it was after a mixture of trials (e.g. A, B, B, A, B, A, A...). These results suggest again that sequential encounters with an object can establish an attentional predisposition to see the object.

Another way to produce attentional priming in search is to provide an advance signal that is associated with the target. For example, if you hear a song sparrow you may be predisposed to detect a song sparrow in a shrub, or among other birds. A number of experiments have reproduced this effect in animal subjects.

Still other experiments have explored nature of stimulus factors that affect the speed and accuracy of visual search. For example, the time taken to find a single target increases as the number of items in the visual field increases. This rise in RT is steep if the distracters are similar to the target, less steep if they are dissimilar, and may not occur if the distracters are very different in from the target in form or color.

Concepts and Categories

Fundamental but difficult to define, the concept of "concept" was discussed for hundreds of years by philosophers before it became a focus of psychological study. Concepts enable humans and animals to organize the world into functional groups; the groups may be composed of perceptually similar objects or events, diverse things that have a common function, relationships such as same versus different, or relations among relations such as analogies. Extensive discussions on these matters together with many references may be found in Shettleworth (2010) Wasserman and Zentall (2006) and in Zentall *et al.* (2008). The latter is freely available online

Methods

Most work on animal concepts has been done with visual stimuli, which can easily be constructed and presented in great variety, but auditory and other stimuli have been used as well. Pigeons have been widely used, for they have excellent vision and are readily conditioned to respond to visual targets; other birds and a number of other animals have been studied as well. In a typical experiment, a bird or other animal confronts a computer monitor on which a large number of pictures appear one by one, and the subject gets a reward for pecking or touching a picture of a category item and no reward for non-category items. Alternatively, a subject may be offered a choice between two or more pictures. Many experiments end with the presentation of items never seen before; successful sorting of these items shows that the animal has not simply learned many specific stimulus-response associations. A related method, sometimes used to study relational concepts, is matching-to-sample. In this task an animal sees one stimulus and then chooses between two or more alternatives, one of which is the same as the first; the animal is then rewarded for choosing the matching stimulus.

Perceptual Categories

Perceptual categorization is said to occur when a person or animal responds in a similar way to a range of stimuli that share common features. For example, a squirrel climbs a tree when it sees Rex, Shep, or Trixie, which suggests that it categorizes all three as something to avoid. This sorting of instances into groups is crucial to survival. Among other things, an animal must categorize if it is to apply learning about one object (e.g. Rex bit me) to new instances of that category (dogs may bite).

Natural Categories

Many animals readily classify objects by perceived differences in form or color. For example, bees or pigeons quickly learn to choose any red object and reject any green object if red leads to reward and green does not. Seemingly much more difficult is an animal's ability to categorize natural objects that vary a great deal in color and form even while belonging to the same group. In a classic study, Richard J. Herrnstein trained pigeons to respond to the presence or absence of human beings in photographs. The birds readily learned to peck photos that contained partial or full views of humans and to avoid pecking photos with no human, despite great differences in the form, size, and color of both the humans displayed and in the non-human pictures. In follow-up studies, pigeons categorized other natural objects (e.g. trees) and after training they were able without reward to sort photos they had not seen before . Similar work has been done with natural auditory categories, for example, bird songs. Honeybees (*Apis mellifera*) are able to form concepts of "up" and "down".

Functional or Associative Categories

Perceptually unrelated stimuli may come to be responded to as members of a class if they have a common use or lead to common consequences. An oft-cited study by Vaughan (1988) provides an example. Vaughan divided a large set of unrelated pictures into two arbitrary sets, A and B. Pigeons got food for pecking at pictures in set A but not for pecks at pictures in set B. After they had learned this task fairly well, the outcome was reversed: items in set B led to food and items in set A did not. Then the outcome was reversed again, and then again, and so on. Vaughan found that after 20 or more reversals, associating reward with a few pictures in one set caused the birds to respond to the other pictures in that set without further reward, as if they were thinking "if these pictures in set A bring food, the others in set A must also bring food." That is, the birds now categorized the pictures in each set as functionally equivalent. Several other procedures have yielded similar results.

Relational or Abstract Categories

When tested in a simple stimulus matching-to-sample task (described above) many animals readily learn specific item combinations, such as "touch red if the sample is red, touch green if the sample is green." But this does not demonstrate that they distinguish between "same" and "different" as general concepts. Better evidence is provided if, after training, an animal successfully makes a choice that matches a novel sample that it has never seen before. Monkeys and chimpanzees do learn to do this, as do pigeons if they are given a great deal of practice with many different stimuli. However, because the sample is presented first, successful matching might mean that the animal is simply choosing the most recently seen "familiar" item rather than the conceptually "same" item. A number of studies have attempted to distinguish these possibilities, with mixed results.

Rule Learning

The use of rules has sometimes been considered an ability restricted to humans, but a number of experiments have shown evidence of simple rule learning in primates and also in other animals. Much of the evidence has come from studies of sequence learning in which the "rule" consists of the order in which a series of events occurs. Rule use is shown if the animal learns to discriminate different orders of events and transfers this discrimination to new events arranged in the same order. For example, Murphy *et al.* (2008) trained rats to discriminate between visual sequences. For one group ABA and BAB were rewarded, where A="bright light" and B="dim light." Other stimulus triplets were not rewarded. The rats learned the visual sequence, although both bright and dim lights were equally associated with reward. More importantly, in a second experiment with auditory stimuli, rats responded correctly to sequences of novel stimuli that were arranged in the same order as those previously learned. Similar sequence learning has been demonstrated in birds and other animals as well.

Memory

The categories that have been developed to analyze human memory (short term memory, long term memory, working memory) have been applied to the study of animal memory, and some of the phenomena characteristic of human short term memory (e.g. the serial position effect) have been detected in animals, particularly monkeys. However most progress has been made in the analysis of spatial memory; some of this work has sought to clarify the physiological basis of spatial memory and the role of the hippocampus; other work has explored the spatial memory of scatter-hoarder animals such as Clark's nutcracker, certain jays, tits and certain squirrels, whose ecological niches require them to remember the locations of thousands of caches, often following radical changes in the environment.

Memory has been widely investigated in foraging honeybees, *Apis mellifera*, which use both transient short-term working memory that is non-feeder specific and a feeder specific long-term reference memory. Memory induced in a free-flying honeybee by a single learning trial lasts for days and, by three learning trials, for a lifetime. Slugs, *Limax flavus*, have a short-term memory of approximately 1 min and long-term memory of 1 month.

Methods

As in humans, research with animals distinguishes between "working" or "short-term" memory from "reference" or long-term memory. Tests of working memory evaluate memory for events that happened in the recent past, usually within the last few seconds or minutes. Tests of reference memory evaluate memory for regularities such as "pressing a lever brings food" or "children give me peanuts."

Habituation

This is one of the simplest tests for memory spanning a short time interval. The test compares an animal's response to a stimulus or event on one occasion to its response on a previous occasion. If the second response differs consistently from the first, the animal must have remembered something about the first, unless some other factor such as motivation, sensory sensitivity, or the test stimulus has changed.

Delayed Response

Delayed response tasks are among the most useful methods used to study short-term memory in animals. Dating from research by Hunter (1913), the animal was shown a stimulus, such as a picture or a colored light, and a few seconds or minutes later the animal had to choose among alternative stimuli. In Hunter's studies, for example, a light appeared briefly in one of three goal boxes and then later the animal was allowed to choose among the boxes, finding food behind the one that had been lighted. Most research has been done with some variation of the "delayed matching-to-sample" task. For example, in the initial study with this task, a pigeon was presented with a flickering or steady light. Then, a few seconds later, two pecking keys were illuminated, one with a steady light and one with a flickering light. The bird got food if it pecked the key that matched the original stimulus.

A commonly-used variation of the matching-to-sample task requires the animal to use the initial stimulus to control a later choice between different stimuli. For example, if the initial stimulus is a black circle, the animal learns to choose "red" after the delay; if it is a black square, the correct choice is "green". Ingenious variations of this method have been used to explore many aspects of memory, including forgetting due to interference and memory for multiple items.

Radial Arm Maze

The radial arm maze is used to test memory for spatial location and to determine the mental processes by which location is determined. In a radial maze test, an animal is placed on a small platform from which paths lead in various directions to goal boxes; the animal finds food in one or more goal boxes. Having found food in a box, the animal must return to the central platform. The maze may be used to test both reference and working memory. Suppose, for example, that over a number of sessions the same 4 arms of an 8-arm maze always lead to food. If in a later test session the animal goes to a box that has never been baited, this indicates a failure of reference memory. On the other hand, if the animal goes to a box that it has already emptied during the same test session, this indicates a failure of working memory. Various confounding factors, such as odor cues, are carefully controlled in such experiments.

Water Maze

The water maze is used to test an animal's memory for spatial location and to discover how an animal is able to determine locations. Typically the maze is circular tank filled with water that has been made milky so that it is opaque. Located somewhere in the maze is small platform placed just below the surface of the water. When placed in the tank, the animal swims around until it finds and climbs up on the platform. With practice the animal finds the platform more and more quickly. Reference memory is assessed by removing the platform and observing the relative amount of time the animal spends swimming in the area where the platform had been located. Visual and other cues in and around the tank may be varied to assess the animal's reliance on landmarks and the geometric relations among them.

Spatial Cognition

Whether an animal ranges over a territory of measured in square kilometers or square meters, its survival typically depends on its ability to do such things as find a food source and then return to its nest. Sometimes such a task can be performed rather simply, for example by following a chemical trail. Typically, however, the animal must somehow acquire and use information about locations, directions, and distances. The following paragraphs outline some of the ways that animals do this.

- Beacons Animals often learn what their nest or other goal looks like, and if it is within sight they may simply move toward it; it is said to serve as a "beacon".
- Landmarks When an animal is unable to see its goal, it may learn the appearance of nearby objects and use these landmarks as guides. Researchers working with birds and bees have demonstrated this by moving prominent objects in the vicinity of nest sites, causing returning foragers to hunt for their nest in a new location.
- Dead reckoning Dead reckoning, also known as "path integration," is the process of computing one's position by starting from a known location and keeping track of the distances and directions subsequently traveled. Classic experiments have shown that the desert ant keeps track of its position in this way as it wanders for many meters searching for food. Though it travels in a randomly twisted path, it heads straight home when it finds food. However, if the ant is picked up and released some meters to the east, for example, it heads for a location displaced by the same amount to the east of its home nest.
- Cognitive maps Some animals appear to construct a cognitive map of their surroundings, meaning that they acquire and use information that enables them to compute how far and in what direction to go to get from one location to another. Such a map-like representation is thought to be used, for example, when an animal goes directly from one food source to

another even though its previous experience has involved only travel between each source and home. Research in this area has also explored such topics as the use of geometric properties of the environment by rats and pigeons, and the ability of rats to represent a spatial pattern in either radial arm mazes or water mazes. Spatial cognition is sometimes explored in visual search experiments in which a human or animal searches the environment for a particular object.

- Detour behaviour Some animals appear to have advanced understanding of their spatial environment and will not take the most direct route if this confers an advantage to them. Some jumping spiders take an indirect route to prey rather than the most direct route, thereby indicating flexibility in behaviour and route planning, and possibly insight learning.

Long-distance Navigation; Homing

Many animals travel hundreds or thousands of miles in seasonal migrations or returns to breeding grounds. They may be guided by the sun, the stars, the polarization of light, magnetic cues, olfactory cues, winds, or a combination of these.

It has been hypothesized that animals such as apes and wolves are good at spatial cognition because this skill is necessary for survival. Some researchers argue that this ability may have diminished somewhat in dogs because humans have provided necessities such as food and shelter during some 15,000 years of domestication.

Timing

Time of Day: Circadian Rhythms

The behavior of most animals is synchronized with the earth's daily light-dark cycle. Thus, many animals are active during the day, others are active at night, still others near dawn and dusk. Though one might think that these "circadian rhythms" are controlled simply by the presence or absence of light, nearly every animal that has been studied has been shown to have a "biological clock" that yields cycles of activity even when the animal is in constant illumination or darkness. Circadian rhythms are so automatic and fundamental to living things — they occur even in plants - that they are usually discussed separately from cognitive processes.

Interval Timing

Survival often depends on an animal's ability to time intervals. For example, rufous hummingbirds feed on the nectar of flowers, and they often return to the same flower, but only after the flower had had enough time to replenish its supply of nectar. In one experiment hummingbirds fed on artificial flowers that quickly emptied of nectar but were refilled at some fixed time (e.g. twenty minutes) later. The birds learned to come back to the flowers at about the right time, learning the refill rates of up to eight separate flowers and remembering how long ago they had visited each one.

The details of interval timing have been studied in a number of species. One of the most common methods is the "peak procedure". In a typical experiment, a rat in an operant chamber presses a lever for food. A light comes on, a lever-press brings a food pellet at a fixed later time, say 10 seconds, and then the light goes off. Timing is measured during occasional test trials on which no food is presented and the light stays on. On these test trials the rat presses the lever more and more until about 10 sec and then, when no food comes, gradually stops pressing. The time at which the rat presses most on these test trials is taken to be its estimate of the payoff time.

Experiments using the peak procedure and other methods have shown that animals can time short intervals quite exactly, can time more than one event at once, and can integrate time with spatial and other cues. Such tests have also been used for quantitative tests of theories of animal timing, such as Gibbon's Scalar Expectancy Theory ("SET"), Killeen's Behavioral Theory of Timing, and Machado's Learning to Time model. No one theory has yet gained unanimous agreement.

Tool and Weapon use

Although tool use was long assumed to be a uniquely human trait, there is now much evidence that many animals use tools, including mammals, birds, fish, cephalopods and insects. Discussions of tool use often involve debate about what constitutes a "tool", and they often consider the relation of tool use to the animal's intelligence and brain size.

Mammals

Series of Photographs Showing a Bonobo Fishing for Termites.

Tool use has been reported many times in both wild and captive primates, particularly the great apes. The use of tools by primates is varied and includes hunting (mammals, invertebrates, fish), collecting honey, processing food (nuts, fruits, vegetables and seeds), collecting water, weapons and shelter. Research in 2007 shows that chimpanzees in the Fongoli savannah sharpen sticks to use as spears when hunting, considered the first evidence of systematic use of weapons in a species other than humans. Other mammals that spontaneously use tools in the wild or in captivity include elephants, bears, cetaceans, sea otters and mongooses.

A bonobo inserting a stick into a termite mound

The bonobo starts "fishing" for the termites

The bonobo withdraws the stick and begins eating the termites

Birds

Several species of birds have been observed to use tools in the wild, including warblers, parrots, Egyptian vultures, brown-headed nuthatches, gulls and owls. Some species, such as the woodpecker finch of the Galapagos Islands, use particular tools as an essential part of their foraging behavior. However, these behaviors are often quite inflexible and cannot be applied effectively in new situations. A great many species of birds build nests with a wide range of complexities, but although nest-building behaviour fulfills the criteria of some definitions of "tool-use", this is not the case with other definitions.

Several species of corvids have been trained to use tools in controlled experiments. One species examined extensively under laboratory conditions is the New Caledonian crow. One individual called "Betty" spontaneously made a wire tool to solve a novel problem. She was being tested to see whether she would select a wire hook rather than a straight wire to pull a little bucket of meat out of a well. Betty tried poking the straight wire at the meat. After a series of failures with this direct approach, she withdrew the wire and began directing it at the bottom of the well, which was secured to its base with duct tape. The wire soon became stuck, whereupon Betty pulled it sideways, bending it and unsticking it. She then inserted the hook into the well and extracted the meat. In all but one of 10 subsequent trials with only straight wire provided, she also made and used a hook in the same manner, but not before trying the straight wire first.

Fish

Several species of wrasses have been observed using rocks as anvils to crack bivalve (scallops, urchins and clams) shells. This behavior was first filmed in an orange-dotted tuskfish (*Choerodon anchorago*) in 2009 by Giacomo Bernardi. The fish fans sand to unearth the bivalve, takes it into its mouth, swims several meters to a rock, which it then uses as an anvil by smashing the mollusc apart with sideward thrashes of the head. This behaviour has also been recorded in a blackspot tuskfish (*Choerodon schoenleinii*) on Australia's Great Barrier Reef, yellowhead wrasse (*Halichoeres garnoti*) in Florida and a six-bar wrasse (*Thalassoma hardwicke*) in an aquarium setting. These species are at opposite ends of the phylogenetic tree in this family, so this behaviour may be a deep-seated trait in all wrasses.

Invertebrates

Some cephalopods are known to use coconut shells for protection or camouflage.

Ants of the species *Conomyrma bicolor* pick up stones and other small objects with their mandibles and drop them down the vertical entrances of rival colonies, allowing workers to forage for food without competition.

Reasoning and problem solving

It is clear that animals of quite a range of species are capable of solving problems that appear to require abstract reasoning; Wolfgang Köhler's (1917) work with chimpanzees is a famous early example. He observed that chimpanzees did not use trial and error to solve problems such as retrieving bananas hung out of reach. Instead, they behaved in a manner that was "unwaveringly purposeful," spontaneously placing boxes so that they could climb to reach the fruit. Modern research has identified similar behavior in animals usually thought of as much less intelligent, if appropriate pre-training is given. Causal reasoning has also been observed in rooks and New Caledonian crows.

It has been shown that Barbados bullfinches (*Loxigilla barbadensis*) from urbanized areas are better at innovative problem-solving tasks than bullfinches from rural environments, but that they did not differ in colour discrimination learning.

Cognitive Bias

A cognitive bias refers to a systematic pattern of deviation from norm or rationality in judgment, whereby inferences about other individuals or situations may be drawn in an illogical fashion.

Cognitive bias is sometimes illustrated by using answers to the question "Is the glass half empty or half full?". Choosing "half empty" is supposed to indicate pessimism whereas choosing "half full" indicates optimism. To test this in animals, an individual is trained to anticipate that stimulus A, e.g. a 100 Hz tone, precedes a positive event, e.g. highly desired food is delivered when a lever is pressed by the animal. The same individual is trained to anticipate that stimulus B, e.g. a 900 Hz tone, precedes a negative event, e.g. bland food is delivered when the animal presses a lever. The animal is then tested by being given an intermediate stimulus C, e.g. a 500 Hz tone, and observing whether the animal presses the lever associated with the positive or negative reward. This has been suggested to indicate whether the animal is in a positive or negative mood.

Is the glass half empty or half full?

In a study that used this approach, rats that were playfully tickled responded differently than rats that were simply handled. The rats that had been tickled were more optimistic than the handled rats. The authors suggested that they had demonstrated "...for the first time a link between the directly measured positive affective state and decision making under uncertainty in an animal model."

There is some evidence for cognitive bias in a number of species, including rats, dogs, rhesus macaques, sheep, chicks, starlings and honeybees.

Language

The modeling of human language in animals is known as animal language research. In addition to the ape-language experiments mentioned above, there have also been more or less successful attempts to teach language or language-like behavior to some non-primate species, including parrots and great spotted woodpeckers. Arguing from his own results with the animal Nim Chimpsky and his analysis of others results, Herbert Terrace criticized the idea that chimps can produce new sentences. Shortly thereafter Louis Herman published research on artificial language comprehension in the bottlenosed dolphin (Herman, Richards, & Wolz, 1984). Though this sort of research has been controversial, especially among cognitive linguists, many researchers agree that many animals can understand the meaning of individual words, and that some may understand simple sentences and syntactic variations, but there is little evidence that any animal can produce new strings of symbols that correspond to new sentences.

Insight

Wolfgang Köhler is usually credited with introducing the concept of insight into experimental psychology. Working with chimpanzees, Köhler came to dispute Edward Thorndike's theory that animals must solve problems gradually, by trial and error. He said that Thorndike's animals could only use trial and error because the situation precluded other forms of problem solving. He provided chimps with a relatively unstructured situation, and he observed sudden "ah-ha!" insightful changes of behavior, as, for example, when a chimp suddenly moved a box into position so that it could retrieve a banana. More recently, Asian elephants (*Elephas maximus*) were shown to exhibit similar insightful problem solving. A male was observed moving a box to a position where it could be stood upon to reach food that had been deliberately hung out of reach.

Numeracy

There a variety of studies indicate that animals are able to use and communicate quantitative information, and that some can count in a rudimentary way. Some examples of this research follow.

Elephants have been known to perform simple arithmetic, and rhesus monkeys and pigeons, in some sense, can count.

Ants are able to use quantitative values and transmit this information. For instance, ants of several species are able to estimate quite precisely numbers of encounters with members of other colonies on their feeding territories. Numeracy has been described in the yellow mealworm beetle (*Tenebrio molitor*) and the honeybee.

Western lowland gorillas given the choice between two food trays demonstrated the ability to choose the tray with more food items at a rate higher than chance after training. In a similar task, chimpanzees chose the option with larger amount of food. Salamanders given a choice between two displays with differing amounts of fruit flies, used as a food reward, reliably choose the display with more flies, as shown in a particular experiment.

Other experiments have been conducted that show animals' abilities to differentiate between non-food quantities. American black bears demonstrated quantity differentiation abilities in a task with a computer screen. The bears were trained to touch a computer monitor with a paw or nose to choose a quantity of dots in one of two boxes on the screen. Each bear was trained with reinforcement to pick a larger or smaller amount. During training, the bears were rewarded with food for a correct response. All bears performed better than what random error predicted on the trials with static, non-moving dots, indicating that they could differentiate between the two quantities. The bears choosing correctly in congruent (number of dots coincided with area of the dots) and incongruent (number of dots did not coincide with area of the dots) trials suggests that they were indeed choosing between quantities that appeared on the screen, not just a larger or smaller retinal image, which would indicate they are only judging size.

Bottlenose dolphins have shown the ability to choose an array with fewer dots compared to one with more dots. Experimenters set up two boards showing various numbers of dots in a poolside setup. The dolphins were initially trained to choose the board with the fewer number of dots. This was done by rewarding the dolphin when it chose the board with the fewer number of dots. In the experimental trials, two boards were set up, and the dolphin would emerge from the water and point to one board. The dolphins chose the arrays with fewer dots at a rate much larger than chance, indicating they can differentiate between quantities. A particular grey parrot, after training, has shown the ability to differentiate between the numbers zero through six using vocalizations. After number and vocalization training, this was done by asking the parrot how many objects there were in a display. The parrot was able to identify the correct amount at a rate higher than chance. Angelfish, when put in an unfamiliar environment will group together with conspecifics, an action named shoaling. Given the choice between two groups of differing size, the angelfish will choose the larger of the two groups. This can be seen with a discrimination ratio of 2:1 or greater, such that, as long as one group has at least twice the fish as another group, it will join the larger one.

Monitor lizards have been shown to be capable of numeracy, and some species can distinguish among numbers up to six.

Intelligence

Cognitive ability and intelligence in non-human animals cannot be measured with verbal scales, as in humans. Instead, intelligence in animals has been measured using a variety of methods that involve such things as habit reversal, social learning, and responses to novelty. However, as with humans, the intelligence attributed to an animal depends on the tasks used to measure that intelligence. In particular, an animal might appear "stupid" when measured with tasks appropriate for a human or an animal of another species, while that same animal might be a marvel of intelligence when given tasks like those it needs to survive in a natural setting.

Principal Component Analysis and factor analytic studies have shown that a single factor of intelligence is responsible for 47% of the individual variance in cognitive ability measures in primates and between 55% and 60% of the variance in mice. These values are similar to the accepted variance in IQ explained by a similar single factor known as the general factor of intelligence in humans (40-50%).

The general factor of intelligence, or *g* factor, is a psychometric construct that summarizes the correlations observed between an individual's scores on various measures of cognitive abilities. It has been suggested that *g* is related to evolutionary life histories and the evolution of intelligence as well as to social learning and cultural intelligence. Non-human models of *g* have been used in genetic and neurological research on intelligence to help understand the mechanisms behind variation in *g*.

Theory of Mind

Theory of mind is the ability to attribute mental states, e.g. intents, desires, pretending, knowledge, to oneself and others and to understand that others have desires, intentions, and perspectives that are different from one's own.

Some research with ravens provides an example of evidence for theory of mind in a non-human species. Ravens are members of the corvidae family, which is widely regarded as having high cognitive abilities. These birds have been observed to hide their food when dominant ravens are visible and audible at the same time. Based on this observation, ravens were tested for their understanding of "seeing" as a mental state. In a first step, the birds protected their cache when dominants were visible but not when they could only be heard from an adjacent room. In the next step, they had access to a small peephole which allowed them to see into the adjacent room. With the peephole open, the ravens guarded their caches against discovery when they could hear dominants in the adjacent room, even when the dominant's sounds were playbacks of recordings.

Consciousness

The sense in which animals can be said to have consciousness or a self-concept has been hotly debated. The best known research technique in this area is the mirror test devised by Gordon G. Gallup, in which an animal's skin is marked in some way while it is asleep or sedated, and it is then allowed to see its reflection in a mirror; if the animal spontaneously directs grooming behavior towards the mark, that is taken as an indication that it is aware of itself. Self-awareness, by this criterion, has been reported for chimpanzees and also for other great apes, the European magpie, some cetaceans and a solitary elephant, but not for monkeys. The mirror test has been criticized researchers because it is entirely focused on vision, the primary sense in humans, while other species rely more heavily on other senses such as the olfactory sense in dogs.

Mirror test with a baboon

It has been suggested that metacognition in some animals provides some evidence for cognitive self-awareness. The great apes, dolphins, and rhesus monkeys have demonstrated the ability to monitor their own mental states and use an "I don't know" response to avoid answering difficult questions. Unlike the mirror test, which reveals awareness of the condition of one's own body, this uncertainty monitoring is thought to reveal awareness of one's internal mental state. A 2007 study has provided some evidence for metacognition in rats, although this interpretation has been questioned. These species might also be aware of the strength of their memories.

Some researchers propose that an animal calls and other vocal behavior provide evidence as to consciousness. This idea arose from research on children's crib talk by Weir (1962) and in investigations of early speech in children by Greenfield and others (1976). Some such research has been done with a macaw.

In July, 2012 during the "Consciousness in Human and Nonhuman Animals" conference in Cambridge a group of scientists announced and signed a declaration with the following conclusions:

Convergent evidence indicates that non-human animals have the neuroanatomical, neurochemical, and neurophysiological substrates of conscious states along with the capacity to exhibit intentional behaviors. Consequently, the weight of evidence indicates that humans are not unique in possessing the neurological substrates that generate consciousness. Non-human animals, including all mammals and birds, and many other creatures, including octopuses, also possess these neurological substrates.

Biological Constraints

Animals differ widely in many learning and cognitive tasks in ways that reflect their evolutionary history and their instinctual behaviors in natural environments. For example, dogs and rats easily learn to avoid an electric shock from the floor by moving to another part of the experimental chamber when they hear a tone preceding the shock; this is an appropriate response to a dangerous situation. However, hedgehogs fail to learn this avoidance behavior. This might seem to show the hedgehog's inability to learn, but the hedgehog's instinctive reaction to a threat is to curl up into a ball, a response that interferes with possible escape behavior in this situation.

Hedgehogs instinctively roll into a ball when threatened, making them unsuitable for studies on aversion avoidance

Instinctive drift is another factor that can influence the interpretation of cognitive research. Instinctive drift is the tendency of an animal to revert to instinctive behaviors that can interfere with learned responses. The concept originated with Keller and Marian Breland when they taught a raccoon to put coins into a box. The raccoon drifted to its instinctive behavior of rubbing the coins with its paws, as it would do when foraging for food.

Animal ability to process and respond to stimuli is correlated with brain size. Small-brain animals tend to show simple behaviors that are less dependent on learning than those of large-brained animals. Vertebrates, particularly mammals, have large brains and complex behavior that changes with experience. A formula called the encephalization quotient (EC) expresses a relationship between brain and body size; it was developed by H.J. Jerison in the late 1960s. When the encephalization quotient is plotted as a curve, an animal with an EC above the curve is expected to show more cognitive ability than the average animal of its size, whereas an animal with an EC below the curve is expected to have less. Various formulas been suggested, but the equation $Ew(brain) = 0.12w(body)^{2/3}$ has been found to fit data from a sample of mammals. The formula is suggestive at best, and should only be applied to non-mammals with extreme caution. For some of the other vertebrate classes, the power of 3/4 rather than 2/3 is sometimes used, and for many groups of invertebrates, the formula may not give meaningful results.

Cognitive faculty by species

A traditionally common image is the *scala naturae*, the ladder of nature on which animals of different species occupy successively higher rungs, with humans typically at the top. However, rather than using such an arbitrary hierarchy, it seems more fruitful to understand cognitive capacities as adaptations to differing ecological niches.

Whether fairly or not, the performance of animals is often compared to that of humans on cognitive tasks. Not surprisingly, our closest biological relatives, the great apes, tend to perform most like humans. Among the birds, corvids and parrots have typically been found to perform well on human-like tasks. Some octopodes have also been shown to exhibit a number of higher-level skills such as tool use, but the amount of research on cephalopod intelligence is still limited.

Baboons have been shown to be capable of recognizing words.

Adaptation

In biology, an adaptation, also called an adaptive trait, is a trait with a current functional role in the life of an organism that is maintained and evolved by means of natural selection. Adaptation refers to both the current state of being adapted and to the dynamic evolutionary process that leads to the adaptation. Adaptations enhance the fitness and survival of individuals. Organisms face a succession of environmental challenges as they grow and develop and are equipped with an adaptive plasticity as the phenotype of traits develop in response to the imposed conditions. The developmental norm of reaction for any given trait is essential to the correction of adaptation as it affords a kind of biological insurance or resilience to varying environments.

General Principles

Adaptation is, first of all, a *process*, rather than a part of a body. An internal parasite (such as a liver fluke) can illustrate the distinction: such a parasite may have a very simple bodily structure, but nevertheless the organism is highly adapted to its specific environment. From this we see that adaptation is not just a matter of visible traits: in such parasites critical adaptations take place in the life cycle, which is often quite complex. However, as a practical term, adaptation is often used for the *product*: those features of a species which result from the process. Many aspects of an animal or plant can be correctly called adaptations, though there are always some features whose function is in doubt. By using the term *adaptation* for the evolutionary *process*, and *adaptive trait* for the bodily part or function (the product), the two senses of the word may be distinguished.

Adaptation is one of the two main processes that explain the diverse species we see in biology, such as the different species of Darwin's finches. The other is speciation (species-splitting or cladogenesis), caused by geographical isolation or some other mechanism. A favorite example used today to study the interplay of adaptation and speciation is the evolution of cichlid fish in African lakes, where the question of reproductive isolation is much more complex.

Adaptation is not always a simple matter, where the ideal phenotype evolves for a given external environment. An organism must be viable at all stages of its development and at all stages of its evolution. This places *constraints* on the evolution of development, behavior and structure of organisms. The main constraint, over which there has been much debate, is the requirement that each genetic and phenotypic change during evolution should be relatively small, because developmental systems are so complex and interlinked. However, it is not clear what "relatively small" should mean, for example polyploidy in plants is a reasonably common large genetic change. The origin of eukaryotic symbiosis is a more dramatic example.

All adaptations help organisms survive in their ecological niches. These adaptive traits may be structural, behavioral or physiological. Structural adaptations are physical features of an organism (shape, body covering, armament; and also the internal organization). Behavioral adaptations are composed of inherited behavior chains and/or the ability to learn: behaviors may be inherited in detail (instincts), or a capacity for learning may be inherited. Examples: searching for food, mating, vocalizations. Physiological adaptations permit the organism to perform special functions (for instance, making venom, secreting slime, phototropism); but also more general functions such as growth and development, temperature regulation, ionic balance and other aspects of homeostasis. Adaptation, then, affects all aspects of the life of an organism.

Definitions

The following definitions are mainly due to Theodosius Dobzhansky.

1. *Adaptation* is the evolutionary process whereby an organism becomes better able to live in its habitat or habitats.

2. *Adaptedness* is the state of being adapted: the degree to which an organism is able to live and reproduce in a given set of habitats.

3. An *adaptive trait* is an aspect of the developmental pattern of the organism which enables or enhances the probability of that organism surviving and reproducing.

Adaptedness and Fitness

From the above definitions, it is clear that there is a relationship between adaptedness and *fitness* (a key population genetics concept). Differences in fitness between genotypes predict the rate of evolution by natural selection. Natural selection changes the relative frequencies of alternative phenotypes, insofar as they are heritable. Although the two are connected, the one does not imply the other: a phenotype with high adaptedness may not have high fitness. Dobzhansky mentioned the example of the Californian redwood, which is highly adapted, but a relict species in danger of extinction. Elliott Sober commented that adaptation was a retrospective concept since it implied something about the history of a trait, whereas fitness predicts a trait's future.

1. Relative fitness. The average contribution to the next generation by a genotype or a class of genotypes, relative to the contributions of other genotypes in the population. This is also known as *Darwinian fitness, selection coefficient*, and other terms.

2. Absolute fitness. The absolute contribution to the next generation by a genotype or a class of genotypes. Also known as the Malthusian parameter when applied to the population as a whole.

3. Adaptedness. The extent to which a phenotype fits its local ecological niche. This can sometimes be tested through a reciprocal transplant.

Brief History

Adaptation is a fact of life that has been accepted by many of the great thinkers who have tackled the world of living organisms. It is their explanations of how adaptation arises that separates these thinkers. A few of the most significant ideas:

- Empedocles did not believe that adaptation required a final cause (~ purpose), but "came about naturally, since such things survived." Aristotle, however, did believe in final causes.

Jean-Baptiste Lamarck (1744–1829)

- In natural theology, adaptation was interpreted as the work of a deity, even as evidence for the existence of God. William Paley believed that organisms were perfectly adapted to the lives they lead, an argument that shadowed Gottfried Wilhelm Leibniz, who had argued that God had brought about "the best of all possible worlds." Voltaire's Dr. Pangloss is a parody of this optimistic idea, and David Hume also argued against design. The *Bridgewater Treatises* are a product of natural theology, though some of the authors managed to present their work in a fairly neutral manner. The series was lampooned by Robert Knox, who held quasi-evolutionary views, as the *Bilgewater Treatises*. Charles Darwin broke with the tradition by emphasising the flaws and limitations which occurred in the animal and plant worlds.

- Lamarckism is a proto-evolutionary hypothesis of the inheritance of acquired characteristics, whose main purpose is to explain adaptations by natural means. Jean-Baptiste Lamarck proposed a tendency for organisms to become more complex, moving up a ladder of progress, plus "the influence of circumstances," usually expressed as *use and disuse*. His evolutionary ideas, and those of Étienne Geoffroy Saint-Hilaire, fail because they cannot be reconciled with heredity. This was known even before Gregor Mendel by medical men interested in human races (William Charles Wells, William Lawrence), and especially by August Weismann.

Many other students of natural history, such as Buffon, accepted adaptation, and some also accepted evolution, without voicing their opinions as to the mechanism. This illustrates the real merit of Darwin and Alfred Russel Wallace, and secondary figures such as Henry Walter Bates, for putting forward a mechanism whose significance had only been glimpsed previously. A century later, experimental field studies and breeding experiments by people such as E. B. Ford and Dobzhansky produced evidence that natural selection was not only the 'engine' behind adaptation, but was a much stronger force than had previously been thought.

Types of Adaptations

Adaptation is the heart and soul of evolution.

—Niles Eldredge, Reinventing Darwin: The Great Debate at the High Table of Evolutionary Theory

Changes in Habitat

Before Charles Darwin, adaptation was seen as a fixed relationship between an organism and its habitat. It was not appreciated that as the climate changed, so did the habitat; and as the habitat changed, so did the biota. Also, habitats are subject to changes in their biota: for example, invasions of species from other areas. The relative numbers of species in a given habitat are always changing. Change is the rule, though much depends on the speed and degree of the change.

When the habitat changes, three main things may happen to a resident population: habitat tracking, genetic change or extinction. In fact, all three things may occur in sequence. Of these three effects only genetic change brings about adaptation.

Habitat Tracking

When a habitat changes, the most common thing to happen is that the resident population moves to another locale which suits it; this is the typical response of flying insects or oceanic organisms, who have wide (though not unlimited) opportunity for movement. This common response is called *habitat tracking*. It is one explanation put forward for the periods of apparent stasis in the fossil record (the punctuated equilibrium theory).

Genetic Change

Genetic change is what occurs in a population when natural selection acts on the genetic variability of the population; moreover, some mutations may create genetic variation that will lead to differing characteristics of offspring and hence abet adaptation. The first pathways of enzyme-based metabolism may have been parts of purine nucleotide metabolism, with previous metabolic pathways being part of the ancient RNA world. By this means, the population adapts genetically to its circumstances. Genetic changes may result in visible structures, or may adjust physiological activity in a way that suits the habitat.

It is now clear that habitats and biota do frequently change. Therefore, it follows that the process of adaptation is never finally complete. Over time, it may happen that the environment changes little, and the species comes to fit its surroundings better and better. On the other hand, it may happen that changes in the environment occur relatively rapidly, and then the species becomes less and less well adapted. Seen like this, adaptation is a genetic *tracking process*, which goes on all the time to some extent, but especially when the population cannot or does not move to another, less hostile area. Also, to a greater or lesser extent, the process affects every species in a particular ecosystem.

Leigh Van Valen thought that even in a stable environment, competing species had to constantly adapt to maintain their relative standing. This became known as the Red Queen hypothesis. One of the manifestations of the Red Queen dynamics can be seen in host-parasite interaction.

Intimate Relationships: Co-adaptations

In coevolution, where the existence of one species is tightly over bound up with the life of another species, new or 'improved' adaptations which occur in one species are often followed by the appearance and spread of corresponding features in the other species. There are many examples of this; the idea emphasises that the life and death of living things is intimately connected, not just with the physical environment, but with the life of other species. These relationships are intrinsically dynamic, and may continue on a trajectory for millions of years, as has the relationship between flowering plants and insects (pollination).

Pollinator constancy: these honey bees selectively visit flowers from only one species, as can be seen by the colour of the pollen in their baskets:

The gut contents, wing structures, and mouthpart morphologies of fossilized beetles and flies suggest that they acted as early pollinators. The association between beetles and angiosperms during the Early Cretaceous period led to parallel radiations of angiosperms and insects into the Late Cretaceous. The evolution of nectaries in Late Cretaceous flowers signals the beginning of the mutualism between hymenopterans and angiosperms.

Mimicry

Bates' work on Amazonian butterflies led him to develop the first scientific account of mimicry, especially the kind of mimicry which bears his name: Batesian mimicry. This is the mimicry by a palatable species of an unpalatable or noxious species. A common example seen in temperate gardens is the hoverfly, many of which—though bearing no sting—mimic the warning colouration of hymenoptera (wasps and bees). Such mimicry does not need to be perfect to improve the survival of the palatable species.

A and B show real wasps; the rest are mimics: three hoverflies and one beetle.

Bates, Wallace and Fritz Müller believed that Batesian and Müllerian mimicry provided evidence for the action of natural selection, a view which is now standard amongst biologists. All aspects of this situation can be, and have been, the subject of research. Field and experimental work on these ideas continues to this day; the topic connects strongly to speciation, genetics and development.

The Basic Machinery: Internal Adaptations

There are some important adaptations to do with the overall coordination of the systems in the body. Such adaptations may have significant consequences. Examples, in vertebrates, would be temperature regulation, or improvements in brain function, or an effective immune system. An example in plants would be the development of the reproductive system in flowering plants. Such adaptations may make the clade (monophyletic group) more viable in a wide range of habitats.

The acquisition of such major adaptations has often served as the spark for adaptive radiation, and huge success over long periods of time for a whole group of animals or plants.

Compromise and Conflict Between Adaptations

It is a profound truth that Nature does not know best; that genetical evolution... is a story of waste, makeshift, compromise and blunder.

—Peter Medawar, The Future of Man

All adaptations have a downside: horse legs are great for running on grass, but they can't scratch their backs; mammals' hair helps temperature, but offers a niche for ectoparasites; the only flying penguins do is under water. Adaptations serving different functions may be mutually destructive. Compromise and makeshift occur widely, not perfection. Selection pressures pull in different directions, and the adaptation that results is some kind of compromise.

Since the phenotype as a whole is the target of selection, it is impossible to improve simultaneously all aspects of the phenotype to the same degree.

—Ernst Mayr, The Growth of Biological Thought: Diversity, Evolution, and Inheritance

Consider the antlers of the Irish elk, (often supposed to be far too large; in deer antler size has an allometric relationship to body size). Obviously, antlers serve positively for defence against predators, and to score victories in the annual rut. But they are costly in terms of resource. Their size during the last glacial period presumably depended on the relative gain and loss of reproductive capacity in the population of elks during that time. Another example: camouflage to avoid detection is destroyed when vivid colors are displayed at mating time. Here the risk to life is counterbalanced by the necessity for reproduction.

Stream-dwelling salamanders, such as Caucasian salamander or Gold-striped salamander have very slender, long bodies, perfectly adapted to life at the banks of fast small rivers and mountain brooks. Elongated body protects their larvae from being washed out by current. However, elongated body increases risk of desiccation and decreases dispersal ability of the salamanders; it also negatively affects their fecundity. As a result, fire salamander, less perfectly adapted to the mountain brook habitats, is in general more successful, have a higher fecundity and broader geographic range.

An Indian peacock's train in full display

The peacock's ornamental train (grown anew in time for each mating season) is a famous adaptation. It must reduce his maneuverability and flight, and is hugely conspicuous; also, its growth costs food resources. Darwin's explanation of its advantage was in terms of sexual selection: "This depends on the advantage which certain individuals have over other individuals of the same sex and species, in exclusive relation to reproduction." The kind of sexual selection represented by the peacock is called 'mate choice,' with an implication that the process selects the more fit over the less fit, and so has survival value. The recognition of sexual selection was for a long time in abeyance, but has been rehabilitated. In practice, the Indian peafowl (*Pavo cristatus*) is a successful species, with a large natural range in India, so the overall outcome of their mating system is quite viable.

The conflict between the size of the human foetal brain at birth, (which cannot be larger than about 400 cm^3, else it will not get through the mother's pelvis) and the size needed for an adult brain (about 1400 cm^3), means the brain of a newborn child is quite immature. The most vital things in human life (locomotion, speech) just have to wait while the brain grows and matures. That is the result of the birth compromise. Much of the problem comes from our upright bipedal stance, without which our pelvis could be shaped more suitably for birth. Neanderthals had a similar problem.

As another example, the long neck of a giraffe is a burden and a blessing. The neck of a giraffe can be up to 2 m (6 ft 7 in) in length. This neck can be used for inter-species competition or for foraging on tall trees where shorter herbivores cannot reach. However, as previously stated, there is always a trade-off. This long neck is heavy and it adds to the body mass of a giraffe, so the giraffe needs an abundance of nutrition to provide for this costly adaptation.

Shifts in Function

Adaptation and function are two aspects of one problem.

—Julian Huxley, Evolution: The Modern Synthesis

Pre-adaptations

This occurs when a species or population has characteristics which (by chance) are suited for conditions which have not yet arisen. For example, the polyploid cordgrass *Spartina townsendii* is better adapted than either of its parent species to their own habitat of saline marsh and mud-flats. White Leghorn chicken are markedly more resistant to vitamin B_1 deficiency than other breeds. On a plentiful diet there is no difference, but on a restricted diet this preadaptation could be decisive.

Pre-adaptation may occur because a natural population carries a huge quantity of genetic variability. In diploid eukaryotes, this is a consequence of the system of sexual reproduction, where mutant alleles get partially shielded, for example, by the selective advantage of heterozygotes. Microorganisms, with their huge populations, also carry a great deal of genetic variability.

The first experimental evidence of the pre-adaptive nature of genetic variants in microorganisms was provided by Salvador Luria and Max Delbrück who developed Fluctuation Test, a method to show the random fluctuation of pre-existing genetic changes that conferred resistance to bacteriophage in the bacterium *Escherichia coli.*

Co-option of Existing Traits: Exaptation

The classic example is the ear ossicles of mammals, which we know from paleontological and embryological studies originated in the upper and lower jaws and the hyoid bone of their synapsid ancestors, and further back still were part of the gill arches of early fish. We owe this esoteric knowledge to the comparative anatomists, who, a century ago, were at the cutting edge of evolutionary studies. The word *exaptation* was coined to cover these shifts in function, which are surprisingly common in evolutionary history. The origin of wings from feathers that were originally used for temperature regulation is a more recent discovery.

Related Issues

Non-adaptive Traits

Some traits do not appear to be adaptive, that is, they appear to have a neutral or even deleterious effect on fitness in the current environment. Because genes have pleiotropic effects, not all traits may be functional (i.e. spandrels). Alternatively, a trait may have been adaptive at some point in an organism's evolutionary history, but a change in habitats caused what used to be an adaptation to become unnecessary or even a hindrance (maladaptations). Such adaptations are termed vestigial.

Vestigial Organs

Many organisms have vestigial organs, which are the remnants of fully functional structures in their ancestors. As a result of changes in lifestyle the organs became redundant, and are either not functional or reduced in functionality. With the loss of function goes the loss of positive selection, and the subsequent accumulation of deleterious mutations. Since any structure represents some kind of cost to the general economy of the body, an advantage may accrue from their elimination once they are not functional. Examples: wisdom teeth in humans; the loss of pigment and functional eyes in cave fauna; the loss of structure in endoparasites.

Fitness Landscapes

Sewall Wright proposed that populations occupy *adaptive peaks* on a fitness landscape. In order to evolve to another, higher peak, a population would first have to pass through a valley of maladaptive intermediate stages. A given population might be "trapped" on a peak that is not optimally adapted.

Extinction

If a population cannot move or change sufficiently to preserve its long-term viability, then obviously, it will become extinct, at least in that locale. The species may or may not survive in other locales. Species extinction occurs when the death rate over the entire species exceeds the birth rate for a long enough period for the species to disappear. It was an observation of Van Valen that groups of species tend to have a characteristic and fairly regular rate of extinction.

Coextinction

Just as we have co-adaptation, there is also coextinction. Coextinction refers to the loss of a species due to the extinction of another; for example, the extinction of parasitic insects following the loss of their hosts. Coextinction can also occur when a flowering plant loses its pollinator, or through the disruption of a food chain. Ecologist Lian Pin Koh and colleagues discuss coextinction, stating, "Species coextinction is a manifestation of the interconnectedness of organisms in complex ecosystems. . . . While coextinction may not be the most important cause of species extinctions, it is certainly an insidious one."

Flexibility, Acclimatization, Learning

Flexibility deals with the relative capacity of an organism to maintain themselves in different habitats: their degree of specialization. *Acclimatization* is a term used for automatic physiological adjustments during life; *learning* is the term used for improvement in behavioral performance during life. In biology these terms are preferred, not adaptation, for changes during life which improve the performance of individuals. These adjustments are not inherited genetically by the next generation.

Generalists, such as birds, are sometimes able to adapt to urban areas.

Adaptation, on the other hand, occurs over many generations; it is a gradual process caused by natural selection which changes the genetic make-up of a population so the collective performs better in its niche.

Flexibility

Populations differ in their phenotypic plasticity, which is the ability of an organism with a given genotype to change its phenotype in response to changes in its habitat, or to move to a different habitat.

To a greater or lesser extent, all living things can adjust to circumstances. The degree of flexibility is inherited, and varies to some extent between individuals. A highly specialized animal or plant lives only in a well-defined habitat, eats a specific type of food, and cannot survive if its needs are not met. Many herbivores are like this; extreme examples are koalas which depend on *eucalyptus*, and giant pandas which require bamboo. A generalist, on the other hand, eats a range of food, and can survive in many different conditions. Examples are humans, rats, crabs and many carnivores. The *tendency* to behave in a specialized or exploratory manner is inherited—it is an adaptation.

Rather different is *developmental flexibility:* “An animal or plant is developmentally flexible if when it is raised in or transferred to new conditions, it changes in structure so that it is better fitted to survive in the new environment,” writes evolutionary biologist John Maynard Smith. Once again, there are huge differences between species, and the *capacities* to be flexible are inherited.

Acclimatization

If humans move to a higher altitude, respiration and physical exertion become a problem, but after spending time in high altitude conditions they *acclimatize* to the pressure by increasing production of Red blood cells. The *ability* to acclimatize is an adaptation, but not the acclimatization itself. Fecundity goes down, but deaths from some tropical diseases also goes down.

Over a longer period of time, some people will reproduce better at these high altitudes than others. They will contribute more heavily to later generations. Gradually the whole population becomes adapted to the new conditions. This we know takes place, because the performance of long-term communities at higher altitude is significantly better than the performance of new arrivals, even when the new arrivals have had time to make physiological adjustments.

Some kinds of acclimatization happen so rapidly that they are better called reflexes. The rapid colour changes in some flatfish, cephalopods, chameleons are examples.

Learning

Social learning is supreme for humans, and is possible for quite a few mammals and birds: of course, that does not involve genetic transmission except to the extent that the capacities are inherited. Similarly, the *capacity to learn* is an inherited adaptation, but not what is learnt; the capacity for human speech is inherited, but not the details of language.

Diversity of Genome DNAs

A large diversity of genome DNAs in a species is the basis for species’ adaptation and for species’ differentiation. A great number of individuals are needed for carrying the great number of different genome DNAs. According to the Misrepair-accumulation aging theory, Misrepair mechanism is important in maintaining the sufficient number of individuals in a species. Misrepair is a way of repair for increasing the surviving chance of an organism when it has severe injuries. Without Misrepairs, no individual could survive to reproduction age. Thus Misrepair mechanism is an essential mechanism for the survival of a species and for maintaining the number of individuals. Although individuals die from aging, genome DNAs are being recopied and transmitted by individuals generation by generation. In addition, the DNA Misrepairs in germ cells contribute also to the diversity of genome DNAs.

Function and Teleonomy

Adaptation raises some issues concerning how biologists use key terms such as *function.*

Function

To say something has a function is to say something about what it does for the organism, obviously. It also says something about its history: how it has come about. A heart pumps blood: that is its function. It also emits sound, which is just an ancillary side-effect. That is not its function. The heart has a history (which may be well or poorly understood), and that history is about how natural selection formed and maintained the heart as a pump. Every aspect of an organism that has a function has a history. Now, an adaptation must have a functional history: therefore we expect it must have undergone selection caused by relative survival in its habitat. It would be quite wrong to use the word adaptation about a trait which arose as a by-product.

It is widely regarded as unprofessional for a biologist to say something like "A wing is for flying," although that is their normal function. A biologist would be conscious that sometime in the remote past feathers on a small dinosaur had the function of retaining heat, and that later many wings were not used for flying (e.g. penguins, ostriches). So, the biologist would rather say that the wings on a bird or an insect usually had the *function* of aiding flight. That would carry the connotation of being an adaptation with a history of evolution by natural selection.

Teleonomy

Teleonomy is a term invented to describe the study of goal-directed functions which are not guided by the conscious forethought of man or any supernatural entity. It is contrasted with Aristotle's teleology, which has connotations of intention, purpose and foresight. Evolution is teleonomic; *adaptation hoards hindsight rather than foresight*. The following is a definition for its use in biology:

> Teleonomy: The hypothesis that adaptations arise without the existence of a prior purpose, but by chance may change the fitness of an organism.

The term may have been suggested by Colin Pittendrigh in 1958; it grew out of cybernetics and self-organising systems. Ernst Mayr, George C. Williams and Jacques Monod picked up the term and used it in evolutionary biology.

Philosophers of science have discussed the concept. Ernest Nagel analysed goal-directedness in biology; and David Hull commented on the use of teleology in biology:

> ...Haldane can be found remarking, 'Teleology is like a mistress to a biologist: he cannot live without her but he's unwilling to be seen with her in public.' Today the mistress has become a lawfully wedded wife. Biologists no longer feel obligated to apologize for their use of teleological language; they flaunt it. The only concession which they make to its disreputable past is to rename it 'teleonomy'.

References

- Chiesa, Mecca (1994). Radical Behaviorism: The Philosophy and the Science. Authors Cooperative, Inc. pp. 1–241. ISBN 0962331147. Retrieved July 31, 2016.
- Cheney, Carl D.; Ferster, Charles B. (1997). Schedules of Reinforcement (B.F. Skinner Reprint Series). Acton, MA: Copley Publishing Group. p. 758. ISBN 0-87411-828-X.
- Skinner, Burrhus Frederick (1957). Verbal link=B.F. Skinner. Acton, Massachusetts: Copley Publishing Group. ISBN 1-58390-021-7.
- Skinner, B.F. (1969). "An operant analysis of problem-solving": 133–57.; chapter in Skinner, B.F. (1969). Contingencies of reinforcement: a theoretical analysis. Appleton-Century-Crofts. p. 283. ISBN 0-13-171728-6.

- Douglas Brown, H (2000). Principles of Language Learning and Teaching (Fourth ed.). White Plains: Longman/Pearson Education. pp. 8–9. ISBN 0-13-017816-0.
- Skinner, B.F. (1969). Contingencies of reinforcement: a theoretical analysis. Appleton-Century-Crofts. p. 283. ISBN 0-13-171728-6.
- Dennett, Daniel (1981). Brainstorms: Philosophical Essays on Mind and Psychology. Bradford Books. MIT Press. p. 53. ISBN 978-0-262-54037-7. LCCN 78013723.
- Cao, Longbing; Yu, Philip (eds) (2012). Behavior Computing: Modeling, Analysis, Mining and Decision. Springer. ISBN 978-1-4471-2969-1.
- King, Dennis & Green, Brian. 1999. Goannas: The Biology of Varanid Lizards. University of New South Wales Press. ISBN 0-86840-456-X, p. 43.
- Brett-Surman, Michael K.; Holtz, Thomas R.; Farlow, James O. (eds.). The complete dinosaur. Illustrated by Bob Walters (2nd ed.). Bloomington, Ind.: Indiana University Press. pp. 191–208. ISBN 978-0-253-00849-7.

3

Evolutionary Basis for Animal Behavior

Behavioral ecology is the study of the development in the change of animal behavior due to the ecological pressures. The features explained in this section are habituation, Lamarckism, natural selection and neuroethology. This chapter is an overview of the subject matter incorporating all the major aspects of the evolutionary basis for animal behavior.

Behavioral Ecology

Behavioral ecology, also spelled behavioural ecology, is the study of the evolutionary basis for animal behavior due to ecological pressures. Behavioral ecology emerged from ethology after Niko Tinbergen outlined four questions to address when studying animal behavior which are the proximate causes, ontogeny, survival value, and phylogeny of behavior.

If an organism has a trait which provides them with a selective advantage (i.e. has an adaptive significance) in its environment, then natural selection can potentially favor it. Adaptive significance therefore refers to the beneficial qualities (such as in terms of increased survival and reproduction), any given modified trait conveys. For example, genetic differences between individuals may lead to behavioral differences, some of which in turn may drive differences in reproductive success, and ultimately over generations, the increased dominance of individuals with those favoured traits, i.e. evolution.

Individuals are always in competition with others for limited resources, including food, territories, and mates. Conflict will occur between predators and prey, between rivals for mates, between siblings, mates, and even between parents and their offspring.

Competing for Resources

The value of a social behavior depends in part on the social behavior of an animal's neighbors. For example, the more likely a rival male is to back down from a threat, the more value a male gets out of making the threat. The more likely, however, that a rival will attack if threatened, the less useful it is to threaten other males. When a population exhibits a number of interacting social behaviors such as this, it can evolve a stable pattern of behaviors known as an evolutionarily stable strategy (or ESS). This term, derived from economic game theory, became prominent after John Maynard Smith (1982) recognized the possible application of the concept of a Nash equilibrium to model the evolution of behavioral strategies.

Evolutionarily Stable Strategy

In short, evolutionary game theory asserts that only strategies that, when common in the population, cannot be "invaded" by any alternative (mutant) strategy will be an ESS, and thus maintained in the population. In other words, at equilibrium every player should play the best strategic response to each other. When the game is two player and symmetric each player should play the strategy which is the best response to itself.

Therefore, the ESS is considered to be the evolutionary end point subsequent to the interactions. As the fitness conveyed by a strategy is influenced by what other individuals are doing (the relative frequency of each strategy in the population), behavior can be governed not only by optimality but the frequencies of strategies adopted by others and are therefore frequency dependent (frequency dependence).

Behavioral evolution is therefore influenced by both the physical environment and interactions between other individuals.

An example of how changes in geography can make a strategy susceptible to alternative strategies is the parasitization of the African honey bee, *A. m. scutellata*.

Resource Defense

The term economic defendability was first introduced by Jerram Brown in 1964. Economic defendability states that defense of a resource have costs, such as energy expenditure or risk of injury, as well as benefits of priority access to the resource. Territorial behavior arises when benefits are greater than the costs.

Studies of the golden-winged sunbird have validated the concept of economic defendability. Comparing the energetic costs a sunbird expends in a day to the extra nectar gained by defending a territory, researchers showed that birds only became territorial when they were making a net energetic profit. When resources are at low density, the gains from excluding others may not be sufficient to pay for the cost of territorial defense. In contrast, when resource availability is high, there may be so many intruders that the defender would have no time to make use of the resources made available by defense.

Sometimes the economics of resource competition favors shared defense. An example is the feeding territories of the white wagtail. The white wagtails feed on insects washed up by the river onto the bank, which acts as a renewing food supply. If any intruders harvested their territory then the prey would quickly become depleted, but sometimes territory owners tolerate a second bird, known as a satellite. The two sharers would then move out of phase with one another, resulting in decreased feeding rate but also increased defense, illustrating advantages of group living.

Ideal Free Distribution

One of the major models used to predict the distribution of competing individuals amongst resource patches is the ideal free distribution model. Within this model, resource patches can be of variable quality, and there is no limit to the number of individuals that can occupy and extract resources from a particular patch. Competition within a particular patch means that the benefit each individual receives from exploiting a patch decreases logarithmically with increasing number of competitors sharing that resource patch. The model predicts that individuals will initially flock to higher-quality patches until the costs of crowding bring the benefits of exploiting them in line with the benefits of being the only individual on the lesser-quality resource patch. After this point has been reached, individuals will alternate between exploiting the higher-quality patches and the lower-quality patches in such a way that the average benefit for all individuals in both patches is the same. This model is *ideal* in that individuals have complete information about the quality of a resource patch and the number of individuals currently exploiting it, and *free* in that individuals are freely able to choose which resource patch to exploit.

An experiment by Manfred Malinski in 1979 demonstrated that feeding behavior in three-spined sticklebacks follows an ideal free distribution. Six fish were placed in a tank, and food items were dropped into opposite ends of the tank at different rates. The rate of food deposition at one end was set at twice that of the other end, and the fish distributed themselves with four individuals at the faster-depositing end and two individuals at the slower-depositing end. In this way, the average feeding rate was the same for all of the fish in the tank.

Mating Strategies and Tactics

As with any competition of resources, species across the animal kingdom may also engage in competitions for mating. If one considers mates or potentials mates as a resource, these sexual partners can be randomly distributed amongst resource pools within a given environment. Following the ideal free distribution model, suitors will distribute themselves amongst the potential mates in an effort to maximize their chances or the number of potential mates that they can consummate. For all competitors, males of a species in most cases, there will be variations in both the strategies and tactics used to obtain matings. Strategies generally refer to the genetically determined behaviors which can be described as conditional. Tactics refer to the subset of behaviors within a given genetic strategy. Thus it is not difficult for a great many variations in mating strategies to exist in a given environment or species.

An experiment conducted by Anthony Arak, where playback of synthetic calls from male natterjack toads was used to manipulate behavior of the males in a chorus, the difference between strategies and tactics is clear. While small and immature, male natterjack toads adopted a satellite tactic to parasitize larger males. Though large males on average still retained greater reproductive success, smaller males were able to intercept matings. When the large males of the chorus were removed, smaller males adopted a calling behavior, no longer competing against the loud calls of larger males. When smaller males got larger and their calls more competitive, then they started calling and competing directly for mates.

Sexual Selection

Mate Choice by Resources

In many sexually reproducing species, such as mammals, birds and amphibians, the females are responsible for bearing the offspring for a certain period of time, during which the males are free to mate with other available females and therefore can father many more offspring, thus continue to pass on their genes. The fundamental difference between male and female reproduction mechanisms determines the different strategies each sex employs to maximize their reproductive success. For males, their reproductive success is limited by access to females, while females are limited by their access to resources. In this sense, females can be way choosier than males because they have to bet on the resources provided by the males to ensure reproductive success.

Resources usually include nest sites, food and protection. In some cases, the males provide all of them (e.g. sedge warblers). The females dwell in their chosen males' territories for access to these resources. The males gain ownership to the territories through male-male competition that often involves physical aggression. Only the largest and strongest males manage to defend the best quality nest sites. Females choose males by inspecting the quality of different territories or by looking at some male traits that can indicate the quality of resources. Sometimes, males leave after mating. The only resource that a male provides is a nuptial gift, such as protection or food. The female can evaluate the quality of the protection or food provided by the male so as to decide whether to mate or not or how long she is willing to copulate.

Mate Choice by Genes

When males' only contribution to offspring is their sperm, females are particularly choosy. With this high level of female choice, sexual ornaments are seen in males, where the ornaments reflect the male's social status. Two hypotheses have been proposed to conceptualize the genetic benefits from female mate choice.

First, the good genes hypothesis suggests that female choice is for higher genetic quality and that this preference is favored because it increases fitness of the offspring. This includes Zahavi's handicap hypothesis and Hamilton and Zuk's host and parasite arms race. Zahavi's handicap hypothesis was proposed within the context of looking at elaborate male sexual displays. He suggested that females favor ornamented traits because they are handicaps and are indicators of the male's genetic quality. Since these ornamented traits are hazards, the male's survival must be indicative of his high genetic quality in other areas. In this way, the degree that a male expresses his sexual display indicates to the female his genetic quality. Zuk and Hamilton proposed a hypothesis after observing disease as a powerful selective pressure on a rabbit population. They suggested that sexual displays were indicators of resistance of disease on a genetic level.

Such 'choosiness' from the female individuals can be seen in wasp species too, especially among *Polistes dominula* wasps. The females tend to prefer males with smaller, more elliptically shaped spots than those with larger and more irregularly shaped spots. Those males would have reproductive superiority over males with irregular spots.

Fisher's hypothesis of runaway sexual selection suggests that female preference is genetically correlated with male traits and that the preference co-evolves with the evolution of that trait, thus the preference is under indirect selection. Fisher suggests that female preference began because the trait indicated the male's quality. The female preference spread, so that the females' offspring now benefited from the higher quality from specific trait but also greater attractiveness to mates. Eventually, the trait will only represent attractiveness to mates and no longer represent increased survival.

An example of mate choice by genes is seen in the cichlid fish *Tropheus moorii* where males provide no parental care. An experiment found that a female *T. moorii* is more likely to choose a mate with the same color morph as her own. In another experiment, females have been shown to share preferences for the same males when given two to choose from, meaning some males get to reproduce more often than others.

Sensory Bias

The sensory bias hypothesis states that the preference for a trait evolves in a non-mating context and is then exploited by one sex in order to obtain more mating opportunities. The competitive sex evolves traits that exploit a pre-existing bias that the choosy sex already possesses. This mechanism is thought to explain remarkable trait differences in closely related species because it produces a divergence in signaling systems which leads to reproductive isolation.

Sensory bias has been demonstrated in guppies, freshwater fish from Trinidad and Tobago. In this mating system, female guppies prefer to mate with males with more orange body coloration. However, outside of a mating context, both sexes prefer animate orange objects which suggests that preference originally evolved in another context, like foraging. Orange fruits are a rare treat that fall into streams where the guppies live. The ability to find these fruits quickly is an adaptive quality that has evolved outside of a mating context. Sometime after the affinity for orange objects arose, male guppies exploited this preference by incorporating large orange spots to attract females.

Another example of sensory exploitation is in the water mite *Neumania papillator*, an ambush predator which hunts copepods (small crustaceans) passing by in the water column. When hunting, *N. papillator* adopts a characteristic stance termed the 'net stance' - their first four legs are held out into the water column, with their four hind legs resting on aquatic vegetation; this allows them to detect vibrational stimuli produced by swimming prey and use this to orient towards and clutch at prey. During courtship, males actively search for females - if a male finds a female, he slowly circles around the female whilst trembling his first and second leg near her. Male leg trembling causes females (who were in the 'net stance') to orient towards often clutch the male. This did not damage the male or deter further courtship; the male then deposited spermatophores and began to vigorously fan and jerk his fourth pair of legs over the spermatophore, generating a current of water that passed over the spermatophores and towards the female. Sperm packet uptake by the female would sometimes follow. Heather Proctor hypothesised that the vibrations trembling male legs made were done to mimic the vibrations that females detect from swimming prey - this would trigger the female prey-detection responses causing females to orient and then clutch at males, mediating courtship. If this was true and males were exploiting female predation responses, then hungry females should be more receptive to male trembling – Proctor found that unfed captive females did orient and clutch at males significantly more than fed captive females did, consistent with the sensory exploitation hypothesis.

Other examples for the sensory bias mechanism include traits in auklets, wolf spiders, and manakins. Further experimental work is required to reach a fuller understanding of the prevalence and mechanisms of sensory bias.

Sexual Conflict

Sexual conflict, in some form or another, may very well be inherent in the ways most animals reproduce. Females invest more in offspring prior to mating, due to the differences in gametes in species that exhibit anisogamy, and often invest more in offspring after mating. This unequal investment leads, on one hand, to intense competition between males for mates and, on the other hand, to females choosing among males for better access to resources and good genes. Because of differences in mating goals, males and females may have very different preferred outcomes to mating.

Sexual conflict occurs whenever the preferred outcome of mating is different for the male and female. This difference, in theory, should lead to each sex evolving adaptations that bias the outcome of reproduction towards its own interests. This sexual competition leads to sexually antagonistic coevolution between males and females, resulting in what has been described as an evolutionary arms race between males and females.

Conflict Over Mating

Males' reproductive successes are often limited by access to mates, whereas females' reproductive successes are more often limited by access to resources. Thus, for a given sexual encounter, it will benefit the male to mate but the female to be choosy and resist. For example, male small tortoiseshell butterfly will compete in order to gain the best territory to mate. Another example of this conflict can be found in the Eastern carpenter bee, *Xylocopa virginica*. Males of this species are limited in reproduction primarily by access to mates, so they will claim a territory and wait for a female to pass through. Big males are, therefore, more successful in mating because they claim territories near the female nesting sites that are more sought after. Smaller males, on the other hand, will monopolize less competitive sites in foraging areas so that they may mate with reduced conflict.

Male scorpionfly

Extreme manifestations of this conflict are seen throughout nature. For example, the male *Panorpa* scorpionflies attempt to force copulation. Male scorpionflies usually acquire mates by presenting them with edible nuptial gifts in the forms of salivary secretions or dead insects. However, some males attempt to force copulation by grabbing females with a specialized abdominal organ without offering a gift. Forced copulation is costly to the female as she does not receive the food from the male and has to search for food herself (costing time and energy), while it is beneficial for the male as he does not need to find a nuptial gift.

In other cases, however, it pays for the female to gain more matings and her social mate to prevent these so as to guard paternity. For example, in many socially monogamous birds, males will follow females closely during her fertile period and attempt to chase away any other males so as to prevent extra-pair matings. The female may attempt to sneak off to achieve these extra matings. In species where males are incapable of constant guarding, the social male will often frequently copulate with the female so as to swamp rival males' sperm.

Female red junglefowl in Thailand

Sexual conflict after mating has also been shown to occur in both males and females. Males employ a diverse array of tactics to increase their success in sperm competition. These can include removing other male's sperm from females, displacing other male's sperm by flushing out prior inseminations with large amounts of their own sperm, creating copulatory plugs in females' reproductive tracts to prevent future matings with other males, spraying females with anti-aphrodisiacs to discourage other males from mating with the female, and producing sterile parasperm to protect fertile eusperm in the female's reproductive tract. Furthermore, males may control the strategic allocation of sperm, producing more sperm when females are more promiscuous. All these methods are meant to ensure that females will be more likely to produce offspring belonging to the males who uses the method.

Females also control the outcomes of matings, and there exists the possibility that females choose sperm (cryptic female choice). A dramatic example of this is the feral fowl *Gallus gallus*. In this species, females prefer to copulate with dominant males, but subordinate males can force matings. In these cases, the female is able to eject the subordinate male's sperm using cloacal contractions.

Parental Care and Family Conflicts

Parental care is the investment a parent will put into their offspring, which includes protecting and feeding the young, preparing burrows or nests, and providing eggs with yolk. There is great variation in parental care in the animal kingdom. In some species, the parents may not care for their offspring at all, while in others the parents exhibit single-parental or even bi-parental care. As with other topics in behavioral ecology, interactions within a family will involve conflicts. These conflicts can be broken down into three general types: sexual (male-female) conflict, parent-offspring conflict, and sibling conflict.

Types of Parental Care

There are many different patterns of parental care in the animal kingdom. The patterns can be explained by physiological constraints or ecological conditions, such as mating opportunities. In invertebrates, there is no parental care in most species because it is more favorable for parents to produce a large number of eggs whose fate is left to chance than to protect a few individual young. For example, female *L. figueresi* die after stocking their larvae's cells with pollen and nectar and before their larvae hatch. In birds, biparental care is the most common, because reproductive success directly depends on the parents' ability to feed their chicks. Two parents can feed twice as many young, so it is more favorable for birds to have both parents delivering food. In mammals, female-only care is the most common. This is most likely because females are internally fertilized and so are holding the young inside for a prolonged period of gestation, which provides males with the opportunity to desert. Females also feed the young through lactation after birth, so males are not required for feeding. Male parental care is only observed in species where they contribute to feeding or carrying of the young, such as in marmosets. In fish there is no parental care in 79% of bony fish. In fish with parental care, it usually limited to selecting, preparing, and defending a nest, as seen in sockeye salmon, for example. Also, parental care in fish, if any, is primarily done by males, as seen in gobies and redlip blennies. The cichlid fish *V. moorii* exhibits biparental care. In species with internal fertilization, the female is usually the one to take care of the young. In cases where fertilization is external the male becomes the main caretaker.

Familial Conflict

Familial conflict is a result of trade-offs as a function of lifetime parental investment. Parental investment was defined by Robert Trivers in 1972 as "any investment by the parent in an individual offspring that increases the offspring's chance of surviving at the cost of the parent's ability to invest in other offspring". Parental investment includes behaviors like guarding and feeding. Each parent has a limited amount of parental investment over the course of their lifetime. Investment trade-offs in offspring quality and quantity within a brood and trade offs between current and future broods leads to conflict over how much parental investment to provide and to whom parents should invest in. There are three major types of familial conflict: sexual, parent-offspring, and sibling-sibling conflict.

Sexual Conflict

There is conflict among parents as to who should provide the care as well as how much care to provide. Each parent must decide whether or not to stay and care for their offspring, or to desert their offspring. This decision is best modeled by game theoretic approaches to evolutionarily stable strategies (ESS) where the best strategy for one parent depends on the strategy adopted by the other parent. Recent research has found response matching in parents who determine how much care to invest in their offspring. Studies found that parent great tits will match their partner's increased care-giving efforts with increased provisioning rates of their own. This cued parental response is a type of behavioral negotiation between parents that leads to stabilized compensation.

Great tit

Parent-offspring Conflict

Blackbird chicks in a nest

According to Robert Trivers's theory on relatedness, each offspring is related to itself by 1, but is only 0.5 related to their parents and siblings. Genetically, offspring are predisposed to behave in their own self-interest while parents are predisposed to behave equally to all their offspring, including both current and future ones. Offspring will selfishly attempt to take more than their fair shares of parental investment while parents will attempt to spread out their parental investment equally amongst their present young and future young. There are many examples of parent-offspring conflict in nature. One manifestation of this is asynchronous hatching in birds. A behavioral ecology hypothesis is known as Lack's brood reduction hypothesis (named after David Lack). Lack's hypothesis posits an evolutionary and ecological explanation as to why birds lay a series of eggs with an asynchronous delay leading to nestlings of mixed age and weights. According to Lack, this brood behavior is an ecological insurance that allows the larger birds to survive in poor years and all birds to survive when food is plentiful. We also see sex-ratio conflict between the queen and her workers in social hymenoptera. Because of haplodiploidy, the workers (offspring) prefer a 3:1 female to male sex allocation while the queen prefers a 1:1 sex ratio. Both the queen and the workers will attempt to bias the sex ratio in their favor. In some species, the workers gain control of the sex ratio, while in other species, like *B. terrestris*, the queen has a considerable amount of control over the colony sex ratio. Lastly, there has been recent evidence regarding genomic imprinting that is a result of parent-offspring conflict. Paternal genes in offspring will demand more maternal resources than maternal genes in the same offspring and vice versa. This has been show in imprinted genes like insulin-like growth factor-II.

Parent-offspring Conflict Resolution

Parents need an honest signal from their offspring indicating their level of hunger or need, so that the parents can distribute resources accordingly. Offspring want to get more than their fair share of resources, so they will want to exaggerate their signals to wheedle more investment from their parents. However, this conflict is resolved by the cost of excessive begging. Not only does excessive begging attract predators, but it also retards chick growth if begging goes unrewarded. Thus, the cost of increased begging will enforce offspring honesty.

Another resolution for parent-offspring conflict is that parental provisioning and offspring demand have actually coevolved, so that there is no obvious underlying conflict. Cross-fostering experiments in great tits (*Parus major*) have shown that offspring beg more when their biological mothers are more generous. Therefore, it seems that the willingness to invest in offspring is co-adapted to offspring demand.

Sibling-sibling Conflict

The lifetime parental investment is the fixed amount of parental resources available for all of a parent's young, and an offspring will want as much of it as possible. Siblings in a brood will often compete for parental resources by trying to gain more than their fair share of what their parents have to offer. There are numerous examples in nature where sibling rivalry is escalated to such an extreme that one sibling will try to kill off his other broodmates in order to maximize his own parental investment. In the Galapagos fur seal, the second pup of a female is usually born when the first pup is still suckling. This competition for the mother's milk is especially fierce during periods of food shortage such as an El Niño year, and this usually results in the older pup directly attacking and killing the younger one.

Galápagos fur seals

In some bird species, sibling rivalry is also abetted by the asynchronous hatching of eggs. In the blue-footed booby, for example, the first egg in a nest is hatched four days before the second one, resulting in the elder chick having a four-day head start in growth. When the elder chick falls 20-25% below its expected weight threshold, it will attack its younger sibling and drive it from the nest.

Sibling relatedness in a brood also influences the level of sibling-sibling conflict. In a study on passerine birds, it was found that chicks begged more loudly in species with higher levels of extra-pair paternity.

Brood Parasitism

Adult reed warbler feeding a common cuckoo chick

Some animals deceive other species into providing all parental care. These brood parasites selfishly exploit their hosts' parents and host offspring. The common cuckoo is a well known example of a brood parasite. Female cuckoos lay a single egg in the nest of the host species and when the cuckoo chick hatches, it ejects all the host eggs and young. Other examples of brood parasites include honeyguides, cowbirds, and the large blue butterfly. Brood parasite offspring have many strategies to induce their host parents to invest parental care. Studies show that the common cuckoo uses vocal mimicry to reproduce the sound of multiple hungry host young to solicit more food. Other cuckoos will use visual deception with their wings to exaggerate the begging display. False gapes from brood parasite offspring cause host parents to collect more food. Another example of a brood parasite is *Phengaris* butterflies such as *Phengaris rebeli* and *Phengaris arion*, which differ from the cuckoo in that the butterflies do not oviposit directly in the nest of the host, an ant species *Myrmica schencki*. Rather, the butterfly larvae release chemicals that deceive the ants into believing that they are ant larvae, causing the ants to bring the butterfly larvae back to their own nests to feed them. Other examples of brood parasites are *Polistes sulcifer*, a paper wasp that has lost the ability to build its own nests so females lay their eggs in the nest of a host species, *Polistes dominula*, and rely on the host workers to take care of their brood, as well as *Bombus bohemicus*, a bumblebee that relies on host workers of various other *Bombus* species. Similarly, in *Eulaema meriana*, some Leucospidae wasps exploit the brood cells and nest for shelter and food from the bees. *Vespula austriaca* is another wasp in which the females force the host workers to feed and take care of the brood. In particular, *Bombus hyperboreus*, an Arctic bee species, is also classified as a brood parasite in that it attacks and enslaves other species within their subgenus, *Alpinobombus* to propagate their population.

Mating Systems

Various types of mating systems include monogamy, polygyny, polyandry, promiscuity, and polygamy. Each is differentiated by the sexual behavior between mates, such as which males mate with certain females. An influential paper by Stephen Emlen and Lewis Oring (1977) argued that two main factors of animal behavior influence the diversity of mating systems: the relative accessibility that each sex has to mates, and the parental desertion by either sex.

Mating Systems with No Male Parental Care

In a system that does not have male parental care, resource Dispersion, predation, and the effects of social living primarily influence female dispersion, which in turn influences male dispersion. Since males' primary concern is female acquisition, the males will either indirectly or directly compete for the females. In direct competition, the males are directly focused on the females. Blue-headed wrasse demonstrate the behavior in which females follow resources—such as good nest sites—and males follow the females. Conversely, species with males that exemplify indirectly competitive behavior tend towards the males' anticipation of the resources desired by females and their subsequent effort to control or acquire these resources, which helps them to achieve success with females. Grey-sided voles demonstrate indirect male competition for females. The males were experimentally observed to home in on the sites with the best food in anticipation of females settling in these areas. Males of *Euglossa imperialis*, a non-social bee species, also demonstrate indirect competitive behavior by forming aggregations of territories, which can be considered leks, to defend fragrant-rich primary territories. The purpose of these aggregations is largely only facultative, since the more suitable fragrant-rich sites there are, the more habitable territories there are to inhabit, giving females of this species a large selection of males with whom to potentially mate. Leks and choruses have also been deemed another behavior among the phenomena of male competition for females. Due to the resource-poor nature of the territories that lekking males often defend, it is difficult to categorize them as indirect competitors. Additionally, it is difficult to classify them as direct competitors seeing as they put a great deal of effort into their defense of their territories before females arrive, and upon female arrival they put for the great mating displays to attract the females to their individual sites. These observations make it difficult to determine whether female or resource dispersion primarily influences male aggregation, especially in lieu of the apparent difficulty that males may have defending resources and females in such densely populated areas. Because the reason for male aggregation into leks is unclear, five hypothesis have been proposed. These postulates propose the following as reasons for male lekking: hotspot, predation reduction, increased female attraction, hotshot males, facilitation of female choice. With all of the mating behaviors discussed, the primary factors influencing differences within and between species are ecology, social conflicts, and life history differences.

In some other instances, neither direct nor indirect competition is seen. Instead, in species like the Edith's checkerspot butterfly, males' efforts are directed at acquisition of females and they exhibit indiscriminate mate location behavior, where, given the low cost of mistakes, they will blindly attempt to mate both correctly with females and incorrectly with other objects.

Mating Systems with Male Parental Care

Monogamy

Monogamy is the mating system in 90% of birds, possibly because each male and female will have a greater number of offspring if they share in raising a brood. In obligate monogamy, males feed females on the nest, or share in incubation and chick-feeding. In some species, males and females form lifelong pair bonds. Monogamy may also arise from limited opportunities for polygamy, due to strong competition among males for mates, females suffering from loss of male help, and female-female aggression.

Polygyny

In birds, polygyny occurs when males indirectly monopolize females by controlling resources. In species where males normally do not contribute much to parental care, females suffer relatively little or not at all. In other species, however, females suffer through the loss of male contribution, and the cost of having to share resources that the male controls, such as nest sites or food. In some cases, a polygynous male may control a high-quality territory so for the female, the benefits of polygyny may outweigh the costs.

Polyandry Threshold

There also seems to be a "polyandry threshold" where males may do better by agreeing to share a female instead of attempting to be in a monogamous mating system. Situations that may lead to cooperation among males include when food is scarce, and when there is intense competition for territories or females. For example, male lions sometimes form coalitions to gain control of a pride of females. In some populations of Galapagos hawks, groups of males would cooperate to defend one breeding territory. The males would share matings with the female and share paternity with the offspring.

Female Desertion and Sex Role Reversal

In birds, desertion often happens when food is abundant, so the remaining partner is better able to raise the young unaided. Desertion also occurs if there is a great chance of a parent to gain another mate, which depends on environmental and populational factors. Some birds, such as the phalaropes, have reversed sex roles where the female is larger and more brightly colored, and compete for males to incubate their clutches. In jacanas, the female is larger than the male and her territory could overlap the multiple territories of up to four males.

Social Behaviors

Animals cooperate with each other in order to increase their own fitness. These altruistic, and sometimes spiteful behaviors can be explained by Hamilton's rule, which states that $rB\text{-}C > 0$ where r= relatedness, B= benefits, and C= costs.

Kin Selection

Kin selection refers to evolutionary strategies where an individual acts to favor the reproductive success of relatives, or kin, even if the action incurs some cost to the organism's own survival and ability to procreate. John Maynard Smith coined the term in 1964, although the concept was referred to by Charles Darwin who cited that helping relatives would be favored by group selection. Mathematical descriptions of kin selection were initially offered by R. A. Fisher in 1930 and J. B. S. Haldane in 1932. and 1955. W. D. Hamilton popularized the concept later, including the mathematical treatment by George Price in 1963 and 1964.

Kin selection predicts that individuals will harbor personal costs in favor of one or multiple individuals because this can maximize their genetic contribution to future generations. For example, an organism may be inclined to expend great time and energy in parental investment to rear offspring since this future generation may be better suited for propagating genes that are highly shared between the parent and offspring. Ultimately, the initial actor performs apparent altruistic actions for kin in order to enhance its own reproductive fitness. In particular, organisms are hypothesized to act in favor of kin depending on their genetic relatedness. So, individuals will be inclined to act altruistically for siblings, grandparents, cousins and other relatives, but to differing degrees.

Inclusive Fitness

Inclusive fitness describes the component of reproductive success in both a focal individual and their relatives. Importantly, the measure embodies the sum of direct and indirect fitness and the change in their reproductive success based on the actor's behavior. That is, the effect an individual's behaviors have on: being personally better-suited to reproduce offspring, and aiding descendent and non-descendent relatives in their reproductive efforts. Natural selection is predicted to push individuals to behave in ways that maximize their inclusive fitness. Studying inclusive fitness is often done using predictions from Hamilton's rule.

Kin Recognition

Genetic cues

One possible method of kin selection is based on genetic cues that can be recognized phenotypically. Genetic recognition has been exemplified in a species that is usually not thought of as a social creature: amoebae. Social amoebae form fruiting bodies when starved for food. These amoebae preferentially formed slugs and fruiting bodies with members of their own lineage, which is clonially related. The genetic cue comes from variable lag genes, which are involved in signaling and adhesion between cells.

Kin can also be recognized a genetically determined odor, as studied in the primitively social sweat bee, *Lasioglossum zephyrum*. These bees can even recognize relatives they have never met and roughly determine relatedness. The Brazilian stingless bee *Schwarziana quadripunctata* uses a distinct combination of chemical hydrocarbons to recognize and locate kin. Each chemical odor, emitted from the organism's epicuticles, is unique and varies according to age, sex, location, and hierarchical position. Similarly, individuals of the stingless bee species *Trigona fulviventris* can distinguish kin from non-kin through recognition of a number of compounds, including hydrocarbons and fatty acids that are present in their wax and floral oils from plants used to construct their nests. In the species, *Osmia rufa,* kin selection has also been associated with mating selection. Females, specifically, will select males for mating with whom they are genetically more related to.

Environmental Cues

There are two simple rules that animals follow to determine who is kin. These rules can be exploited, but exist because they are generally successful.

The first rule is 'treat anyone in my home as kin.' This rule is readily seen in the reed warbler, a bird species that will only focus on chicks in their own nest. Interestingly, if its own kin is placed outside of the nest, a parent bird will ignore that chick. This rule can sometimes lead to odd results, especially if there is a parasitic bird that lays eggs in the reed warbler nest. For example, an adult cuckoo may sneak its egg into the nest. Once the cuckoo hatches, the reed warbler parent will feed the invading bird like its own child. Even with the risk for exploitation, the rule generally proves successful.

The second rule, named by Konrad Lorenz as 'imprinting,' states that those who you grow up with are kin. Several species exhibit this behavior, including, but not limited to the Belding's ground squirrel. Experimentation with these squirrels showed that regardless of true genetic relatedness, those that were reared together rarely fought. Further research suggests that there is partially some genetic recognition going on as well, as siblings that were raised apart were less aggressive toward one another compared to non-relatives reared apart.

Cooperation

Cooperation is broadly defined as behavior that provides a benefit to another individual that specifically evolved for that benefit. This excludes behavior that has not been expressly selected for to provide a benefit for another individual, because there are many commensal and parasitic relationships where the behavior one individual (which has evolved to benefit that individual and no others) is taken advantage of by other organisms. For cooperative behavior to be stable, it must provide a benefit to both the actor and recipient, although the benefit to the actor can take many different forms.

Within Species

Within species cooperation occurs among members of the same species. Examples of intraspecific cooperation include cooperative breeding (such as in weeper capuchins) and cooperative foraging (such as in wolves). There are also forms of cooperative defense mechanisms, such as the "fighting swarm" behavior used by the stingless bee *Tetragonula carbonaria*. Much of this behavior occurs due to kin selection. Kin selection allows cooperative behavior to evolve where the actor receives no direct benefits from the cooperation.

Cooperation (without kin selection) must evolve to provide benefits to both the actor and recipient of the behavior. This includes reciprocity, where the recipient of the cooperative behavior repays the actor at a later time. This may occur in vampire bats but it is uncommon in non-human animals. Cooperation can occur willingly between individuals when both benefit directly as well. Cooperative breeding, where one individual cares for the offspring of another, occurs in several species, including wedge-capped capuchin monkeys.

Cooperative behavior may also be enforced, where there failure to cooperate results in negative consequences. One of the best examples of this is worker policing, which occurs in social insect colonies.

Between Species

Cooperation can occur between members of different species. For interspecific cooperation to be evolutionarily stable, it must benefit individuals in both species. Examples include pistol shrimp and goby fish, nitrogen fixing microbes and legumes, ants and aphids. In ants and aphids, aphids secrete a sugary liquid called honeydew, which that ants eat. The ants provide protection to the aphids against predators, and, in some instances, raise the aphid eggs and larvae inside the ant colony. This behavior is analogous to human domestication. The genus of goby fish, *Elacatinus* also demonstrate cooperation by removing and feeding on ectoparasites of their clients. The species of wasp *Polybia rejecta* and ants *Azteca chartifex* show a cooperative behavior protecting one another's nests from predators.

Spite

Hamilton's rule can also predict spiteful behaviors between non-relatives. A spiteful behavior is one that is harmful to both the actor and to the recipient. Spiteful behavior will be favored if the actor is less related to the recipient than to the average member of the population making r negative and if rB-C is still greater than zero. Spite can also be thought of as a type of altruism because harming a non-relative, by taking his resources for example, could also benefit a relative, by allowing him access to those resources. Furthermore, certain spiteful behaviors may provide harmful short term consequences to the actor but also give long term reproductive benefits. Many behaviors that are commonly thought of as spiteful are actually better explained as being selfish, that is benefiting the actor and harming the recipient, and true spiteful behaviors are rare in the animal kingdom.

An example of spite is the sterile soldiers of the polyembryonic parasitoid wasp. A female wasp will lay a male and a female egg in a caterpillar. The eggs will divide asexually creating many genetically identical male and female larvae. Sterile soldier wasps will also develop and attack the relatively unrelated brother larvae so that the genetically identical sisters will have more access to food.

Another example is bacteria that release bacteriocins. The bacteria that releases the bacteriocin may have to die in order to do so; however most of the harm will be done to unrelated individuals who will be killed by the bacteriocin. This is because the ability to produce and release the bacteriocin is linked with immunity to it. Therefore, close relatives to the releasing cell will be less likely to die than non-relatives.

Altruism and Conflict in Social Insects

Many insect species of the order Hymenoptera (bees, ants, wasps) are eusocial. Within the nests or hives of social insects, individuals engage in specialized tasks to ensure the survival of the colony. Dramatic examples of these specializations include changes in body morphology or unique behaviors, such as the engorged bodies of the honeypot ant *Myrmecocystus mexicanus* or the waggle dance of honey bees and a wasp species, *Vespula vulgaris*.

Honeypot ant

In many, but not all social insects, reproduction is monopolized by the queen of the colony. Due to the effects of a haplodiploid mating system, in which unfertilized eggs become male drones and fertilized eggs become worker females, average relatedness values between sister workers can be higher than those seen in humans or other eutherian mammals. This has led to the suggestion that kin selection may be a driving force in the evolution of eusociality, as individuals could provide cooperative care that establishes a favorable benefit to cost ratio ($rB-c > 0$). However, not all social insects follow this rule. In the social wasp *Polistes dominula*, 35% of the nest mates are unrelated. In many other species, unrelated individuals only help the queen when no other options are present. In this case, subordinates work for unrelated queens even when other options may be present. No other social insect submits to unrelated queens in this way. This seemingly unfavorable behavior parallels some vertebrate systems. It is thought that this unrelated assistance is evidence of altruism in *P. dominula*.

Cooperation in social organisms has numerous ecological factors that can determine the benefits and costs associated with this form of organization. One suggested benefit is a type of "life insurance" for individuals who participate in the care of the young. In this instance, individuals may have a greater likelihood of transmitting genes to the next generation when helping in a group compared to individual reproduction. Another suggested benefit is the possibility of "fortress defense", where soldier castes will threaten or attack intruders, thus protecting related individuals inside the territory. Such behaviors are seen in the snapping shrimp *Synalpheus regalis* and gall-forming aphid *Pemphigus spyrothecae*. A third ecological factor that is posited to promote eusociality is the distribution of resources: when food is sparse and concentrated in patches, eusociality is favored. Evidence supporting this third factor comes from studies of naked mole-rats and Damaraland mole-rats, which have communities containing a single pair of reproductive individuals.

Naked mole-rats

Conflicts in Social Insects

Although eusociality has been shown to offer many benefits to the colony, there is also potential for conflict. Examples include the sex-ratio conflict and worker policing seen in certain species of social Hymenoptera such as *Dolichovespula media, Dolichovespula sylvestris, Dolichovespula norwegica* and *Vespula vulgaris*. The queen and the worker wasps either indirectly kill the laying-workers' offspring by neglecting them or directly condemn them by cannibalizing and scavenging.

The sex-ratio conflict arises from a relatedness asymmetry, which is caused by the haplodiploidy nature of Hymenoptera. For instance, workers are most related to each other because they share half of the genes from the queen and inherit all of the father's genes. Their total relatedness to each other would be 0.5+ (0.5 x 0.5) = 0.75. Thus, sisters are three-fourths related to each other. On the other hand, males arise from unfertilized larva, meaning they only inherit half of the queen's genes and none from the father. As a result, a female is related to her brother by 0.25, because 50% of her genes that come from her father have no chance of being shared with a brother. Her relatedness to her brother would therefore be 0.5 x 0.5=0.25.

According to Trivers and Hare's population-level sex-investment ratio theory, the ratio of relatedness between sexes determines the sex investment ratios. As a result, it has been observed that there is a tug-of-war between the queen and the workers, where the queen would prefer a 1:1 female to male ratio because she is equally related to her sons and daughters (r=0.5 in each case). However, the workers would prefer a 3:1 female to male ratio because they are 0.75 related to each other and only 0.25 related to their brothers. Allozyme data of a colony may indicate who wins this conflict.

Conflict can also arise between workers in colonies of social insects. In some species, worker females retain their ability to mate and lay eggs. The colony's queen will be related to her sons by half of her genes and a quarter to the sons of her worker daughters. Workers, however, are related to their sons by half of their genes and to their brothers by a quarter. Thus, the queen and her worker daughters would compete for reproduction to maximize their own reproductive fitness. Worker reproduction is limited by other workers who are more related to the queen than their sisters, a situation occurring in many polyandrous hymenopteran species. Workers police the egg-laying females by engaging in oophagy or directed acts of aggression.

The Monogamy Hypothesis

The monogamy hypothesis states that the presence of monogamy in insects is crucial for eusociality to occur. This is thought to be true because of Hamilton's rule that states that rB-C>0. By having a monogamous mating system, all of the offspring have high relatedness to each other. This means that it is equally beneficial to help out a sibling, as it is to help out an offspring. If there were many fathers the relatedness of the colony would be lowered.

This monogamous mating system has been observed in insects such as termites, ants, bees and wasps. In termites the queen commits to a single male when founding a nest. In ants, bees and wasps the queens have a functional equivalent to lifetime monogamy. The male can even die before the founding of the colony. The queen can store and use the sperm from a single male throughout their lifetime, sometimes up to 30 years.

In an experiment looking at the mating of 267 hymenopteran species, the results were mapped onto a phylogeny. It was found that monogamy was the ancestral state in all the independent transitions to eusociality. This indicates that monogamy is the ancestral, likely to be crucial state for the development of eusociality. In species where queens mated with multiple mates, it was found that these were developed from lineages where sterile castes already evolved, so the multiple mating was secondary. In these cases, multiple mating is likely to be advantageous for reasons other than those important at the origin of eusociality. Most likely reasons are that a diverse worker pool attained by multiple mating by the queen increases disease resistance and may facilitate a division of labor among workers

Communication and Signalling

Communication is varied at all scales of life, from interactions between microscopic organisms to those of large groups of people. Nevertheless, the signals used in communication abide by a fundamental property: they must be a quality of the receiver that can transfer information to a receiver that is capable of interpreting the signal and modifying its behavior accordingly. Signals are distinct from cues in that evolution has selected for signalling between both parties, whereas cues are merely informative to the observer and may not have originally been used for the intended purpose. The natural world is replete with examples of signals, from the luminescent flashes of light from fireflies, to chemical signaling in red harvester ants to prominent mating displays of birds such as the Guianan cock-of-the-rock, which gather in leks, the pheromones released by the corn earworm moth, the dancing patterns of the blue-footed booby, or the alarm sound *Synoeca cyanea* make by rubbing their mandibles against their nest.

The nature of communication poses evolutionary concerns, such as the potential for deceit or manipulation on the part of the sender. In this situation, the receiver must be able to anticipate the interests of the sender and act appropriately to a given signal. Should any side gain advantage in the short term, evolution would select against the signal or the response. The conflict of interests between the sender and the receiver results in an evolutionarily stable state only if both sides can derive an overall benefit.

Although the potential benefits of deceit could be great in terms of mating success, there are several possibilities for how dishonesty is controlled, which include indices, handicaps, and common interests. Indices are reliable indicators of a desirable quality, such as overall health, fertility, or fighting ability of the organism. Handicaps, as the term suggests, place a restrictive cost on the organisms that own them, and thus lower quality competitors experience a greater relative cost compared to their higher quality counterparts. In the common interest situation, it is beneficial to both sender and receiver to communicate honestly such that the benefit of the interaction is maximized.

Signals are often honest, but there are exceptions. Prime examples of dishonest signals include the luminescent lure of the anglerfish, which is used to attract prey, or the mimicry of non-poisonous butterfly species, like the Batesian mimic *Papilio polyxenes* of the poisonous model *Battus philenor*. Although evolution should normally favor selection against the dishonest signal, in these cases it appears that the receiver would benefit more on average by accepting the signal.

Altruism (Biology)

In biology, altruism refers to behaviour by an individual that increases the fitness of another individual while decreasing the fitness of the actor. Altruism in this sense is different from the philosophical concept of altruism, in which an action would only be called "altruistic" if it was done with the conscious intention of helping another. In the behavioural sense, there is no such requirement. As such, it is not evaluated in moral terms—it is the consequences of an action for reproductive fitness that determine whether the action is considered altruistic, not the intentions, if any, with which the action is performed.

The term altruism was coined by the French philosopher Auguste Comte in French, as *altruisme*, for an antonym of egoism. He derived it from an Italian *altrui*, which in turn was derived from Latin *alteri*, meaning "other people" or "somebody else".

Altruistic behaviours appear most obviously in kin relationships, such as in parenting, but may also be evident among wider social groups, such as in social insects. They allow an individual to increase the success of its genes by helping relatives that share those genes. Obligate altruism is the permanent loss of direct fitness (with potential for indirect fitness gain). For example, honey bee workers may forage for the colony. Facultative altruism is temporary loss of direct fitness (with potential for indirect fitness gain followed by personal reproduction) example: Florida scrub jay helping at the nest, then gaining parental territory.

Overview

In the science of ethology (the study of behavior), and more generally in the study of social evolution, on occasion, some animals do behave in ways that reduce their individual fitness but increase the fitness of other individuals in the population; this is a functional definition of altruism. Research in evolutionary theory has been applied to social behaviour, including altruism. Cases of animals helping individuals to whom they are closely related can be explained by kin selection, and are not considered true altruism. Beyond the physical exertions that in some species mothers and in some species fathers undertake to protect their young, extreme examples of sacrifice may occur. One example is matriphagy (the consumption of the mother by her offspring) in the spider *Stegodyphus*; another example is a male spider allowing a female fertilized by him to eat him. Hamilton's rule describes the benefit of such altruism in terms of Wright's coefficient of relationship to the beneficiary and the benefit granted to the beneficiary minus the cost to the sacrificer. Should this sum be greater than zero a fitness gain will result from the sacrifice.

When apparent altruism is not between kin, it may be based on reciprocity. A monkey will present its back to another monkey, who will pick out parasites; after a time the roles will be reversed. Such reciprocity will pay off, in evolutionary terms, as long as the costs of helping are less than the benefits of being helped and as long as animals will not gain in the long run by "cheating" – that is to say, by receiving favours without returning them. This is elaborated on in evolutionary game theory and specifically the prisoner's dilemma as social theory.

Altruism is an evolutionary enigma, not because it cannot arise by mutation (as all new features in biology ultimately arise through mutation), but because it is evolutionarily unstable. It is even possible for altruism to spread to all the members of a population, through a variety of mechanisms (e.g. founder effects or population bottlenecks). However, should a mutation for selfishness arise, the carriers of that altered gene will almost always be fitter than the rest of the population, resulting, ultimately, in the replacement of altruism by selfishness. This is the conundrum that has exercised biologists since Darwin initially recognized the problem.

Implications in Evolutionary Theory

Cooperative hunting by wolves allows them to tackle much larger and more nutritious prey than any individual wolf could handle. However, such cooperation could, potentially, be exploited by selfish individuals who do not expose themselves to the dangers of the hunt, but nevertheless share in the spoils.

The existence of altruism in nature is at first sight puzzling, because altruistic behaviour reduces the likelihood that an individual will reproduce. The idea that *group selection* might explain the evolution of altruism was first broached by Darwin himself in The Descent of Man, and Selection in Relation to Sex, (1871). The concept of group selection has had a chequered and controversial history in evolutionary biology but the uncritical 'good of the species' tradition came to an abrupt halt in the 1960s, due largely to the work of George C. Williams, and John Maynard Smith as well as Richard Dawkins. These evolutionary theorists pointed out that natural selection acts on the individual, and that it is the individual's fitness (number of offspring and grand-offspring produced compared to the rest of the population) that drives evolution. A group advantage (e.g. hunting in a pack) that is disadvantageous to the individual (who might be harmed during the hunt, when it could avoid injury by hanging back from the pack but still share in the spoils) cannot evolve, because the selfish individual will leave, on average, more offspring than those who join the pack and suffer injuries as a result. If the selfishness is hereditary, this will ultimately result in the population consisting entirely of selfish individuals. However, in the 1960s and 1970s an alternative to the "group selection" theory emerged. This was the kin selection theory, due originally to W. D. Hamilton. Kin selection is an instance of inclusive fitness, which is based on the notion that an individual shares only half its genes with each offspring, but also with each full-sib. From an evolutionary genetic point of view it is therefore as advantageous to help with the upbringing of full sibs as it is to produce and raise one's own offspring. The two activities are evolutionarily entirely equivalent. Co-operative breeding (i.e. helping one's parents raise sibs—provided they are full sibs) could thus evolve without the need for group-level selection. This quickly gained prominence among biologists interested in the evolution of social behaviour.

Olive baboons grooming

In 1971 Robert Trivers introduced his reciprocal altruism theory to explain the evolution of helping at the nest of an unrelated breeding pair of birds. He argued that an individual might act as a helper if there was a high probabilistic expectation of being helped by the recipients at some later date. If, however, the recipients did not reciprocate when it was possible to do so, the altruistic interaction with these recipients would be permanently terminated. But if the recipients did not cheat then the reciprocal altruism would continue indefinitely to both parties' advantage. This model was considered by many (e.g. West-Eberhard and Dawkins) to be evolutionarily unstable because it is prone to invasion by cheats for the same reason that cooperative hunting can be invaded and replaced by cheats. However, Trivers did make reference to the Prisoner's Dilemma Game which, 10 years later, would restore interest in Trivers' reciprocal altruism theory, but under the title of "tit-for-tat".

In its original form the Prisoner's Dilemma Game (PDG) described two awaiting trial prisoners, A and B, each faced with the choice of betraying the other or remaining silent. The "game" has four possible outcomes: (a) they both betray each other, and are both sentenced to two years in prison; (b) A betrays B, which sets A free and B is sentenced to four years in prison; (c) B betrays A, with the same result as (b) except that it is B who is set free and the other spends four years in jail; (d) both remain silent, resulting in a six-month sentence each. Clearly (d) ("cooperation") is the best mutual strategy, but from the point of view of the individual betrayal is unbeatable (resulting in being set free, or getting only a two-year sentence). Remaining silent results in a four-year or six-month sentence. This is exemplified by a further example of the PDG: two strangers attend a restaurant together and decide to split the bill. The mutually best ploy would be for both parties to order the cheapest items on the menu (mutual cooperation). But if one member of the party exploits the situation by ordering the most expensive items, then it is best for the other member to do likewise. In fact, if the fellow diner's personality is completely unknown, and the two diners are unlikely ever to meet again, it is always in one's own best interests to eat as expensively as possible. Situations in nature that are subject to the same dynamics (rewards and penalties) as the PDG define cooperative behaviour: it is never in the individual's fitness interests to cooperate, even though mutual cooperation rewards the two contestants (together) more highly than any other strategy. Cooperation cannot evolve under these circumstances.

However, in 1981 Axelrod and Hamilton noted that if the same contestants in the PDG meet repeatedly (the so-called Iterated Prisoner's Dilemma game, IPD) then tit-for-tat (foreshadowed by Robert Triver's reciprocal altruism theory) is a robust strategy which promotes altruism. In "tit-for-tat" both players' opening moves are cooperation. Thereafter each contestant repeats the other player's last move, resulting in a seemingly endless sequence of mutually cooperative moves. However, mistakes severely undermine tit-for-tat's effectiveness, giving rise to prolonged sequences of betrayal, which can only be rectified by another mistake. Since these initial discoveries, all the other possible IPD game strategies have been identified (16 possibilities in all, including, for instance, "generous tit-for-tat", which behaves like "tit-for-tat", except that it cooperates with a small probability when the opponent's last move was "betray".), but all can be outperformed by at least one of the other strategies, should one of the players switch to such a strategy. The result is that none is evolutionarily stable, and any prolonged series of the iterated prisoner's dilemma game, in which alternative strategies arise at random, gives rise to a chaotic sequence of strategy changes that never ends.

The Handicap Principle

A male peacock with its beautiful but clumsy, aerodynamically unsound tail—a handicap, comparable to a race horse's handicap.

The best horses in a handicap race carry the largest weights, so the size of the handicap is a measure of the animal's quality

In the light of the Iterated Prisoner's Dilemma Game failing to provide a full answer to the evolution of cooperation or altruism, several alternative explanations have been proposed.

There are striking parallels between altruistic acts and exaggerated sexual ornaments displayed by some animals, particularly certain bird species, such as, amongst others, the peacock. Both are costly in fitness terms, and both are generally conspicuous to other members of the population or species. This led Amotz Zahavi to suggest that both might be fitness signals rendered evolutionarily stable by his handicap principle. If a signal is to remain reliable, and generally resistant to falsification, the signal has to be evolutionarily costly. Thus, if a (low fitness) liar were to use the highly costly signal, which seriously eroded its real fitness, it would find it difficult to maintain a semblance or normality. Zahavi borrowed the term "handicap principle" from sports handicapping systems. These systems are aimed at reducing disparities in performance, thereby making the outcome of contests less predictable. In a horse handicap race, provenly faster horses are given heavier weights to carry under their saddles than inherently slower horses. Similarly, in amateur golf, better golfers have fewer strokes subtracted from their raw scores than the less talented players. The handicap therefore correlates with unhandicapped performance, making it possible, if one knows nothing about the horses, to predict which unhandicapped horse would win an open race. It would be the one handicapped with the greatest weight in the saddle. The handicaps in nature are highly visible, and therefore a peahen, for instance, would be able to deduce the health of a potential mate by comparing its handicap (the size of the peacock's tail) with those of the other males. The loss of the male's fitness caused by the handicap is offset by its increased access to females, which is as much of a fitness concern as is its health. An altruistic act is, by definition, similarly costly. It would therefore also signal fitness, and is probably as attractive to females as a physical handicap. If this is the case altruism is evolutionarily stabilized by sexual selection.

There is an alternate strategy for identifying fit mates which does not rely on one gender having exaggerated sexual ornaments or other handicaps, but is generally applicable to most, if not all sexual creatures. It derives from the concept that the change in appearance and functionality caused by a non-silent mutation will generally stand out in a population. This is because that altered appearance and functionality will be unusual, peculiar, and different from the norm within that population. The norm against which these unusual features are judged is made up of fit attributes that have attained their plurality through natural selection, while less adaptive attributes will be in the minority or frankly rare. Since the overwhelming majority of mutant features are maladaptive, and it is impossible to predict evolution's future direction, sexual creatures would be expected to prefer mates with the least number of unusual or minority features. This will have the effect of a sexual population rapidly shedding peripheral phenotypic features and canalizing the entire outward appearance and behaviour so that all the members of that population will begin to look remarkably similar in every detail, as illustrated in the accompanying photograph of the African pygmy kingfisher, *Ispidina picta*. Once a population has become as homogeneous in appearance as is typical of most species, its entire repertoire of behaviours will also be rendered evolutionarily stable, including any altruistic, cooperative and social characteristics. Thus, in the example of the selfish individual who hangs back from the rest of the hunting pack, but who nevertheless joins in the spoils, that individual will be recognized as being different from the norm, and will therefore find it difficult to attract a mate. Its genes will therefore have only a very small probability of being passed on to the next generation, thus evolutionarily stabilizing cooperation and social interactions at whatever level of complexity is the norm in that population.

African pygmy kingfisher, showing details of appearance and colouration that are shared by all African pygmy kingfishers to a high degree of fidelity.

Reciprocity Mechanisms

Altruism in animals describes a range of behaviors performed by animals that may be to their own disadvantage but which benefit others. The costs and benefits are measured in terms of reproductive fitness, or expected number of offspring. So by behaving altruistically, an organism reduces the number of offspring it is likely to produce itself, but boosts the likelihood that other organisms are to produce offspring. There are other forms of altruism in nature other than risk-taking behavior, such as reciprocal altruism. This biological notion of altruism is not identical to the everyday human concept. For humans, an action would only be called 'altruistic' if it was done with the conscious intention of helping another. Yet in the biological sense there is no such requirement. Instead, until we can communicate directly with other species, an accurate theory to describe altruistic acts between species is Biological Market Theory. Humans and other animals exchange benefits in several ways, known technically as reciprocity mechanism. No matter what the mechanism, the common thread is that benefits find their way back to the original giver.

Symmetry-based

Also known as the "buddy-system", mutual affection between two parties prompts similar behavior in both directions without need to track of daily give-and-take, so long as the overall relationship remains satisfactory. This is one of the most common mechanism of reciprocity in nature, this kind is present in humans, primates, and many other mammals.

Attitudinal

Also known as, "If you're nice, I'll be nice too." This mechanism of reciprocity is similar to the heuristic of the golden rule, "Treat others how you would like to be treated." Parties mirror one another's attitudes, exchanging favors on the spot. Instant attitudinal reciprocity occurs among monkeys, and people often rely on it with strangers and acquaintances.

Calculated

Also known as, "what have you done for me lately?" Individuals keep track of the benefits they exchange with particular partners, which helps them decide to whom to return favors. This mechanism is typical of chimpanzees and very common among human relationships. Yet some opposing experimental research suggests that calculated or contingent reciprocity does not spontaneously arise in laboratory experimental settings, despite patterns of behavior.

Biological Market Theory

Biological market theory is an extension of the idea of reciprocal altruism, as a mechanism to explain altruistic acts between unrelated individuals in a more flexible system of exchanging commodities. The term 'biological market' was first used by Ronald Noe and Hammerstein in 1994 to refer to all the interactions between organisms in which different organisms function as 'traders' that exchange goods and services such as food and water, grooming, warning calls, shelter, etc. Biological market theory consists of five formal characteristics which present a basis for altruism.

1. Commodities are exchanged between individuals that differ in the degree of control over those commodities.
2. Trading partners are chosen from a number of potential partners.
3. There is competition among the members of the chosen class to be the most attractive partner. This competition by 'outbidding' causes an increase in the value of the commodity offered.
4. Supply and demand determine the bartering value of commodities exchanged.
5. Commodities on offer can be advertised. As in commercial advertisements there is a potential for false information.

The applicability of biological market theory with its emphasis on partner choice is evident in the interactions between the cleaner wrasse and its "client" reef fish. Cleaners have small territories, which the majority of reef fish species actively visit to invite inspection of their surface, gills, and mouth. Clients benefit from the removal of parasites while cleaners benefit from the access to a food source. Some particularly choosy client species have large home ranges that cover several cleaning stations, whereas other clients have small ranges and have access to one cleaning station only (resident clients). Field observations, field manipulations, and laboratory experiments revealed that whether or not a client has choice options influences several aspects of both cleaner and client behaviour. Cleaners give choosy clients priority of access. Choosy clients switch partners if cheated by a cleaner by taking a bite of out of the cleaner, whereas resident clients punish cheats. Cleaners and resident clients, but not choosy clients, build up relationships before normal cleaning interactions take place. Cleaners are particularly cooperative if choosy clients are bystanders of an interaction but less so when resident clients are bystanders.

Two bluestreak cleaner wrasses cleaning a potato cod

Researchers tested whether wild white-handed gibbon males from Khao Yai National Park, Thailand, increased their grooming activity when the female partner was fertile. Adult females and males of our study population are codominant (in terms of aggression), they live in pairs or small multi male groups and mate promiscuously. They found that males groomed females more than vice versa and more grooming was exchanged when females were cycling than during pregnancy or lactation. The number of copulations/day was elevated when females were cycling, and females copulated more frequently with males on days when they received more grooming. When males increased their grooming efforts, females also increased their grooming of males, perhaps to equalize give and take. Although grooming might be reciprocated because of intrinsic benefits of receiving grooming, males also interchange grooming as a commodity for sexual opportunities during a female's fertile period.

Examples in Vertebrates

Mammals

- Wolves and wild dogs bring meat back to members of the pack not present at the kill.
- Mongooses support elderly, sick, or injured animals.
- Meerkats often have one standing guard to warn while the rest feed in case of predator attack.

- Raccoons inform conspecifics about feeding grounds by droppings left on commonly shared latrines. A similar information system has been observed to be used by common ravens.
- Male baboons threaten predators and cover the rear as the troop retreats.
- Gibbons and chimpanzees with food will, in response to a gesture, share their food with others of the group. Chimpanzees will help humans and conspecifics without any reward in return.
- Bonobos have been observed aiding injured or handicapped bonobos.
- Vampire bats commonly regurgitate blood to share with unlucky or sick roost mates that have been unable to find a meal, often forming a buddy system.
- Vervet monkeys give alarm calls to warn fellow monkeys of the presence of predators, even though in doing so they attract attention to themselves, increasing their personal chance of being attacked.
- Lemurs of all ages and of both sexes will take care of infants unrelated to them.
- Dolphins support sick or injured animals, swimming under them for hours at a time and pushing them to the surface so they can breathe.
- Walruses have been seen adopting orphans who lost their parents to predators.
- African buffalo will rescue a member of the herd captured by predators. (Battle at Kruger)

Birds

- In numerous bird species, a breeding pair receives support in raising its young from other "helper" birds, including help with the feeding of its fledglings. Some will even go as far as protecting an unrelated bird's young from predators.

Fish

- *Harpagifer bispinis*, a species of fish, live in social groups in the harsh environment of the Antarctic Peninsula. If the parent guarding the nest of eggs is removed, a usually male replacement unrelated to the parents guards the nest from predators and prevents fungal growth that would kill off the brood. There is no clear benefit to the male so the act may be considered altruistic.

Examples in Invertebrates

- Some termites and ants release a sticky secretion by fatally rupturing a specialized gland. This autothysis altruistically aids the colony at the expense of the individual insect. For example, defending against invading ants by creating a tar baby effect. This can be attributed to the fact that ants share their genes with the entire colony, and so this behaviour is evolutionarily beneficial (not necessarily for the individual ant but for the continuation of its specific genetic make-up).

- Synalpheus regalis is a species of eusocial marine snapping shrimp that lives in sponges in coral reefs. They live in colonies of about 300 individuals with one reproductive female. Other colony members defend the colony against intruders, forage, and care for the young. Eusociality in this system entails an adaptive division of labor which results in enhanced reproductive output of the breeders and inclusive fitness benefits for the nonbreeding helpers. *S. regalis* are exceptionally tolerant of conspecifics within their colonies due to close genetic relatedness among nestmates. Allozyme data reveals that relatedness within colonies is high, which is an indication that colonies in this species represent close kin groups. The existence of such groups is an important prerequisite of explanations of social evolution based on kin selection.

Examples in Protists

An interesting example of altruism is found in the cellular slime moulds, such as *Dictyostelium mucoroides*. These protists live as individual amoebae until starved, at which point they aggregate and form a multicellular fruiting body in which some cells sacrifice themselves to promote the survival of other cells in the fruiting body.

Habituation

Habituation is a form of learning in which an organism decreases or ceases to respond to a stimulus after repeated presentations. Essentially, the organism learns to stop responding to a stimulus which is no longer biologically relevant. For example, organisms may habituate to repeated sudden loud noises when they learn these have no consequences. Habituation usually refers to a reduction in innate behaviours, rather than behaviours developed during conditioning in which the process is termed "extinction". A progressive decline of a behavior in a habituation procedure may also reflect nonspecific effects such as fatigue, which must be ruled out when the interest is in habituation as a learning process.

The habituation process is a form of adaptive behavior (or neuroplasticity) that is classified as non-associative learning. Non-associative learning is a change in a response to a stimulus that does not involve associating the presented stimulus with another stimulus or event such as a reward or punishment. (Examples of associative learning include classical conditioning and operant conditioning). Habituation is the decrease of a response to a repeated eliciting stimulus that is not due to sensory adaption or motor fatigue. Sensory adaptation (or neural adaptation) occurs when an organism can no longer detect the stimulus as efficiently as when first presented and motor fatigue occurs when an organism is able to detect the stimulus but can no longer respond efficiently. In contrast, habituation is a learned adaption to the repeated presentation of a stimulus, not a reduction in sensory or motor ability.

Some related phenomena to habituation include sensitization and stimulus generalization/discrimination. Sensitization is the opposite process to habituation, i.e. an increase in the elicited behavior from repeated presentation of a stimulus. There may also be an initial increase in response immediately prior to the decline (a sensitization process followed by a habituation process). Another related phenomenon is stimulus generalization, when habituation occurs in response to other stimuli that are similar to the original stimulus. The opposing process, stimulus discrimination, is when habituation does not occur to other stimuli that are dissimilar to the original stimulus.

Drug Habituation

There is an additional connotation to the term habituation which applies to psychological dependency on drugs, and is included in several online dictionaries. A team of specialist from the World Health Organization (WHO) assembled in 1957 to address the problem of drug addiction and adopted the term "drug habituation" to distinguish some drug-use behaviors from drug addiction. According to the WHO lexicon of alcohol and drug terms, habituation is defined as "becoming accustomed to any behavior or condition, including psychoactive substance use". By 1964 the America Surgeon's General report on smoking and health included four features that characterize drug habituation according to WHO: 1) "a desire (but not a compulsion) to continue taking the drug for the sense of improved well-being which it engenders"; 2) "little or no tendency to increase the dose"; 3) "some degree of psychic dependence on the effect of the drug, but absence of physical dependence and hence of an abstinence syndrome"; 4)"detrimental effects, if any, primarily on the individual". However, also in 1964, a committee from the World Health Organization once again convened and decided the definitions of drug habituation and drug addiction were insufficient, replacing the two terms with "drug dependence". Substance dependence is the preferred term today when describing drug-related disorders, whereas the use of the term drug habituation has declined substantially.

Characteristics

Habituation as a form of non-associative learning can be distinguished from other behavioral changes (e.g., sensory adaption, fatigue) by considering the characteristics of habituation that have been identified over several decades of research. The characteristics first described by Thompson and Spencer have recently been updated and include the following:

Repeated presentation of a stimulus will cause a decrease in reaction to the stimulus. Habituation is also proclaimed to be a form of implicit learning, which is commonly the case with continually repeated stimuli. This characteristic is consistent with the definition of habituation as a procedure, but to confirm habituation as a process, additional characteristics must be demonstrated. Also observed is spontaneous recovery. That is, a habituated response to a stimulus recovers (increases in magnitude) when a significant amount of time (hours, days, weeks) passes between stimulus presentations.

"Potentiation of habituation" is observed when tests of spontaneous recovery are given repeatedly. In this phenomenon, the decrease in responding that follows spontaneous recovery becomes more rapid with each test of spontaneous recovery. Also noted was that an increase in the frequency of stimulus presentation (i.e., shorter interstimulus interval) will increase the rate of habituation. Furthermore, continued exposure to the stimulus after the habituated response has plateaued (i.e., show no further decrement) may have additional effects on subsequent tests of behavior such as delaying spontaneous recovery. The concepts of stimulus generalization and stimulus discrimination will be observed. Habituation to an original stimulus will also occur to other stimuli that are similar to the original stimulus (*stimulus generalization*). The more similar the new stimulus is to the original stimulus, the greater the habituation that will be observed. When a subject shows habituation to a new stimulus that is similar to the original stimulus but not to a stimulus that is different from the original stimulus, then the subject is showing *stimulus discrimination*. (For example, if one was habituated to the taste of lemon, their responding would increase significantly when presented with the taste of lime). Stimulus discrimination can be used to rule out sensory adaptation and fatigue as an alternative explanation of the habituation process.

Another observation mentioned is when a single introduction of a different stimulus late in the habituation procedure when responding to the eliciting stimulus has declined can cause an increase in the habituated response. This increase in responding is temporary and is called "dishabituation" and always occurs to the original eliciting stimulus (not to the added stimulus). Researchers also use evidence of dishabituation to rule out sensory adaptation and fatigue as alternative explanations of the habituation process. Habituation of dishabituation can occur. The amount of dishabituation that occurs as a result of the introduction of a different stimulus can decrease after repeated presentation of the "dishabituating" stimulus.

Some habituation procedures appear to result in a habituation process that last days or weeks. This is considered long-term habituation. It persists over long durations of time (i.e., shows little or no spontaneous recovery). Long-term habituation can be distinguished from short-term habituation which is identified by the nine characteristics listed above.

Biological Mechanisms

Habituation can refer to a decrease in behavior, subjective experience, or synaptic transmission. The changes in synaptic transmission that occur during habituation have been well-characterized in the *Aplysia* gill and siphon withdrawal reflex.

Habituation has been shown in essentially every species of animal. The experimental investigation of simple organisms such as the large protozoan *Stentor coeruleus* provides an understanding of the cellular mechanisms that are involved in the habituation process.

Neuroimaging

Within psychology, habituation has been studied through different forms of neuroimaging like PET scan and FMRI. Habituation is observed after repeated presentations of stimuli. Within fMRI, the stimuli's effect is measured using blood oxygen level-dependent (BOLD) signals, where long-term decreases of the BOLD signal are interpreted as habituation and increases of the BOLD signal are considered sensitization.

The amygdala is one of the most-studied areas of the brain when looking at habituation. One of the most common ways to study this is to observe the visual processing of facial expressions. A study by Breiter and colleagues used fMRI scans to identify which areas of the brain habituate and at what rate this happens. The results of this study showed that the human amygdala responds and rapidly habituates preferentially to fearful facial expressions over neutral. They also observed significant amygdala signal changes in response to happy faces over neutral faces.

One of the most recent studies that has looked at habituation was run by Blackford, Allen, Cowan and Avery. In 2012 they looked at the effect extreme inhibited temperaments verses extreme uninhibited temperaments has on habituation. Their study found that individuals with uninhibited temperament demonstrated habituation in both the amygdala and hippocampus regions of the brain. Whereas, participants with inhibited temperaments these regions of the brain failed to habituate over repeated presentations. The researchers suggest that this failure to habituate reflects a social learning deficit in individuals with an extreme inhibited temperament, which may provide a possible mechanism for higher risk of social anxiety.

History

Early studies relied on the demonstration of dishabituation (the brief recovery of the response to the eliciting stimulus when another stimulus is added) to distinguish habituation from sensory adaptation and fatigue. More recently, stimulus specificity and frequency-dependent spontaneous recovery have been identified as experimental evidence for the habituation process. Sensitization is also conceptualized as a nonassociative process because it involves an increase in responding with repeated presentations to a single stimulus. Much less is understood about sensitization than habituation, but the sensitization process is often observed along with the habituation process.

Theories

In an article written 20 years after his initial research with Groves, renowned authority on the behavioral phenomenon of habituation, Richard F. Thompson, reviews several theories of the process of habituation. The Stimulus-Model Comparator theory formulated by Evgeny Sokolov, and the Groves and Thompson dual-process theory are two examples.

The stimulus-model comparator theory emerged from the research of Sokolov who used the orienting response as the cornerstone of his studies, and operationally defining the orienting response as EEG activity. Orienting responses are heightened sensitivity experienced by an organism when exposed to a new or changing stimulus. Orienting responses can result in overt, observable behaviors as well as psychophysiological responses such as EEG activity and undergo habituation with repeated presentation of the eliciting stimulus. The Sokolov model assumes that when a stimulus is experienced several times the nervous system creates a model of the expected stimulus (a stimulus model). With additional presentations of the stimulus the experienced stimulus is compared with the stimulus model. If the experienced stimulus matches the stimulus model responding is inhibited. At first the stimulus model is not a very good representation of the presented stimulus and therefore there is a mismatch and responding continues, but with additional presentations the stimulus model is improved; consequently there is no longer a mismatch and responding is inhibited causing habituation. However, if the stimulus is changed so that it no longer matches the stimulus model the inhibition of the orienting response is weakened, and an orienting response returns. Sokolov places the location of the Stimulus-Model within the cerebral cortex of the brain.

The Groves and Thompson dual-process theory of habituation posits that two separate processes exist in the central nervous system that interacts to produce habituation. The two distinct processes are a habituation process and a sensitization process. The dual-process theory argues that all noticeable stimuli will elicit both of these processes and that the behavioral output will reflect a summation of both processes. The habituation process is decremental, whereas the sensitization process is incremental enhancing the tendency to respond. Thus when the habituation process exceeds the sensitization process behavior shows habituation, but if the sensitization process exceeds the habituation process, then behavior shows sensitization. Groves and Thompson hypothesize the existence of two neural pathways: an "S-R pathway" involved with the habituation process, and a "state pathway" involved with sensitization. The state system is seen as equivalent to a general state of arousal.

Examples of the Habituation Process in Animals and Humans

Habituation has been observed in an enormously wide range of species from motile single-celled organisms such as the amoeba and *Stentor coeruleus* to sea slugs to humans. Habituation processes are adaptive, allowing animals to adjust their innate behaviors to changes in their natural world. A natural animal instinct, for example, is to protect themselves and their territory from any danger and potential predators. It is obvious that an animal needs to respond quickly to the sudden appearance of a predator. What may be less obvious is the importance of defensive responses to the sudden appearance of any new, unfamiliar stimulus, whether it is dangerous or not. An initial defensive response to a new stimulus is important because if an animal fails to respond to a potentially dangerous unknown stimulus, the results could be deadly. Despite this initial, innate defensive response to an unfamiliar stimulus, the response becomes habituated if the stimulus repeatedly occurs but causes no harm. An example of this is the prairie dog habituating to humans. Prairie dogs give alarm calls when they detect a potentially dangerous stimulus. This defensive call occurs when any mammal, snake, or large bird approaches them. However, they habituate to noises, such as human footsteps, that occur repeatedly but result in no harm to them. If prairie dogs never habituate to nonthreatening stimuli, they would be constantly sending out alarm calls and wasting their time and energy. However, the habituation process in prairie dogs may depend on several factors including the particular defensive response. In one study that measured several different responses to the repeated presence of humans, the alarm calls of prairie dogs showed habituation whereas the behavior of escaping into their burrows showed sensitization.

Another example of the importance of habituation in the animal world is provided by a study with harbor seals. In one study researchers measured the responses of harbor seals to underwater calls of different types of killer whales. The seals showed a strong response when they heard the calls of mammal-eating killer whales. However, they did not respond strongly when hearing familiar calls of the local fish-eating population. The seals, therefore, are capable of habituating to the calls of harmless predators, in this case harmless killer whales. While some researchers prefer to simply describe the adaptive value of observable habituated behavior others find it useful to infer psychological processes from the observed behavior change. For example, habituation of aggressive responses in male bullfrogs has been explained as "an attentional or learning process that allows animals to form enduring mental representations of the physical properties of a repeated stimulus and to shift their focus of attention away from sources of irrelevant or unimportant stimulation".

Habituation of innate defensive behaviors is also adaptive in humans, such as habituation of a startle response to a sudden loud noise. But habituation is much more ubiquitous even in humans. An example of habituation that is an essential element of everyone's life is the changing response to food as it is repeatedly experienced during a meal. When people eat the same food during a meal, they begin to respond less to the food as they become habituated to the motivating properties of the food and decrease their consumption. Eating less during a meal is usually interpreted as reaching satiety or "getting full", but experiments suggest that habituation also plays an important role. Many experiments with animals and humans have shown that providing variety in a meal increases the amount that is consumed in a meal, most likely because habituation is stimulus specific and because variety may introduce dishabituation effects. Food variety also slows the rate of habituation in children and may be an important contributing factor to the recent increases in obesity.

We also find that habituation is found in our emotional responses, called the Opponent-Process Theory, proposed by researchers Richard Solomon and John Corbit (1974). It is known that responses by the subject tends to change by repetitively presenting certain stimuli. But concerning the opponent-process theory, some emotional reactions to the stimuli weaken (decrease) while others reactions are strengthened (increase). Take for example, that is the end of the semester at your university. You have been worried about your grade for the entire semester and you need to make an A on the final in order to pass the course. You study efficiently for the test and after taking it, you feel that you will receive a very high grade. But once you check the gradebook, you see that you did not get an A on your exam. Instead you received a C+. Now you are distraught and know that there is no other way to pass the course for the semester. After a few minutes you begin to calm down and by the next hour you are back to your normal emotional state. This is an example of an emotional response explained by the opponent-process theory. It begins with an outside stimulus provoking an emotional reaction that increases rapidly until it is at its most intense (presumably after you learned that you did not receive a high letter grade). Gradually, your emotional state declines to a level lower than normal and eventually returns to neutral. This pattern coincides with two internal processes referred to as the a-process and b-process. The a-process, or "affective" response to a stimulus, is the initial emotional response one has and can be pleasant or unpleasant. The b-process is the after reaction and has a lower intensity than the a-process. The a-process is very fast-acting and ends as soon as the stimulus ends or is removed. Unlike the a-process, b-process is much slower in returning to baseline. Concerning the definition of the opponent process theory—repeated presentations present habituation—the a-process does not necessarily change. It is the b-process that is strengthened instead and rises more quickly to reach the highest intensity, and much slower in attempting to return to baseline after the stimulus is removed. To sum it all up,with the opponent-process theory, repeated presentations of the same stimulus will result in habituation, where subjects show little to no reaction. It is the after-reaction that is much larger and prolonged, than if an initial reaction to a stimulus occurred.

Uses of the Habituation Procedure

Habituation procedures are used by researchers for many reasons. For example, in a study on aggression in female chimpanzees from a group known as the "*Kasakela Chimpanzee Community*", researchers habituated the chimpanzees by repeatedly exposing them to the presence of human beings. Their efforts to habituate the chimpanzees before the field researchers studied the animal's behavior was necessary in order for them to eventually be able to note the natural behavior of the chimpanzees, instead of simply noting chimpanzee behavior as a response to the presence of the researchers. In another study, Mitumba chimpanzees in the Gombe National Park were habituated for at least four years before the introduction of systematic data collection.

Researchers also use habituation and dishabituation procedures in the laboratory to study the perceptual and cognitive capabilities of human infants. The presentation of a visual stimulus to an infant elicits looking behavior that habituates with repeated presentations of the stimulus. When changes to the habituated stimulus are made (or a new stimulus is introduced) the looking behavior returns (dishabituates). A recent fMRI study revealed that the presentation of a dishabituating stimulus has an observable, physical effect upon the brain. In one study the mental spatial representations of infants were assessed using the phenomenon of dishabituation. Infants were presented repeatedly with an object in the same position on a table. Once the infants habituated to the object (i.e., spent less time looking at it) either the object was spatially moved while the infant remained at the same place near the table or the object was left in the same place but the infant was moved to the opposite side of the table. In both cases the spatial relationship between the object and the infant had changed, but only in the former case did the object itself move. Would the infants know the difference? Or would they treat both cases as if the object itself moved? The results revealed a return of looking behavior (dishabituation) when the object's position was changed, but not when the infant's position was changed. Dishabituation indicates that infants perceived a significant change in the stimulus. Therefore, the infants understood when the object itself moved and when it did not. Only when the object itself moved were they interested in it again (dishabituation), When the object remained in the same position as before it was perceived as the same old boring thing (habituation). In general, habituation/dishabituation procedures help researchers determine the way infants perceive their environments.

The habituation/dishabituation procedure is also used to discover the resolution of perceptual systems. For instance, by habituating someone to one stimulus, and then observing responses to similar ones, one can detect the smallest degree of difference that is detectable.

Lamarckism

Lamarckism (or Lamarckian inheritance) is the idea that an organism can pass on characteristics that it has acquired during its lifetime to its offspring (also known as heritability of acquired characteristics or soft inheritance). It is named after the French biologist Jean-Baptiste Lamarck (1744–1829), who incorporated the action of soft inheritance into his evolutionary theories as a supplement to his concept of an inherent progressive tendency driving organisms continuously towards greater complexity, in parallel but separate lineages with no extinction. Lamarck did not originate the idea of soft inheritance, which proposes that individual efforts during the lifetime of the organisms were the main mechanism driving species to adaptation, as they supposedly would acquire adaptive changes and pass them on to offspring.

When Charles Darwin published his theory of evolution by natural selection in *On the Origin of Species* (1859), he continued to give credence to what he called "use and disuse inheritance," but rejected other aspects of Lamarck's theories. Later, Mendelian genetics supplanted the notion of inheritance of acquired traits, eventually leading to the development of the modern evolutionary synthesis, and the general abandonment of the Lamarckian theory of evolution in biology. Despite this abandonment, interest in Lamarckism has continued as studies in the field of epigenetics have highlighted the possible inheritance of behavioral traits acquired by the previous generation. However, this remains controversial as science historians have asserted that it is inaccurate to describe transgenerational epigenetic inheritance as a form of Lamarckism.

History

Between 1794 and 1796 Erasmus Darwin wrote *Zoonomia* suggesting "that all warm-blooded animals have arisen from one living filament... with the power of acquiring new parts" in response to stimuli, with each round of "improvements" being inherited by successive generations. Subsequently, Jean-Baptiste Lamarck repeated in his *Philosophie Zoologique* of 1809 the folk wisdom that characteristics which were "needed" were acquired (or diminished) during the lifetime of an organism then passed on to the offspring. He incorporated this mechanism into his thoughts on evolution, seeing it as resulting in the adaptation of life to local environments.

Jean-Baptiste Lamarck

Lamarck founded a school of French *Transformationism* which included Étienne Geoffroy Saint-Hilaire, and which corresponded with a radical British school of anatomy based in the extramural anatomy schools in Edinburgh, Scotland, which included the surgeon Robert Knox and the comparative anatomist Robert Edmond Grant. In addition, the Regius Professor of Natural History at the University of Edinburgh, Robert Jameson, was the probable author of an anonymous paper in 1826 praising "Mr. Lamarck" for explaining how the higher animals had "evolved" from the "simplest worms"—this was the first use of the word "evolved" in a modern sense. As a young student, Charles Darwin was tutored by Grant, and worked with him on marine creatures.

The *Vestiges of the Natural History of Creation*, authored by Robert Chambers in St Andrews, Scotland, and published anonymously in England in 1844, proposed a theory which combined radical phrenology with Lamarckism, causing political controversy for its radicalism and unorthodoxy, but exciting popular interest and preparing a huge and prosperous audience for Darwin.

Darwin's *On the Origin of Species* proposed natural selection as the main mechanism for development of species, but did not rule out a variant of Lamarckism as a supplementary mechanism. Darwin called his Lamarckian hypothesis pangenesis, and explained it in the final chapter of his book *The Variation of Animals and Plants under Domestication* (1868), after describing numerous examples to demonstrate what he considered to be the inheritance of acquired characteristics. Pangenesis, which he emphasised was a hypothesis, was based on the idea that somatic cells would, in response to environmental stimulation (use and disuse), throw off 'gemmules' or 'pangenes' which travelled around the body (though not necessarily in the bloodstream). These pangenes were microscopic particles that supposedly contained information about the characteristics of their parent cell, and Darwin believed that they eventually accumulated in the germ cells where they could pass on to the next generation the newly acquired characteristics of the parents. Darwin's half-cousin, Francis Galton, carried out experiments on rabbits, with Darwin's cooperation, in which he transfused the blood of one variety of rabbit into another variety in the expectation that its offspring would show some characteristics of the first. They did not, and Galton declared that he had disproved Darwin's hypothesis of pangenesis, but Darwin objected, in a letter to the scientific journal *Nature*, that he had done nothing of the sort, since he had never mentioned blood in his writings. He pointed out that he regarded pangenesis as occurring in Protozoa and plants, which have no blood.

Lamarck's Theory

The identification of Lamarckism with the inheritance of acquired characteristics is regarded by some as an artifact of the subsequent history of evolutionary thought, repeated in textbooks without analysis. American paleontologist and historian of science Stephen Jay Gould wrote that in the late 19th century, evolutionists "re-read Lamarck, cast aside the guts of it ... and elevated one aspect of the mechanics—inheritance of acquired characters—to a central focus it never had for Lamarck himself." He argued that "the restriction of 'Lamarckism' to this relatively small and non-distinctive corner of Lamarck's thought must be labelled as more than a misnomer, and truly a discredit to the memory of a man and his much more comprehensive system." Gould advocated defining "Lamarckism" more broadly, in line with Lamarck's overall evolutionary theory.

The long neck of the giraffe is often used as an example in explanations of Lamarckism.

Lamarck incorporated two ideas into his theory of evolution, in his day considered to be generally true. The first was the idea of use versus disuse; he theorized that individuals lose characteristics they do not require, or use, and develop characteristics that are useful. His second point was to argue that the acquired traits were heritable. Examples of what is traditionally called "Lamarckism" would include the idea that when giraffes stretch their necks to reach leaves high in trees (especially Acacias), they strengthen and gradually lengthen their necks. These giraffes have offspring with slightly longer necks (also known as "soft inheritance"). Similarly, a blacksmith, through his work, strengthens the muscles in his arms, and thus his sons will have similar muscular development when they mature.

Lamarck stated the following two laws:

1. *Première Loi: Dans tout animal qui n' a point dépassé le terme de ses développemens, l' emploi plus fréquent et soutenu d' un organe quelconque, fortifie peu à peu cet organe, le développe, l' agrandit, et lui donne une puissance proportionnée à la durée de cet emploi ; tandis que le défaut constant d' usage de tel organe, l'affoiblit insensiblement, le détériore, diminue progressivement ses facultés, et finit par le faire disparoître.*
2. *Deuxième Loi: Tout ce que la nature a fait acquérir ou perdre aux individus par l' influence des circonstances où leur race se trouve depuis long-temps exposée, et, par conséquent, par l' influence de l' emploi prédominant de tel organe, ou par celle d' un défaut constant d' usage de telle partie ; elle le conserve par la génération aux nouveaux individus qui en proviennent, pourvu que les changemens acquis soient communs aux deux sexes, ou à ceux qui ont produit ces nouveaux individus.*

English translation:

1. First Law: In every animal which has not passed the limit of its development, a more frequent and continuous use of any organ gradually strengthens, develops and enlarges that organ, and gives it a power proportional to the length of time it has been so used; while the permanent disuse of any organ imperceptibly weakens and deteriorates it, and progressively diminishes its functional capacity, until it finally disappears.
2. Second Law: All the acquisitions or losses wrought by nature on individuals, through the influence of the environment in which their race has long been placed, and hence through the influence of the predominant use or permanent disuse of any organ; all these are preserved by reproduction to the new individuals which arise, provided that the acquired modifications are common to both sexes, or at least to the individuals which produce the young.

In essence, a change in the environment brings about change in "needs" (*besoins*), resulting in change in behavior, bringing change in organ usage and development, bringing change in form over time—and thus the gradual transmutation of the species.

However, as historians of science such as Michael Ghiselin and Stephen Jay Gould have pointed out, none of these views were original to Lamarck. On the contrary, Lamarck's contribution was a systematic theoretical framework for understanding evolution. He saw evolution as comprising two processes;

1. *Le pouvoir de la vie* (the complexifying force) - in which the natural, alchemical movements of fluids would etch out organs from tissues, leading to ever more complex construc-

tion regardless of the organ's use or disuse. This would drive organisms from simple to complex forms.

2. *L'influence des circonstances* (the adaptive force) - in which the use and disuse of characters led organisms to become more adapted to their environment. This would take organisms sideways off the path from simple to complex, specialising them for their environment.

Weismann's Experiment

The idea that germline cells contain information that passes to each generation unaffected by experience and independent of the somatic (body) cells, came to be referred to as the Weismann barrier, and is frequently quoted as putting a final end to Lamarckism and theory of inheritance of acquired characteristics.

August Weismann

August Weismann conducted the experiment of removing the tails of 68 white mice, repeatedly over five generations, and reporting that no mice were born in consequence without a tail or even with a shorter tail. He stated that "901 young were produced by five generations of artificially mutilated parents, and yet there was not a single example of a rudimentary tail or of any other abnormality in this organ."

However, the experiment has been questioned in relationship to Lamarck's hypothesis as it did not address the *use and disuse* of characteristics in response to the environment. Biologist Peter Gauthier noted that:

Can Weismann's experiment be considered a case of disuse? Lamarck proposed that when an organ was not used, it slowly, and very gradually atrophied. In time, over the course of many generations, it would gradually disappear as it was inherited in its modified form in each successive generation. Cutting the tails off mice does not seem to meet the qualifications of disuse, but rather falls in a category of accidental misuse... Lamarck's hypothesis has never been proven experimentally and there is no known mechanism to support the idea that somatic change, however acquired, can in some way induce a change in the germplasm. On the other hand it is difficult to disprove Lamarck's idea experimentally, and it seems that Weismann's experiment fails to provide the evidence to deny the Lamarckian hypothesis, since it lacks a key factor, namely the willful exertion of the animal in overcoming environmental obstacles.

Science historian Michael Ghiselin also considers the Weismann tail-chopping experiment to have no bearing on the Lamarckian hypothesis:

The acquired characteristics that figured in Lamarck's thinking were changes that resulted from an individual's own drives and actions, not from the actions of external agents. Lamarck was not concerned with wounds, injuries or mutilations, and nothing that Lamarck had set forth was tested or "disproven" by the Weismann tail-chopping experiment.

Neo-Lamarckism

The period of the history of evolutionary thought between Darwin's death in the 1880s, and the foundation of population genetics in the 1920s and beginnings of modern evolutionary synthesis in the 1930s, is called the eclipse of Darwinism by some historians of science. During that time many scientists and philosophers accepted the reality of evolution but doubted whether natural selection was the main evolutionary mechanism.

Edward Drinker Cope

Among the most popular alternatives were theories involving the inheritance of characteristics acquired during an organism's lifetime. Scientists who felt that such Lamarckian mechanisms were the key to evolution were called neo-Lamarckians and included the British botanist George Henslow (1835–1925), who studied the effects of environmental stress on the growth of plants, in the belief that such environmentally-induced variation might explain much of plant evolution, and the American entomologist Alpheus Spring Packard, Jr., who studied blind animals living in caves and wrote a book in 1901 about Lamarck and his work.

Also included were a number of paleontologists like Edward Drinker Cope and Alpheus Hyatt, who felt that the fossil record showed orderly, almost linear, patterns of development that they felt were better explained by Lamarckian mechanisms than by natural selection. Some people, including Cope and the Darwin critic Samuel Butler, felt that inheritance of acquired characteristics would let organisms shape their own evolution, since organisms that acquired new habits would change the use patterns of their organs, which would kick-start Lamarckian evolution. They considered this philosophically superior to Darwin's mechanism of random variation acted on by selective pressures. Lamarckism also appealed to those, like the philosopher Herbert Spencer and the German anatomist Ernst Haeckel, who saw evolution as an inherently progressive process. The German zoologist Theodor Eimer combined Larmarckism with ideas about orthogenesis.

With the development of the modern synthesis of the theory of evolution and a lack of evidence for a mechanism for acquiring and passing on new characteristics, or even their heritability, Lamarckism largely fell from favor. Unlike neo-Darwinism, the term neo-Lamarckism refers more to a loose grouping of largely heterodox theories and mechanisms that emerged after Lamarck's time, than to any coherent body of theoretical work.

Experiments

In a series of experiments from 1869 to 1891, Charles-Édouard Brown-Séquard cut the sciatic nerve of the leg and spinal cord in the dorsal regions of guinea pigs, causing an abnormal nervous condition resembling epilepsy; these were then bred and produced epileptic offspring. Although some scientists considered this evidence for Lamarckian inheritance, the experiments were not Lamarckian, as they did not address the *use and disuse* of characteristics in response to the environment. The results from the experiment were not duplicated by other scientists. One explanation for the results was that they show a transmitted disease, and not evidence for the inheritance of an acquired characteristic. Brown-Séquard's experiments are now considered anomalous and alternative explanations have been suggested.

Charles-Édouard Brown-Séquard

The French botanist Gaston Bonnier, conducting experiments in the French Alps in 1884 and the Pyrenees in 1886, studied structural changes induced by growing plants at various altitudes and transplanting them to others. Bonnier believed he had proven acquired adaptive characteristics; however, he did not weed, cultivate, fertilize or protect his plant specimens from native vegetation. In the 1920s his experiments were analysed and attributed to genetic contamination rather than Lamarckian inheritance.

In a series of experiments (in 1891, 1893 and 1895) on the action of light on the coloration of flatfish, the British marine biologist Joseph Thomas Cunningham (1859–1935) directed light upon the lower sides of flatfishes by means of a glass-bottomed tank placed over a mirror. He discovered the influence of light in producing pigments on the lower sides of flatfishes and gave his results a Lamarckian interpretation. Other scientists wrote that Cunningham had received some definite results, but that they were open to more than one interpretation. The geneticist William Bateson was not convinced that the cause of the increase in pigmentation was from the illumination. George Romanes wrote approvingly of Cunningham's interpretation. Thomas Hunt Morgan criticized the experiments and did not believe the results were evidence for Lamarckism.

In 1906, the philosopher Eugenio Rignano wrote a book, *Sur La Transmissibilité Des Caractères Acquis*, that argued for the inheritance of acquired characteristics. He advanced a moderated Lamarckian hypothesis of inheritance known as "centro-epigenesis." However, his views were controversial and not accepted by the majority in the scientific community.

In a series of experiments from 1907 to 1910, William Lawrence Tower performed experiments on potato beetles which were said by Ernest MacBride to have provided evidence for the inheritance of acquired characteristics. These were heavily criticized by William Bateson. It was later suggested that his research may have been faked. Tower claimed that the records of his experimental results had been lost in a fire. The geneticist William E. Castle who visited Tower's laboratory was not impressed by the experimental conditions. He later concluded that Tower had faked his data. Castle found the fire suspicious and also Tower's claim that a steam leak in his greenhouse had destroyed all his beetle stocks.

Experiments conducted by Gustav Tornier from 1907 to 1918 on goldfish and embryos of frogs and newts were supported by neo-Lamarckians such as Cunningham and MacBride as demonstrating the inheritance of acquired characteristics. The abnormalities were interpreted as the result of an osmotic effect by other researchers.

In the late 19th century, Frederick Merrifield exposed caterpillars and chrysalids to significantly high and low temperatures, and discovered permanent changes in some offspring's wing patterns. Swiss biologist Maximilian Rudolph Standfuss (1854–1917) led 30 years of intensive breeding experiments with European butterflies and after several generations, found similar preserved variations even generations after the cessation of exposing them to low temperatures. Standfuss was a neo-Lamarckian and attributed the results of his experiments as direct changes to the environment. In 1940, Richard Goldschmidt interpreted these results without invoking Lamarckian inheritance, and in 1998 Ernst Mayr wrote that results reported by Standfuss and others on the effects of abnormal temperatures on Lepidoptera are difficult to interpret.

In 1910, the American zoologist Charles Rupert Stockard (1879–1939) tested the effects of alcohol intoxication on the offspring of pregnant guinea pigs. Stockard discovered that repeated alcohol intoxication in the guinea pigs produced defects and malformations in their offspring that was passed down to two or more generations. His results were challenged by the biologist Raymond Pearl who performed the same experiments with chickens. Pearl discovered that the offspring of the chickens that had been exposed to alcohol were not defected but were healthy. He attributed his findings to the detrimental effects of alcohol only on the eggs and sperm which were already weak, the strong eggs and sperm were unaffected by alcohol intoxication. Pearl argued that his results had a Darwinian, not a Lamarckian explanation.

The French zoologist Yves Delage in his book *The Theories of Evolution* (1912) reviewed experiments into Lamarckism concluded the evidence "is not of uniform value and is more or less open to criticism; very little of it is convincing... [due to] difficulties of experimentation and, above all, of interpretation."

Paul Kammerer

In a series of experiments, Francis Bertody Sumner (1874–1945) reared several generations of white mice under different conditions of temperature and relative humidity. Sumner discovered that the white mice at 20 °C to 30 °C developed longer bodies, tails and hind feet which were also transmitted to their offspring over a number of generations, however, later results were not entirely consistent and the experiments ended in uncertainty.

Between 1918 and 1924, two American scientists Michael F. Guyer and Elizabeth A. Smith performed experiments in which fowl serum antibodies for rabbit lens-protein were injected into pregnant rabbits which resulted in defects in the eyes of some of their offspring that were inherited through eight generations. Their experiments were criticized and were not repeated by other scientists.

In the 1920s, experiments by Paul Kammerer on amphibians, particularly the midwife toad, appeared to find evidence supporting Lamarckism. However, his specimens with supposedly acquired black foot-pads were found to have been tampered with. In *The Case of the Midwife Toad* (1971), author and journalist Arthur Koestler surmised that the tampering had been done by a Nazi sympathiser to discredit Kammerer for his political views, and that his research might actually have been valid. However, most biologists believe that Kammerer was a fraud, and even among those who believe he was honest, most believe that he misinterpreted the results of his experiments.

During the 1920s, Harvard University researcher William McDougall studied the abilities of rats to correctly solve mazes. He found that offspring of rats that had learned the maze were able to run it faster. The first rats would get it wrong 165 times before being able to run it perfectly each time, but after a few generations it was down to 20. McDougall attributed this to some sort of Lamarckian evolutionary process. Oscar Werner Tiegs and Wilfred Eade Agar later showed McDougall's results to be incorrect, caused by poor experimental controls. Peter Medawar wrote that "careful and extensive repetitions of McDougall's research failed altogether to confirm it. His work therefore becomes an exhibit in the capacious ill-lit museum of unreproducible phenomena."

In the 1920s, John William Heslop-Harrison conducted experiments on the peppered moth, claiming to have evidence for the inheritance of acquired characteristics. Other scientists failed to replicate his results. The Russian physiologist Ivan Pavlov claimed to have observed a similar phenomenon in white mice being subject to a conditioned reflex experiment involving food and the sound of a bell. He wrote that with each generation, the mice became easier to condition. In 1926, Pavlov announced that there had been a fatal flaw in his experiment and retracted his claim to have demonstrated Lamarckian inheritance. Other researchers were also unable to replicate his results.

In other experiments, Coleman Griffith (1920, 1922) and John Detlefson (1923, 1925) reared rats in cages on a rotating table for three months. The rats adapted to the rotating condition to such an extent that when the rotation was stopped they showed signs of disequilibration and other physiological conditions which were inherited for several generations. In 1933, Roy Dorcus replicated their experiments but obtained different results as the rotated rats did not manifest any abnormalities of posture described by Griffith and Detlefson. Other studies revealed that the same abnormalities could occur in rats without rotation if they were suffering from an ear infection thus the results were interpreted as a case of infection, not Lamarckian inheritance.

In the 1930s, the German geneticist Victor Jollos (1887–1941) in a series of experiments claimed evidence for inherited changes induced by heat treatment in *Drosophila melanogaster*. His experiments were described as Lamarckian. However, Jollos was not an advocate of Lamarckian evolution and attributed the results from his experiments as evidence for directed mutagenesis. American scientists were unable to replicate his results.

The British anthropologist Frederic Wood Jones and the South African paleontologist Robert Broom supported a neo-Lamarckian view of human evolution as opposed to the Darwinian view. The German anthropologist Hermann Klaatsch relied on a neo-Lamarckian model of evolution to try and explain the origin of bipedalism. Neo-Lamarckism remained influential in biology until the 1940s when the role of natural selection was reasserted in evolution as part of the modern evolutionary synthesis.

Herbert Graham Cannon, a British zoologist, defended Lamarckism in his 1959 book *Lamarck and Modern Genetics*.

In the 1960s, "biochemical Lamarckism" was advocated by the embryologist Paul Wintrebert.

Pierre-Paul Grassé

In the 1970s, Australian immunologist Edward J. Steele and colleagues proposed a neo-Lamarckian mechanism to try to explain why homologous DNA sequences from the VDJ gene regions of parent mice were found in their germ cells and seemed to persist in the offspring for a few generations. The mechanism involved the somatic selection and clonal amplification of newly acquired antibody gene sequences that were generated via somatic hypermutation in B-cells. The messenger RNA (mRNA) products of these somatically novel genes were captured by retroviruses endogenous to the B-cells and were then transported through the bloodstream where they could breach the soma-germ barrier and retrofect (reverse transcribe) the newly acquired genes into the cells of the germ line. Although Steele was advocating this theory for the better part of two decades, little more than indirect evidence was ever acquired to support it. An interesting attribute of this idea is that it strongly resembles Darwin's own theory of pangenesis, except in the soma to germ line feedback theory, pangenes are replaced with realistic retroviruses. Regarding Steele's research, historian of biology Peter J. Bowler wrote, "his work was bitterly criticized at the time by biologists who doubted his experimental results and rejected his hypothetical mechanism as implausible."

Neo-Lamarckism was dominant in French biology for more than a century. French scientists who supported neo-Lamarckism included Edmond Perrier (1844–1921), Alfred Giard (1846–1908), Gaston Bonnier (1853–1922) and Pierre-Paul Grassé (1895–1985).

In 1987, Ryuichi Matsuda coined the term "pan-environmentalism" for his evolutionary theory which he saw as a fusion of Darwinism with neo-Lamarckism. He held that heterochrony is a main mechanism for evolutionary change and that novelty in evolution can be generated by genetic assimilation. His views were criticized by Arthur M. Shapiro for providing no solid evidence for his theory. Shapiro noted that "Matsuda himself accepts too much at face value and is prone to wish-fulfilling interpretation."

Within the discipline of history of technology, Lamarckism has been used in linking cultural development to human evolution by classifying artefacts as extensions of human anatomy: in other words, as the acquired cultural characteristics of human beings. Ben Cullen has shown that a strong element of Lamarckism exists in sociocultural evolution.

Ideological Neo-Lamarckism

A form of Lamarckism was revived in the Soviet Union of the 1930s when Trofim Lysenko promoted Lysenkoism which suited the ideological opposition of Joseph Stalin to genetics. This ideologically driven research influenced Soviet agricultural policy which in turn was later blamed for crop failures.

Neo-Lamarckian versions of evolution were widespread in the late 19th century. The idea that living things could to some degree *choose* the characteristics that would be inherited allowed them things to be in charge of their own destiny as opposed to the Darwinian view, which made them puppets at the mercy of the environment. Such ideas were more popular than natural selection in the late 19th century as it made it possible for biological evolution to fit into a framework of a divine or naturally willed plan, thus the neo-Lamarckian view of evolution was often advocated by proponents of orthogenesis. According to Peter J. Bowler:

One of the most emotionally compelling arguments used by the neo-Lamarckians of the late nineteenth century was the claim that Darwinism was a mechanistic theory which reduced living things to puppets driven by heredity. The selection theory made life into a game of Russian roulette, where life or death was predetermined by the genes one inherited. The individual could do nothing to mitigate bad heredity. Lamarckism, in contrast, allowed the individual to choose a new habit when faced with an environmental challenge and shape the whole future course of evolution.

Supporters of neo-Lamarckism such as George Bernard Shaw and Arthur Koestler claimed that Lamarckism is more humane and optimistic than Darwinism.

Critique

George Gaylord Simpson in his book *Tempo and Mode in Evolution* (1944) claimed that experiments in heredity have failed to corroborate any Lamarckian process. Simpson noted that neo-Lamarckism "stresses a factor that Lamarck rejected: inheritance of direct effects of the environment" and neo-Lamarckism is closer to Darwin's pangenesis than Lamarck's views. Simpson wrote, "the inheritance of acquired characters, failed to meet the tests of observation and has been almost universally discarded by biologists."

Botanist Conway Zirkle pointed out that Lamarck did not originate the hypothesis that acquired characters were heritable, therefore it is incorrect to refer to it as Lamarckism:

What Lamarck really did was to accept the hypothesis that acquired characters were heritable, a notion which had been held almost universally for well over two thousand years and which his contemporaries accepted as a matter of course, and to assume that the results of such inheritance were cumulative from generation to generation, thus producing, in time, new species. His individual contribution to biological theory consisted in his application to the problem of the origin of species of the view that acquired characters were inherited and in showing that evolution could be inferred logically from the accepted biological hypotheses. He would doubtless have been greatly astonished to learn that a belief in the inheritance of acquired characters is now labeled "Lamarckian," although he would almost certainly have felt flattered if evolution itself had been so designated.

Peter Medawar wrote regarding Lamarckism, "very few professional biologists believe that anything of the kind occurs—or can occur—but the notion persists for a variety of nonscientific reasons." Medawar stated there is no known mechanism by which an adaption acquired in an individual's lifetime can be imprinted on the genome and Lamarckian inheritance is not valid unless it excludes the possibility of natural selection but this has not been demonstrated in any experiment.

Martin Gardner wrote in his book *Fads and Fallacies in the Name of Science* (1957):

A host of experiments have been designed to test Lamarckianism. All that have been verified have proved negative. On the other hand, tens of thousands of experiments— reported in the journals and carefully checked and rechecked by geneticists throughout the world— have established the correctness of the gene-mutation theory beyond all reasonable doubt... In spite of the rapidly increasing evidence for natural selection, Lamarck has never ceased to have loyal followers.... There is indeed a strong emotional appeal in the thought that every little effort an animal puts forth is somehow transmitted to his progeny.

According to Ernst Mayr, any Lamarckian theory involving the inheritance of acquired characters has been refuted as "DNA does not directly participate in the making of the phenotype and that the phenotype, in turn, does not control the composition of the DNA." Peter J. Bowler has written that although many early scientists took Lamarckism seriously, it was discredited by genetics in the early twentieth century.

Epigenetic Lamarckism

Forms of 'soft' or epigenetic inheritance within organisms have been suggested as neo-Lamarckian in nature by such scientists as Eva Jablonka and Marion J. Lamb. In addition to 'hard' or genetic inheritance, involving the duplication of genetic material and its segregation during meiosis, there are other hereditary elements that pass into the germ cells also. These include things like methylation patterns in DNA and chromatin marks, both of which regulate the activity of genes. These are considered Lamarckian in the sense that they are responsive to environmental stimuli and can differentially affect gene expression adaptively, with phenotypic results that can persist for many generations in certain organisms.

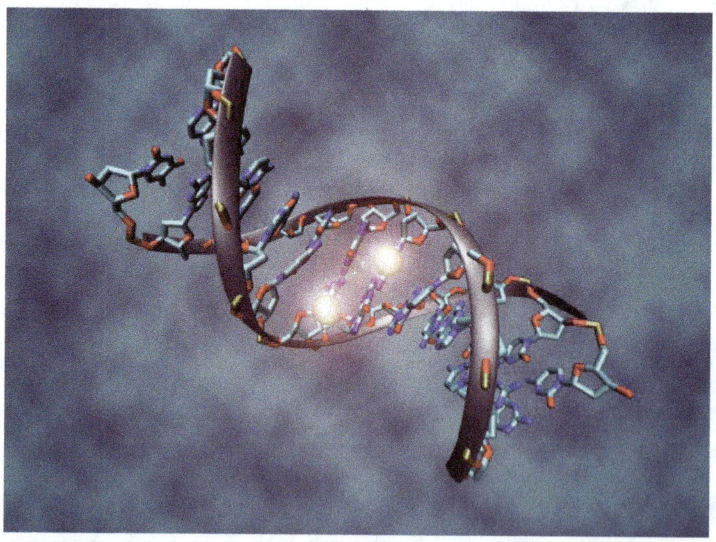

Illustration of a DNA molecule that is methylated at the two center cytosines. DNA methylation plays an important role for epigenetic gene regulation in development and disease.

Jablonka and Lamb have called for an extended evolutionary synthesis. They have argued that there is evidence for Lamarckian epigenetic control systems causing evolutionary changes and the mechanisms underlying epigenetic inheritance can also lead to saltational changes that reorganize the epigenome.

Interest in Lamarckism has increased, as studies in the field of epigenetics have highlighted the possible inheritance of behavioral traits acquired by the previous generation. A 2009 study examined foraging behavior in chickens as a function of stress:

Transmissions of information across generations which does not involve traditional inheritance of DNA-sequence alleles is often referred to as soft inheritance or "Lamarckian inheritance."

The study concluded:

Our findings suggest that unpredictable food access caused seemingly adaptive responses in feeding behavior, which may have been transmitted to the offspring by means of epigenetic mechanisms, including regulation of immune genes. This may have prepared the offspring for coping with an unpredictable environment.

The evolution of acquired characteristics has also been shown in human populations who have experienced starvation, resulting in altered gene function in both the starved population and their offspring. The process of DNA methylation is thought to be behind such changes.

In October 2010, further evidence linking food intake to traits inherited by the offspring were shown in a study of rats conducted by several Australian universities. The study strongly suggested that fathers can transfer a propensity for obesity to their daughters as a result of the fathers' food intake, and not their genetics (or specific genes), prior to the conception of the daughter. A "paternal high-fat diet" was shown to cause cell dysfunction in the daughter, which in turn led to obesity for the daughter. Felicia Nowak, *et al.* reported at the Endocrine Society meeting in June 2013 that obese male rats passed on the tendency to obesity to their male offspring.

Several studies, one conducted by researchers at Massachusetts Institute of Technology and another by researchers at the Tufts University School of Medicine, have rekindled the debate once again. As reported in *MIT Technology Review* in February 2009, "The effects of an animal's environment during adolescence can be passed down to future offspring ... The findings provide support for a 200-year-old theory of evolution that has been largely dismissed: Lamarckian evolution, which states that acquired characteristics can be passed on to offspring." A report investigating the inheritance of resistance to viral infection in the nematode *Caenorhabditis elegans* suggests that small RNA molecules may be inherited in a non-Mendelian fashion and provide resistance to infection. More recent studies in C. elegans have revealed that progeny may inherit information regarding environmental challenges that the parent experienced, such as starvation, and that this epigenetic effect may persist for multiple generations.

A study (Akimoto *et al.* 2007) on epigenetic inheritance in rice plants came to the conclusion that "gene expression is flexibly tuned by methylation, allowing plants to gain or lose particular traits which are heritable as far as methylation patterns of corresponding genes are maintained. This is in support of the concept of Lamarckian inheritance, suggesting that acquired traits are heritable." Another study (Sano, 2010) wrote that observations suggest that acquired traits are heritable in plants as far as the acquired methylation pattern is stably transmitted which is consistent with Lamarckian evolution. Handel and Ramagopalan found that there is evidence that epigenetic alterations such as DNA methylation and histone modifications are transmitted transgenerationally as a mechanism for environmental influences to be passed from parents to offspring. According to Handel and Romagopalan "epigenetics allows the peaceful co-existence of Darwinian and Lamarckian evolution."

In their book *An Introduction to Zoology* (2013), Joseph Springer and Dennis Holley wrote:

Lamarck and his ideas were ridiculed and discredited. In a strange twist of fate, Lamarck may have the last laugh. Epigenetics, an emerging field of genetics, has shown that Lamarck may have been at least partially correct all along. It seems that reversible and heritable changes can occur without a change in DNA sequence (genotype) and that such changes may be induced spontaneously or in response to environmental factors—Lamarck's "acquired traits." Determining which observed phenotypes are genetically inherited and which are environmentally induced remains an important and ongoing part of the study of genetics, developmental biology, and medicine.

Eugene Koonin has written that the prokaryotic CRISPR system and Piwi-interacting RNA could be classified as Lamarckian and came to the conclusion that "Both Darwinian and Lamarckian modalities of evolution appear to be important, and reflect different aspects of the interaction between populations and the environment."

A study in 2013 reported that mutations caused by a father's lifestyle can be inherited by his children through multiple generations. A study from Lund University in Sweden showed that exercise changes the epigenetic pattern of genes that affect fat storage in the body.

Commenting on this, Charlotte Ling explained:

The cells of the body contain DNA, which contains genes. We inherit our genes and they cannot be changed. The genes, however, have 'methyl groups' attached which affect what is known as 'gene expression' – whether the genes are activated or deactivated. The methyl groups can be influenced in various ways, through exercise, diet and lifestyle, in a process known as 'DNA methylation'.

A 2013 study published in *Nature Neuroscience* reported that mice trained to fear the smell of a chemical called acetophenone passed their fear onto at least two generations. The science magazine *New Scientist* commented on the study saying, "While it needs to be corroborated, this finding seems consistent with Lamarckian inheritance. It is, however, based on epigenetics: changes that tweak the action of genes, not the genes themselves. So it fits with natural selection – and may yet give Lamarck's name a sheen of respectability."

Guy Barry wrote that Darwin's hypothesis pangenesis coupled with "Lamarckian somatic cell-derived epigenetic modifications" and *de novo* RNA and DNA mutations can explain the evolution of the human brain.

Lamarckian elements also appear in the hologenome theory of evolution.

Contrary Views

The significance of epigenetic inheritance to the evolutionary process is uncertain. Critics assert that epigenetic inheritance modifications are not inherited past two or three generations, so are not a stable basis for evolutionary change. According to a recent review in 2015, "there are no reported epigenetic marks transmitted via the male germ line during more than three generations."

Jerry Coyne has criticized Lamarckian claims of epigenetic inheritance.

The evolutionary biologist T. Ryan Gregory contends that epigenetic inheritance should not be considered Lamarckian. According to Gregory, Lamarck did not claim the environment imposed direct effects on organisms. Instead, Lamarck "argued that the environment created needs to which organisms responded by using some features more and others less, that this resulted in those features being accentuated or attenuated, and that this difference was then inherited by offspring." Gregory has stated that Lamarckian evolution in the context of epigenetics is actually closer to the view held by Darwin rather than by Lamarck.

In a paper titled *Weismann Rules! OK? Epigenetics and the Lamarckian Temptation* (2007), David Haig writes that research into epigenetic processes does allow a Lamarckian element in evolution but the processes do not challenge the main tenets of the modern evolutionary synthesis as modern Lamarckians have claimed. Haig argued for the primary of DNA and evolution of epigenetic switches by natural selection. Haig has also written there is a "visceral attraction" to Lamarckian evolution from the public and some scientists as it posits the world with a meaning, in which organisms can shape their own evolutionary destiny.

American biologist Jerry Coyne has stated that "lots of studies show us that Lamarckian inheritance doesn't operate" and epigenetic changes are rarely passed on to future generations, thus do not serve as the basis of evolutionary change. Coyne has also written:

Lamarckism is not a "heresy," but simply a hypothesis that hasn't held up... If "epigenetics" in the second sense is so important in evolution, let us have a list of, say, a hundred adaptations of organisms that evolved in this Larmackian way as opposed to the old, boring, neo-Darwinian way involving inherited changes in DNA sequence... I can't think of a single entry for that list.

Thomas Dickens and Qazi Rahman (2012) have written epigenetic mechanisms such as DNA methylation and histone modification are genetically inherited under the control of natural selection and do not challenge the modern synthesis. Dickens and Rahman have taken issue with the claims of Eva Jablonka and Marion J. Lamb on Lamarckian epigenetic processes.

Edith Heard and Robert Martienssen (2014) in a *Cell* review were not convinced that epigenetics has revived Lamarckism as there is no evidence epigenetic changes are passed on to successive generations in mammals. They concluded the characteristics that are thought to be the result of epigenetic inheritance may be caused by other factors such as behavioral changes, undetected mutations, microbiome alterations or the transmission of metabolites.

In 2015, Khursheed Iqbal and colleagues discovered that although "endocrine disruptors exert direct epigenetic effects in the exposed fetal germ cells, these are corrected by reprogramming events in the next generation." Molecular biologist Emma Whitelaw has cited this study as an example of evidence disputing Lamarckian epigenetic inheritance. Another critic recently argued that bringing back Lamarck in the context of epigenetics is misleading, commenting, "We should remember [Lamarck] for the good he contributed to science, not for things that resemble his theory only superficially. Indeed, thinking of CRISPR and other phenomena as Lamarckian only obscures the simple and elegant way evolution really works."

Natural Selection

Natural selection is the differential survival and reproduction of individuals due to differences in phenotype. It is a key mechanism of evolution, the change in heritable traits of a population over time. Charles Darwin popularised the term "natural selection", and compared it with artificial selection.

Variation exists within all populations of organisms. This occurs partly because random mutations arise in the genome of an individual organism, and offspring can inherit such mutations. Throughout the lives of the individuals, their genomes interact with their environments to cause variations in traits. The environment of a genome includes the molecular biology in the cell, other cells, other individuals, populations, species, as well as the abiotic environment. Individuals with certain variants of the trait may survive and reproduce more than individuals with other, less successful, variants. Therefore, the population evolves. Factors that affect reproductive success are also important, an issue that Darwin developed in his ideas on sexual selection (now often included in natural selection) and fecundity selection.

Natural selection acts on the phenotype, or the observable characteristics of an organism, but the genetic (heritable) basis of any phenotype that gives a reproductive advantage may become more common in a population. Over time, this process can result in populations that specialise for particular ecological niches (microevolution) and may eventually result in the emergence of new species (macroevolution). In other words, natural selection is a key process in the evolution of a population. Natural selection can be contrasted with artificial selection, in which humans intentionally choose specific traits, whereas in natural selection there is no intentional choice.

Natural selection is one of the cornerstones of modern biology. The concept, published by Darwin and Alfred Russel Wallace in a joint presentation of papers in 1858, was elaborated in Darwin's influential 1859 book *On the Origin of Species by Means of Natural Selection, or the Preservation of Favoured Races in the Struggle for Life*, which described natural selection as analogous to artificial selection, a process by which animals and plants with traits considered desirable by human breeders are systematically favoured for reproduction. The concept of natural selection originally developed in the absence of a valid theory of heredity; at the time of Darwin's writing, science had yet to develop modern theories of genetics. The union of traditional Darwinian evolution with subsequent discoveries in classical genetics formed the modern evolutionary synthesis of the mid-20th century. The addition of molecular genetics has led to evolutionary developmental biology, which explains evolution at the molecular level. While genotypes can slowly change by random genetic drift, natural selection remains the primary explanation for adaptive evolution.

Historical Development

Pre-Darwinian Theories

Several philosophers of the classical era, including Empedocles and his intellectual successor, the Roman poet Lucretius, expressed the idea that nature produces a huge variety of creatures, randomly, and that only those creatures that manage to provide for themselves and reproduce successfully persist. Empedocles' idea that organisms arose entirely by the incidental workings of causes such as heat and cold was criticised by Aristotle in Book II of *Physics*. He posited natural teleology in its place, and believed that form was achieved for a purpose, citing the regularity of heredity in species as proof. Nevertheless, he accepted in his biology that new types of animals, monstrosities, can occur in very rare instances (*Generation of Animals*, Book IV). As quoted in Darwin's 1872 edition of *The Origin of Species*, Aristotle considered whether different forms (e.g., of teeth) might have appeared accidentally, but only the useful forms survived:

Aristotle considered whether different forms could have appeared, only the useful ones surviving.

So what hinders the different parts [of the body] from having this merely accidental relation in nature? as the teeth, for example, grow by necessity, the front ones sharp, adapted for dividing, and the grinders flat, and serviceable for masticating the food; since they were not made for the sake of this, but it was the result of accident. And in like manner as to the other parts in which there appears to exist an adaptation to an end. Wheresoever, therefore, all things together (that is all the parts of one whole) happened like as if they were made for the sake of something, these were preserved, having been appropriately constituted by an internal spontaneity, and whatsoever things were not thus constituted, perished, and still perish.

—Aristotle, Physics, Book II, Chapter 8

But Aristotle rejected this possibility in the next paragraph, making clear that he is talking about the development of animals as embryos with the phrase "either invariably or normally come about", not the origin of species:

... Yet it is impossible that this should be the true view. For teeth and all other natural things either invariably or normally come about in a given way; but of not one of the results of chance or spontaneity is this true. We do not ascribe to chance or mere coincidence the frequency of rain in winter, but frequent rain in summer we do; nor heat in the dog-days, but only if we have it in winter. If then, it is agreed that things are either the result of coincidence or for an end, and these cannot be the result of coincidence or spontaneity, it follows that they must be for an end; and that such things are all due to nature even the champions of the theory which is before us would agree. Therefore action for an end is present in things which come to be and are by nature.

—Aristotle, Physics, Book II, Chapter 8

The struggle for existence was later described by the Islamic writer Al-Jahiz in the 9th century.

The classical arguments were reintroduced in the 18th century by Pierre Louis Maupertuis and others, including Darwin's grandfather, Erasmus Darwin.

Until the early 19th century, the prevailing view in Western societies was that differences between individuals of a species were uninteresting departures from their Platonic ideals (or typus) of created kinds. However, the theory of uniformitarianism in geology promoted the idea that simple, weak forces could act continuously over long periods of time to produce radical changes in the Earth's landscape. The success of this theory raised awareness of the vast scale of geological time and made plausible the idea that tiny, virtually imperceptible changes in successive generations could produce consequences on the scale of differences between species.

The early 19th-century zoologist Jean-Baptiste Lamarck suggested the inheritance of acquired characteristics as a mechanism for evolutionary change; adaptive traits acquired by an organism during its lifetime could be inherited by that organism's progeny, eventually causing transmutation of species. This theory, Lamarckism, was an influence on the Soviet biologist Trofim Lysenko's antagonism to mainstream genetic theory as late as the mid 20th century.

Between 1835 and 1837, the zoologist Edward Blyth worked on the area of variation, artificial selection, and how a similar process occurs in nature. Darwin acknowledged Blyth's ideas in the first chapter on variation of *On the Origin of Species.*

Darwin's Theory

In 1859, Charles Darwin set out his theory of evolution by natural selection as an explanation for adaptation and speciation. He defined natural selection as the "principle by which each slight variation [of a trait], if useful, is preserved." The concept was simple but powerful: individuals best adapted to their environments are more likely to survive and reproduce. As long as there is some variation between them and that variation is heritable, there will be an inevitable selection of individuals with the most advantageous variations. If the variations are heritable, then differential reproductive success leads to a progressive evolution of particular populations of a species, and populations that evolve to be sufficiently different eventually become different species.

The modern theory of natural selection derives from Charles Darwin's work in the nineteenth century.

Darwin's ideas were inspired by the observations that he had made on the second voyage of HMS *Beagle* (1831–1836), and by the work of a political economist, Thomas Robert Malthus, who, in *An Essay on the Principle of Population* (1798), noted that population (if unchecked) increases exponentially, whereas the food supply grows only arithmetically; thus, inevitable limitations of resources would have demographic implications, leading to a "struggle for existence." When Darwin read Malthus in 1838 he was already primed by his work as a naturalist to appreciate the "struggle for existence" in nature and it struck him that as population outgrew resources, "favourable variations would tend to be preserved, and unfavourable ones to be destroyed. The result of this would be the formation of new species." Darwin wrote:

If during the long course of ages and under varying conditions of life, organic beings vary at all in the several parts of their organisation, and I think this cannot be disputed; if there be, owing to the high geometrical powers of increase of each species, at some age, season, or year, a severe struggle for life, and this certainly cannot be disputed; then, considering the infinite complexity of the relations of all organic beings to each other and to their conditions of existence, causing an infinite diversity in structure, constitution, and habits, to be advantageous to them, I think it would be a most extraordinary fact if no variation ever had occurred useful to each being's own welfare, in the same way as so many variations have occurred useful to man. But if variations useful to any organic being do occur, assuredly individuals thus characterised will have the best chance of being preserved in the struggle for life; and from the strong principle of inheritance they will tend to produce offspring similarly characterised. This principle of preservation, I have called, for the sake of brevity, Natural Selection.

—Darwin summarising natural selection in the fourth chapter of On the Origin of Species

Once he had his theory, Darwin was meticulous about gathering and refining evidence before making his idea public. He was in the process of writing his "big book" to present his research when the naturalist Alfred Russel Wallace independently conceived of the principle and described it in an essay he sent to Darwin to forward to Charles Lyell. Lyell and Joseph Dalton Hooker decided (without Wallace's knowledge) to present his essay together with unpublished writings that Darwin had sent to fellow naturalists, and *On the Tendency of Species to form Varieties; and on the Perpetuation of Varieties and Species by Natural Means of Selection* was read to the Linnean Society of London announcing co-discovery of the principle in July 1858. Darwin published a detailed account of his evidence and conclusions in *On the Origin of Species* in 1859. In the 3rd edition of 1861 Darwin acknowledged that others—like William Charles Wells in 1813, and Patrick Matthew in 1831—had proposed similar ideas, but had neither developed them nor presented them in notable scientific publications.

Darwin thought of natural selection by analogy to how farmers select crops or livestock for breeding, which he called "artificial selection"; in his early manuscripts he referred to a "Nature" which would do the selection. At the time, other mechanisms of evolution such as evolution by genetic drift were not yet explicitly formulated, and Darwin believed that selection was likely only part of the story: "I am convinced that Natural Selection has been the main but not exclusive means of modification." In a letter to Charles Lyell in September 1860, Darwin regretted the use of the term "Natural Selection," preferring the term "Natural Preservation."

For Darwin and his contemporaries, natural selection was in essence synonymous with evolution by natural selection. After the publication of *On the Origin of Species*, educated people generally accepted that evolution had occurred in some form. However, natural selection remained controversial as a mechanism, partly because it was perceived to be too weak to explain the range of observed characteristics of living organisms, and partly because even supporters of evolution balked at its "unguided" and non-progressive nature, a response that has been characterised as the single most significant impediment to the idea's acceptance. However, some thinkers enthusiastically embraced natural selection; after reading Darwin, Herbert Spencer introduced the phrase *survival of the fittest*, which became a popular summary of the theory. The fifth edition of *On the Origin of Species* published in 1869 included Spencer's phrase as an alternative to natural selection, with credit given: "But the expression often used by Mr. Herbert Spencer of the Survival of the Fittest is more accurate, and is sometimes equally convenient." Although the phrase is still often used by non-biologists, modern biologists avoid it because it is tautological if "fittest" is read to mean "functionally superior" and is applied to individuals rather than considered as an averaged quantity over populations.

The "Modern Evolutionary Synthesis" of the Mid-20th Century

Natural selection relies crucially on the idea of heredity, but developed before the basic concepts of genetics. Although the Moravian monk Gregor Mendel, the father of modern genetics, was a contemporary of Darwin's, his work lay in obscurity, only being rediscovered in 1900. Only after the mid-20th century integration of evolution with Mendel's laws of inheritance, the so-called modern evolutionary synthesis, did scientists generally come to accept natural selection. The synthesis grew from advances in different fields. Ronald Fisher developed the required mathematical language and wrote *The Genetical Theory of Natural Selection* (1930). J. B. S. Haldane introduced the concept of the "cost" of natural selection. Sewall Wright elucidated the nature of selection and adaptation. Theodosius Dobzhansky established the idea that mutation, by creating genetic diversity, supplied the raw material for natural selection, in his book *Genetics and the Origin of Species* (1937). Ernst Mayr recognised the key importance of reproductive isolation for speciation in his *Systematics and the Origin of Species* (1942). W. D. Hamilton conceived of kin selection in 1964. This synthesis cemented natural selection as the foundation of evolutionary theory, where it remains today. A second synthesis was brought about at the end of the 20th century by molecular genetics, creating the field of evolutionary developmental biology (evo-devo), which seeks to explain evolution at a molecular level.

Terminology

The term *natural selection* is most often defined to operate on heritable traits, because these directly participate in evolution. However, natural selection is "blind" in the sense that changes in phenotype can give a reproductive advantage regardless of whether or not the trait is heritable. Following Darwin's primary usage, the term is used to refer both to the evolutionary consequence of blind selection and to its mechanisms. It is sometimes helpful to explicitly distinguish between selection's mechanisms and its effects; when this distinction is important, scientists define "(phenotypic) natural selection" specifically as "those mechanisms that contribute to the selection of individuals that reproduce", without regard to whether the basis of the selection is heritable. Traits that cause greater reproductive success of an organism are said to be *selected for*, while those that reduce success are *selected against*.

Mechanism

Heritable Variation, Differential Reproduction

During the industrial revolution, pollution killed many lichens, leaving tree trunks dark. A dark (melanic) morph of the peppered moth largely replaced the formerly usual light morph. Since the moths are subject to predation by birds hunting by sight, the colour change offers better camouflage against the changed background, suggesting natural selection at work.

Natural variation occurs among the individuals of any population of organisms. Some differences may improve an individual's chances of surviving and reproducing, so that its lifetime reproductive rate is increased: it leaves more offspring. If the traits that give these individuals a reproductive advantage are also heritable, that is, passed from parent to offspring, then there will be differential reproduction, that is, a slightly higher proportion of fast rabbits or efficient algae in the next generation. Even if the reproductive advantage is very slight, over many generations any advantageous heritable trait becomes dominant in the population. In this way the natural environment of an organism "selects for" traits that confer a reproductive advantage, causing evolutionary change, as Darwin described. This gives the appearance of purpose, but in natural selection there is no intentional choice. Artificial selection is purposive where natural selection is not, though biologists often use teleological language to describe it.

The peppered moth exists in both light and dark colours in Great Britain, but during the industrial revolution, many of the trees on which the moths rested became blackened by soot, giving the dark-coloured moths an advantage in hiding from predators. This gave dark-coloured moths a better chance of surviving to produce dark-coloured offspring, and in just fifty years from the first dark moth being caught, nearly all of the moths in industrial Manchester were dark. The balance was reversed by the effect of the Clean Air Act 1956, and the dark moths became rare again, demonstrating the influence of natural selection on peppered moth evolution.

Fitness

The concept of fitness is central to natural selection. In broad terms, individuals that are more "fit" have better potential for survival, as in the well-known phrase "survival of the fittest", but the precise meaning of the term is much more subtle. Modern evolutionary theory defines fitness not by how long an organism lives, but by how successful it is at reproducing. If an organism lives half as long as others of its species, but has twice as many offspring surviving to adulthood, its genes become more common in the adult population of the next generation. Though natural selection acts on individuals, the effects of chance mean that fitness can only really be defined "on average" for the individuals within a population. The fitness of a particular genotype corresponds to the average effect on all individuals with that genotype.

Competition

In biology, competition is an interaction between organisms in which the fitness of one is lowered by the presence of another. This may be because both rely on a limited supply of a resource such as food, water, or territory. Competition may be within or between species, and may be direct or indirect. Species less suited to compete should in theory either adapt or die out, since competition plays a powerful role in natural selection, but according to the "room to roam" theory it may be less important than expansion among larger clades.

Competition is modelled by *r/K* selection theory, which is based on Robert MacArthur and E. O. Wilson's work on island biogeography. In this theory, selective pressures drive evolution in one of two stereotyped directions: *r*- or *K*-selection. These terms, *r* and *K*, can be illustrated in a logistic model of population dynamics:

where *r* is the growth rate of the population (*N*), and *K* is the carrying capacity of its local environmental setting. Typically, *r*-selected species exploit empty niches, and produce many offspring, each with a relatively low probability of surviving to adulthood. In contrast, *K*-selected species are strong competitors in crowded niches, and invest more heavily in much fewer offspring, each with a relatively high probability of surviving to adulthood.

Types of Selection

Natural selection can act on any heritable phenotypic trait, and selective pressure can be produced by any aspect of the environment, including sexual selection and competition with members of the same or other species. However, this does not imply that natural selection is always directional and results in adaptive evolution; natural selection often results in the maintenance of the status quo by eliminating less fit variants.

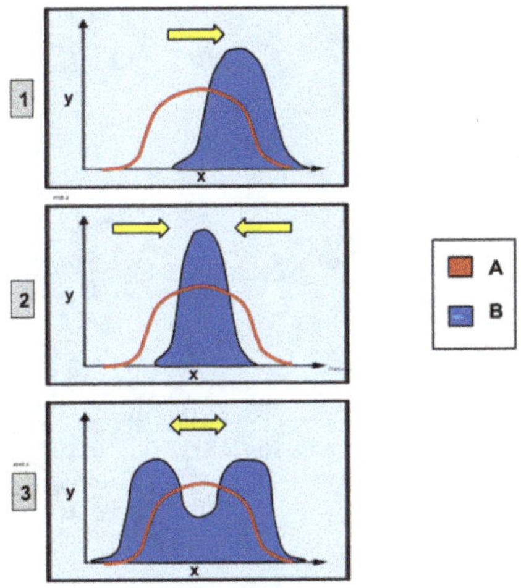

1: directional selection: a single extreme phenotype favoured.
2, stabilizing selection: intermediate favoured over extremes.
3: disruptive selection: extremes favoured over intermediate.
X-axis: phenotypic trait
Y-axis: number of organisms
Group A: original population
Group B: after selection

Selection can be classified in several different ways, such as by its effect on a trait, on genetic diversity, by the life cycle stage where it acts, by the unit of selection, or by the resource being competed for.

Selection has different effects on traits. Stabilizing selection acts to hold a trait at a stable optimum, and in the simplest case all deviations from this optimum are selectively disadvantageous. Directional selection favours extreme values of a trait. The uncommon disruptive selection also acts during transition periods when the current mode is sub-optimal, but alters the trait in more than one direction. In particular, if the trait is quantitative and univariate then both higher and lower trait levels are favoured. Disruptive selection can be a precursor to speciation.

Alternatively, selection can be divided according to its effect on genetic diversity. Purifying or negative selection acts to remove genetic variation from the population (and is opposed by *de novo* mutation, which introduces new variation. In contrast, balancing selection acts to maintain genetic variation in a population, even in the absence of *de novo* mutation, by negative frequency-dependent selection. One mechanism for this is heterozygote advantage, where individuals with two different alleles have a selective advantage over individuals with just one allele. The polymorphism at the human ABO blood group locus has been explained in this way.

Another option is to classify selection by the life cycle stage at which it acts. Some biologists recognise just two types: viability (or survival) selection, which acts to increase an organism's probability of survival, and fecundity (or fertility or reproductive) selection, which acts to increase the rate of reproduction, given survival. Others split the life cycle into further components of selection. Thus viability and survival selection may be defined separately and respectively as acting to improve the probability of survival before and after reproductive age is reached, while fecundity selection may be split into additional sub-components including sexual selection, gametic selection, acting on gamete survival, and compatibility selection, acting on zygote formation.

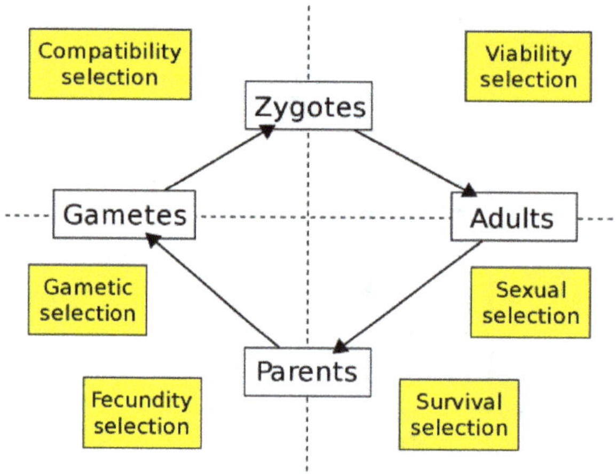

Different types of selection act at each life cycle stage of a sexually reproducing organism.

Selection can also be classified by the level or unit of selection. Individual selection acts on the individual, in the sense that adaptations are "for" the benefit of the individual, and result from selection among individuals. Gene selection acts directly at the level of the gene. In kin selection and intragenomic conflict, gene-level selection provides a more apt explanation of the underlying process. Group selection, if it occurs, acts on groups of organisms, on the assumption that groups replicate and mutate in an analogous way to genes and individuals. There is an ongoing debate over the degree to which group selection occurs in nature.

Finally, selection can be classified according to the resource being competed for. Sexual selection results from competition for mates. Sexual selection typically proceeds via fecundity selection, sometimes at the expense of viability. Ecological selection is natural selection via any means other than sexual selection, such as kin selection, competition, and infanticide. Following Darwin, natural selection is sometimes defined as ecological selection, in which case sexual selection is considered a separate mechanism.

Sexual Selection

Sexual selection refers specifically to competition for mates, which can be *intrasexual*, between individuals of the same sex, that is male–male competition, or *intersexual*, where one gender chooses mates. However, some species exhibit sex-role reversed behaviour in which it is males that are most selective in mate choice; such as in some fishes of the family Syngnathidae, though likely examples have also been found in sexual selection in amphibians, birds, mammals (including humans) and reptiles.

The peacock's tail is a classic example of sexual selection.

Alligator pipefish males select females.

Phenotypic traits can be displayed in one sex and desired in the other sex, causing a positive feedback loop called a Fisherian runaway, for example, the extravagant plumage of some male birds. An alternate theory proposed by the same Ronald Fisher in 1930 is the sexy son hypothesis, that mothers want promiscuous sons to give them large numbers of grandchildren and so choose promiscuous fathers for their children. Aggression between members of the same sex is sometimes associated with very distinctive features, such as the antlers of stags, which are used in combat with other stags. More generally, intrasexual selection is often associated with sexual dimorphism, including differences in body size between males and females of a species.

Natural Selection in Action

Natural selection is seen in action in the development of antibiotic resistance in microorganisms. Since the discovery of penicillin in 1928, antibiotics have been used to fight bacterial diseases. The widespread misuse of antibiotics has selected for microbial resistance to antibiotics in clinical use, to the point that the methicillin-resistant *Staphylococcus aureus* (MRSA) has been described as a "superbug" because of the threat it poses to health and its relative invulnerability to existing drugs. Response strategies typically include the use of different, stronger antibiotics; however, new strains of MRSA have recently emerged that are resistant even to these drugs. This is an evolutionary arms race, in which bacteria develop strains less susceptible to antibiotics, while medical researchers attempt to develop new antibiotics that can kill them. A similar situation occurs with pesticide resistance in plants and insects. Arms races are not necessarily induced by man; a well-documented example involves the spread of a gene in the butterfly *Hypolimnas bolina* suppressing male-killing activity by *Wolbachia* bacteria parasites on the island of Samoa, where the spread of the gene is known to have occurred over a period of just five years

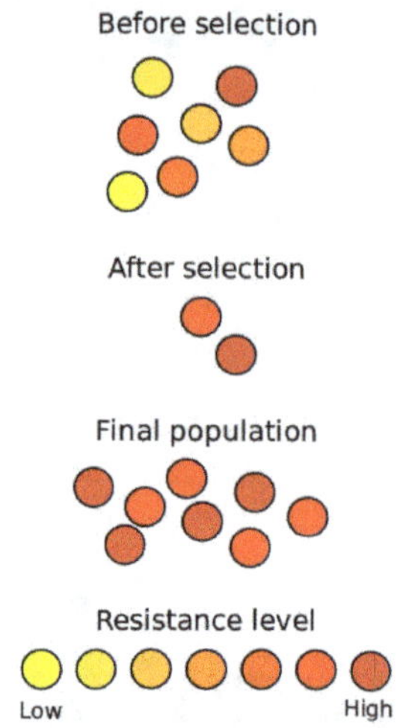

Selection in action: resistance to antibiotics grows though the survival of individuals less affected by the antibiotic. Their offspring inherit the resistance.

Evolution by Means of Natural Selection

A prerequisite for natural selection to result in adaptive evolution, novel traits and speciation is the presence of heritable genetic variation that results in fitness differences. Genetic variation is the result of mutations, genetic recombinations and alterations in the karyotype (the number, shape, size and internal arrangement of the chromosomes). Any of these changes might have an effect that is highly advantageous or highly disadvantageous, but large effects are rare. In the past, most changes in the genetic material were considered neutral or close to neutral because they occurred in noncoding DNA or resulted in a synonymous substitution. However, many mutations in non-coding DNA have deleterious effects. Although both mutation rates and average fitness effects of mutations are dependent on the organism, a majority of mutations in humans are slightly deleterious.

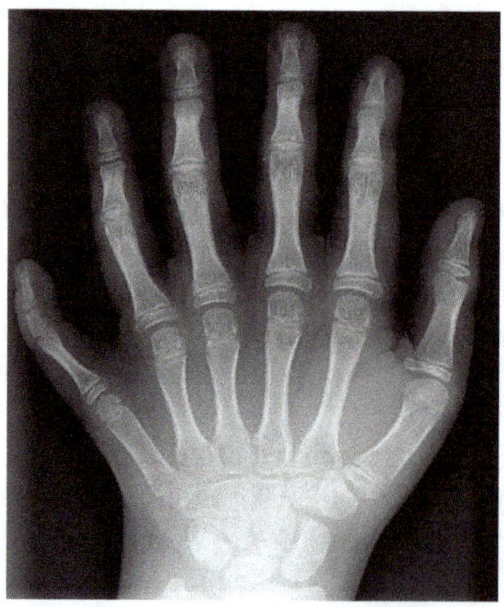

X-ray of the left hand of a ten-year-old boy with polydactyly, caused by a mutant *Hox* gene

Some mutations occur in "toolkit" or regulatory genes. Changes in these often have large effects on the phenotype of the individual because they regulate the function of many other genes. Most, but not all, mutations in regulatory genes result in non-viable embryos. Some nonlethal regulatory mutations occur in HOX genes in humans, which can result in a cervical rib or polydactyly, an increase in the number of fingers or toes. When such mutations result in a higher fitness, natural selection favours these phenotypes and the novel trait spreads in the population. Established traits are not immutable; traits that have high fitness in one environmental context may be much less fit if environmental conditions change. In the absence of natural selection to preserve such a trait, it becomes more variable and deteriorate over time, possibly resulting in a vestigial manifestation of the trait, also called evolutionary baggage. In many circumstances, the apparently vestigial structure may retain a limited functionality, or may be co-opted for other advantageous traits in a phenomenon known as preadaptation. A famous example of a vestigial structure, the eye of the blind mole-rat, is believed to retain function in photoperiod perception.

Speciation

Speciation requires a degree of reproductive isolation—that is, a reduction in gene flow. However, it is intrinsic to the concept of a species that hybrids are selected against, opposing the evolution of reproductive isolation, a problem that was recognised by Darwin. The problem does not occur in allopatric speciation with geographically separated populations, which can diverge with different sets of mutations. E. B. Poulton realized in 1903 that reproductive isolation could evolve through divergence, if each lineage acquired a different, incompatible allele of the same gene. Selection against the heterozygote would then directly create reproductive isolation, leading to the Bateson–Dobzhansky–Muller model, further elaborated by H. Allen Orr and Michael Turelli.

Genetic Basis

Genotype and Phenotype

Natural selection acts on an organism's phenotype, or physical characteristics. Phenotype is determined by an organism's genetic make-up (genotype) and the environment in which the organism lives. When different organisms in a population possess different versions of a gene for a certain trait, each of these versions is known as an allele. It is this genetic variation that underlies differences in phenotype. An example is the ABO blood type antigens in humans, where three alleles govern the phenotype.

Some traits are governed by only a single gene, but most traits are influenced by the interactions of many genes. A variation in one of the many genes that contributes to a trait may have only a small effect on the phenotype; together, these genes can produce a continuum of possible phenotypic values.

Directionality of Selection

When some component of a trait is heritable, selection alters the frequencies of the different alleles, or variants of the gene that produces the variants of the trait. Selection can be divided into three classes, on the basis of its effect on allele frequencies: directional, stabilizing, and purifying selection. Directional selection occurs when an allele has a greater fitness than others, so that it increases in frequency, gaining an increasing share in the population. This process can continue until the allele is fixed and the entire population shares the fitter phenotype. Far more common is stabilizing selection, which lowers the frequency of alleles that have a deleterious effect on the phenotype – that is, produce organisms of lower fitness. This process can continue until the allele is eliminated from the population. Purifying selection conserves functional genetic features, such as protein-coding genes or regulatory sequences, over time by selective pressure against deleterious variants.

Some forms of balancing selection do not result in fixation, but maintain an allele at intermediate frequencies in a population. This can occur in diploid species (with pairs of chromosomes) when heterozygous individuals (with just one copy of the allele) have a higher fitness than homozygous individuals (with two copies). This is called heterozygote advantage or over-dominance, of which the best-known example is the resistance to malaria in humans heterozygous for sickle-cell anaemia. Maintenance of allelic variation can also occur through disruptive or diversifying selection, which favours genotypes that depart from the average in either direction (that is, the opposite of over-dominance), and can result in a bimodal distribution of trait values. Finally, balancing selection can occur through frequency-dependent selection, where the fitness of one particular phenotype depends on the distribution of other phenotypes in the population. The principles of game theory have been applied to understand the fitness distributions in these situations, particularly in the study of kin selection and the evolution of reciprocal altruism.

Selection, Genetic Variation, and Drift

A portion of all genetic variation is functionally neutral, producing no phenotypic effect or significant difference in fitness. Motoo Kimura's neutral theory of molecular evolution by genetic drift proposes that this variation accounts for a large fraction of observed genetic diversity. Neutral events can radically reduce genetic variation through population bottlenecks. which among other things can cause the founder effect in initially small new populations. When genetic variation does not result in differences in fitness, selection cannot directly affect the frequency of such variation. As a result, the genetic variation at those sites is higher than at sites where variation does influence fitness. However, after a period with no new mutations, the genetic variation at these sites is eliminated due to genetic drift. Natural selection reduces genetic variation by eliminating maladapted individuals, and consequently the mutations that caused the maladaptation. At the same time, new mutations occur, resulting in a mutation–selection balance. The exact outcome of the two processes depends both on the rate at which new mutations occur and on the strength of the natural selection, which is a function of how unfavourable the mutation proves to be.

Genetic linkage occurs when the loci of two alleles are in close proximity on a chromosome. During the formation of gametes, recombination reshuffles the alleles. The chance that such a reshuffle occurs between two alleles is inversely related to the distance between them. Selective sweeps occur when an allele becomes more common in a population as a result of positive selection. As the prevalence of one allele increases, closely linked alleles can also become more common by "genetic hitchhiking", whether they are neutral or even slightly deleterious. A strong selective sweep results in a region of the genome where the positively selected haplotype (the allele and its neighbours) are in essence the only ones that exist in the population. Selective sweeps can be detected by measuring linkage disequilibrium, or whether a given haplotype is overrepresented in the population.

Since a selective sweep also results in selection of neighbouring alleles, the presence of a block of strong linkage disequilibrium might indicate a 'recent' selective sweep near the centre of the block.

Background selection is the opposite of a selective sweep. If a specific site experiences strong and persistent purifying selection, linked variation tends to be weeded out along with it, producing a region in the genome of low overall variability. Because background selection is a result of deleterious new mutations, which can occur randomly in any haplotype, it does not produce clear blocks of linkage disequilibrium, although with low recombination it can still lead to slightly negative linkage disequilibrium overall.

Impact

Darwin's ideas, along with those of Adam Smith and Karl Marx, had a profound influence on 19th century thought, including his radical claim that "elaborately constructed forms, so different from each other, and dependent on each other in so complex a manner" evolved from the simplest forms of life by a few simple principles. This inspired some of Darwin's most ardent supporters—and provoked the strongest opposition. Natural selection had the power, according to Stephen Jay Gould, to "dethrone some of the deepest and most traditional comforts of Western thought", such as the belief that humans have a special place in the world.

In the words of the philosopher Daniel Dennett, "Darwin's dangerous idea" of evolution by natural selection is a "universal acid," which cannot be kept restricted to any vessel or container, as it soon leaks out, working its way into ever-wider surroundings. Thus, in the last decades, the concept of natural selection has spread from evolutionary biology to other disciplines, including evolutionary computation, quantum Darwinism, evolutionary economics, evolutionary epistemology, evolutionary psychology, and cosmological natural selection. This unlimited applicability has been called universal Darwinism.

Origin of Life

How life originated from inorganic matter remains an unresolved problem in biology. One prominent hypothesis is that life first appeared in the form of short self-replicating RNA polymers. On this view, life may have come into existence when RNA chains first experienced the basic conditions, as conceived by Charles Darwin, for natural selection to operate. These conditions are: heritability, variation of type, and competition for limited resources. The fitness of an early RNA replicator would likely have been a function of adaptive capacities that were intrinsic (i.e., determined by the nucleotide sequence) and the availability of resources. The three primary adaptive capacities could logically have been: (1) the capacity to replicate with moderate fidelity (giving rise to both heritability and variation of type), (2) the capacity to avoid decay, and (3) the capacity to acquire and process resources. These capacities would have been determined initially by the folded configurations (including those configurations with ribozyme activity) of the RNA replicators that, in turn, would have been encoded in their individual nucleotide sequences.

Cell and Molecular Biology

In 1881, the embryologist Wilhelm Roux published *Der Kampf der Theile im Organismus* (*The Struggle of Parts in the Organism*) in which he suggested that the development of an organism results from a Darwinian competition between the parts of the embryo, occurring at all levels, from molecules to organs. In recent years, a modern version of this theory has been proposed by Jean-Jacques Kupiec. According to this cellular Darwinism, random variation at the molecular level generates diversity in cell types whereas cell interactions impose a characteristic order on the developing embryo.

Social and Psychological Theory

The social implications of the theory of evolution by natural selection also became the source of continuing controversy. Friedrich Engels, a German political philosopher and co-originator of the ideology of communism, wrote in 1872 that "Darwin did not know what a bitter satire he wrote on mankind, and especially on his countrymen, when he showed that free competition, the struggle for existence, which the economists celebrate as the highest historical achievement, is the normal state of the *animal kingdom*." Herbert Spencer and the eugenics advocate Francis Galton's interpretation of natural selection as necessarily progressive, leading to supposed advances in intelligence and civilisation, became a justification for colonialism, eugenics, and social Darwinism. For example, in 1940, Konrad Lorenz, in writings that he subsequently disowned, used the theory as a justification for policies of the Nazi state. He wrote "... selection for toughness, heroism, and social utility ... must be accomplished by some human institution, if mankind, in default of selective factors, is not to be ruined by domestication-induced degeneracy. The racial idea as the basis of our state has already accomplished much in this respect." Others have developed ideas that human societies and culture evolve by mechanisms analogous to those that apply to evolution of species.

More recently, work among anthropologists and psychologists has led to the development of sociobiology and later of evolutionary psychology, a field that attempts to explain features of human psychology in terms of adaptation to the ancestral environment. The most prominent example of evolutionary psychology, notably advanced in the early work of Noam Chomsky and later by Steven Pinker, is the hypothesis that the human brain has adapted to acquire the grammatical rules of natural language. Other aspects of human behaviour and social structures, from specific cultural norms such as incest avoidance to broader patterns such as gender roles, have been hypothesised to have similar origins as adaptations to the early environment in which modern humans evolved. By analogy to the action of natural selection on genes, the concept of memes—"units of cultural transmission," or culture's equivalents of genes undergoing selection and recombination—has arisen, first described in this form by Richard Dawkins in 1976 and subsequently expanded upon by philosophers such as Daniel Dennett as explanations for complex cultural activities, including human consciousness.

Information and Systems Theory

In 1922, Alfred J. Lotka proposed that natural selection might be understood as a physical principle that could be described in terms of the use of energy by a system, a concept later developed by Howard T. Odum as the maximum power principle in thermodynamics, whereby evolutionary systems with selective advantage maximise the rate of useful energy transformation.

The principles of natural selection have inspired a variety of computational techniques, such as "soft" artificial life, that simulate selective processes and can be highly efficient in 'adapting' entities to an environment defined by a specified fitness function. For example, a class of heuristic optimisation algorithms known as genetic algorithms, pioneered by John Henry Holland in the 1970s and expanded upon by David E. Goldberg, identify optimal solutions by simulated reproduction and mutation of a population of solutions defined by an initial probability distribution. Such algorithms are particularly useful when applied to problems whose energy landscape is very rough or has many local minima.

Neuroethology

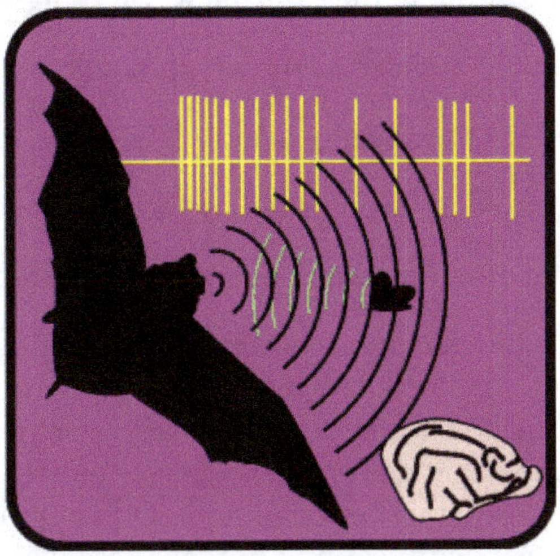

Echolocation in bats is one model system in neuroethology

Neuroethology is the evolutionary and comparative approach to the study of animal behavior and its underlying mechanistic control by the nervous system. This interdisciplinary branch of behavioral neuroscience endeavors to understand how the central nervous system translates biologically relevant stimuli into natural behavior. For example, many bats are capable of echolocation which is used for prey capture and navigation. The auditory system of bats is often cited as an example for how acoustic properties of sounds can be converted into a sensory map of behaviorally relevant features of sounds. Neuroethologists hope to uncover general principles of the nervous system from the study of animals with exaggerated or specialized behaviors.

As its name implies, neuroethology is a multidisciplinary field composed of neurobiology (the study of the nervous system) and ethology (the study of behavior in natural conditions). A central theme of the field of neuroethology, delineating it from other branches of neuroscience, is this focus on natural behavior. Natural behaviors may be thought of as those behaviors generated through means of natural selection (i.e. finding mates, navigation, locomotion, predator avoidance) rather than behaviors in disease states, or behavioral tasks that are particular to the laboratory.

Philosophy

Neuroethology is an integrative approach to the study of animal behavior that draws upon several disciplines. Its approach stems from the theory that animals' nervous systems have evolved to address problems of sensing and acting in certain environmental niches and that their nervous systems are best understood in the context of the problems they have evolved to solve. In accordance with Krogh's principle, neuroethologists often study animals that are "specialists" in the behavior the researcher wishes to study e.g. honeybees and social behavior, bat echolocation, owl sound localization, etc.

The scope of neuroethological inquiry might be summarized by Jörg-Peter Ewert, a pioneer of neuroethology, when he considers the types of questions central to neuroethology in his 1980 introductory text to the field:

1. How are stimuli detected by an organism?
2. How are environmental stimuli in the external world represented in the nervous system?
3. How is information about a stimulus acquired, stored and recalled by the nervous system?
4. How is a behavioral pattern encoded by neural networks?
5. How is behavior coordinated and controlled by the nervous system?
6. How can the ontogenetic development of behavior be related to neural mechanisms?

Often central to addressing questions in neuroethology are comparative methodologies, drawing upon knowledge about related organisms' nervous systems, anatomies, life histories, behaviors and environmental niches. While it is not unusual for many types of neurobiology experiments to give rise to behavioral questions, many neuroethologists often begin their research programs by observing a species' behavior in its natural environment. Other approaches to understanding nervous systems include the systems identification approach, popular in engineering. The idea is to stimulate the system using a non-natural stimulus with certain properties. The system's response to the stimulus may be used to analyze the operation of the system. Such an approach is useful for linear systems, but the nervous system is notoriously nonlinear, and neuroethologists argue that such an approach is limited. This argument is supported by experiments in the auditory system. These experiments show that neural responses to complex sounds, like social calls, can not be predicted by the knowledge gained from studying the responses due to pure tones (one of the non-natural stimuli favored by auditory neurophysiologists). This is because of the non-linearity of the system.

Modern neuroethology is largely influenced by the research techniques used. Neural approaches are necessarily very diverse, as is evident through the variety of questions asked, measuring techniques used, relationships explored, and model systems employed. Techniques utilized since 1984 include the use of intracellular dyes, which make maps of identified neurons possible, and the use of brain slices, which bring vertebrate brains into better observation through intracellular electrodes (Hoyle 1984). Currently, other fields toward which neuroethology may be headed include computational neuroscience, molecular genetics, neuroendocrinology and epigenetics. The existing field of neural modeling may also expand into neuroethological terrain, due to its practical uses in robotics. In all this, neuroethologists must use the right level of simplicity to effectively guide research towards accomplishing the goals of neuroethology.

Critics of neuroethology might consider it a branch of neuroscience concerned with 'animal trivia'. Though neuroethological subjects tend not to be traditional neurobiological model systems (i.e. *Drosophila*, *C. elegans*, or *Danio rerio*), neuroethological approaches emphasizing comparative methods have uncovered many concepts central to neuroscience as a whole, such as lateral inhibition, coincidence detection, and sensory maps. The discipline of neuroethology has also discovered and explained the only vertebrate behavior for which the entire neural circuit has been described: the electric fish jamming avoidance response. Beyond its conceptual contributions, neuroethology makes indirect contributions to advancing human health. By understanding simpler nervous systems, many clinicians have used concepts uncovered by neuroethology and other branches of neuroscience to develop treatments for devastating human diseases.

History

The field of neuroethology owes part of its existence to the establishment of ethology as a unique discipline within the discipline of Zoology. Although animal behavior had been studied since the time of Aristotle (384-342 BC), it was not until the early twentieth century that ethology finally became distinguished from natural science (a strictly descriptive field) and ecology. The main catalysts behind this new distinction were the research and writings of Konrad Lorenz and Niko Tinbergen.

Konrad Lorenz was born in Austria in 1903, and is widely known for his contribution of the theory of fixed action patterns (FAPs): endogenous, instinctive behaviors involving a complex sequence of movements that are triggered ("released") by a certain kind of stimulus. This sequence always proceeds to completion, even if the original stimulus is removed. It is also species-specific and performed by nearly all members. Lorenz constructed his famous "hydraulic model" to help illustrate this concept, as well as the concept of action specific energy, or drives.

Niko Tinbergen was born in the Netherlands in 1907 and worked closely with Lorenz in the development of the FAP theory; their studies focused on the egg retrieval response of nesting geese. Tinbergen performed extensive research on the releasing mechanisms of particular FAPs, and used the bill-pecking behavior of baby herring gulls as his model system. This led to the concept of the supernormal stimulus. Tinbergen is also well known for his four questions that he believed ethologists should be asking about any given animal behavior; among these is that of the mechanism of the behavior, on a physiological, neural and molecular level, and this question can be thought of in many regards as the keystone question in neuroethology. Tinbergen also emphasized the need for ethologists and neurophysiologists to work together in their studies, a unity that has become a reality in the field of neuroethology.

Unlike behaviorism, which studied animals' reactions to non-natural stimuli in artificial, laboratory conditions, ethology sought to categorize and analyze the natural behaviors of animals in a field setting. Similarly, neuroethology asks questions about the neural bases of *naturally occurring* behaviors, and seeks to mimic the natural context as much as possible in the laboratory.

Although the development of ethology as a distinct discipline was crucial to the advent of neuroethology, equally important was the development of a more comprehensive understanding of Neuroscience. Contributors to this new understanding were the Spanish Neuroanatomist, Ramon y Cajal (born in 1852), and physiologists Charles Sherrington, Edgar Adrian, Alan Hodgkin, and Andrew Huxley. Charles Sherrington, who was born in Great Britain in 1857, is famous for his work on the nerve synapse as the site of transmission of nerve impulses, and for his work on reflexes in the spinal cord. His research also led him to hypothesize that every muscular activation is coupled to an inhibition of the opposing muscle. He was awarded a Nobel Prize for his work in 1932 along with Lord Edgar Adrian who made the first physiological recordings of neural activity from single nerve fibers.

Alan Hodgkin and Andrew Huxley (born 1914 and 1917, respectively, in Great Britain), are known for their collaborative effort to understand the production of action potentials in giant squid neurons. The pair also proposed the existence of ion channels to facilitate action potential initiation, and were awarded the Nobel Prize in 1963 for their efforts.

As a result of this pioneering research, many scientists then sought to connect the physiological aspects of the nervous and sensory systems to specific behaviors. These scientists – Karl von Frisch, Erich von Holst, and Theodore Bullock – are frequently referred to as the "fathers" of neuroethology. Neuroethology did not really come into its own, though, until the 1970s and 1980s, when new, sophisticated experimental methods allowed researchers such as Masakazu Konishi, Walter Heiligenberg, Jörg-Peter Ewert, and others to study the neural circuits underlying verifiable behavior.

Modern Neuroethology

The International Society for Neuroethology represents the present discipline of neuroethology, which was founded on the occasion of the NATO-Advanced Study Institute "Advances in Vertebrate Neuroethology" (August 13–24, 1981) organized by J.-P. Ewert, D.J. Ingle and R.R. Capranica, held at the University of Kassel in Hofgeismar, Germany (cf. report Trends in Neurosci. 5:141-143,1982). Its first president was Theodore H. Bullock. The society has met every three years since its first meeting in Tokyo in 1986.

Its membership draws from many research programs around the world; many of its members are students and faculty members from medical schools and neurobiology departments from various universities. Modern advances in neurophysiology techniques have enabled more exacting approaches in an ever-increasing number of animal systems, as size limitations are being dramatically overcome. Survey of the most recent (2007) congress of the ISN meeting symposia topics gives some idea of the field's breadth:

- Comparative aspects of spatial memory (rodents, birds, humans, bats)
- Influences of higher processing centers in active sensing (primates, owls, electric fish, rodents, frogs)
- Animal signaling plasticity over many time scales (electric fish, frogs, birds)
- Song production and learning in passerine birds
- Primate sociality
- Optimal function of sensory systems (flies, moths, frogs, fish)
- Neuronal complexity in behavior (insects, computational)
- Contributions of genes to behavior (Drosophila, honeybees, zebrafish)
- Eye and head movement (crustaceans, humans, robots)

- Hormonal actions in brain and behavior (rodents, primates, fish, frogs, and birds)
- Cognition in insects (honeybee)

Application to Technology

Neuroethology can help create advancements in technology through an advanced understanding of animal behavior. Model systems were generalized from the study of simple and related animals to humans. For example, the neuronal cortical space map discovered in bats, a specialized champion of hearing and navigating, elucidated the concept of a computational space map. In addition, the discovery of the space map in the barn owl led to the first neuronal example of the Jeffress model. This understanding is translatable to understanding spatial localization in humans, a mammalian relative of the bat. Today, knowledge learned from neuroethology are being applied in new technologies. For example, Randall Beer and his colleagues used algorithms learned from insect walking behavior to create robots designed to walk on uneven surfaces (Beer et al.). Neuroethology and technology contribute to one another bidirectionally.

Neuroethologists seek to understand the neural basis of a behavior as it would occur in an animal's natural environment but the techniques for neurophysiological analysis are lab-based, and cannot be performed in the field setting. This dichotomy between field and lab studies poses a challenge for neuroethology. From the neurophysiology perspective, experiments must be designed for controls and objective rigor, which contrasts with the ethology perspective—that the experiment be applicable to the animal's natural condition, which is uncontrolled, or subject to the dynamics of the environment. An early example of this is when Walter Rudolf Hess developed focal brain stimulation technique to examine a cat's brain controls of vegetative functions in addition to other behaviors. Even though this was a breakthrough in technological abilities and technique, it was not used by many neuroethologists originally because it compromised a cat's natural state, and, therefore, in their minds, devalued the experiments' relevance to real situations.

When intellectual obstacles like this were overcome, it led to a golden age of neuroethology, by focusing on simple and robust forms of behavior, and by applying modern neurobiological methods to explore the entire chain of sensory and neural mechanisms underlying these behaviors (Zupanc 2004). New technology allows neuroethologists to attach electrodes to even very sensitive parts of an animal such as its brain while it interacts with its environment. The founders of neuroethology ushered this understanding and incorporated technology and creative experimental design. Since then even indirect technological advancements such as battery-powered and waterproofed instruments have allowed neuroethologists to mimic natural conditions in the lab while they study behaviors objectively. In addition, the electronics required for amplifying neural signals and for transmitting them over a certain distance have enabled neuroscientists to record from behaving animals performing activities in naturalistic environments. Emerging technologies can complement neuroethology, augmenting the feasibility of this valuable perspective of natural neurophysiology.

Another challenge, and perhaps part of the beauty of neuroethology, is experimental design. The value of neuroethological criteria speak to the reliability of these experiments, because these discoveries represent behavior in the environments in which they evolved. Neuroethologists foresee future advancements through using new technologies and techniques, such as computational neuroscience, neuroendocrinology, and molecular genetics that mimic natural environments.

Case Studies

Jamming Avoidance Response

In 1963, Akira Watanabe and Kimihisa Takeda discovered the behavior of the jamming avoidance response in the knifefish *Eigenmannia* sp. In collaboration with T.H. Bullock and colleagues, the behavior was further developed. Finally, the work of W. Heiligenberg expanded it into a full neuroethology study by examining the series of neural connections that led to the behavior. *Eigenmannia* is a weakly electric fish that can generate electric discharges through electrocytes in its tail. Furthermore, it has the ability to electrolocate by analyzing the perturbations in its electric field. However, when the frequency of a neighboring fish's current is very close (less than 20 Hz difference) to that of its own, the fish will avoid having their signals interfere through a behavior known as Jamming Avoidance Response. If the neighbor's frequency is higher than the fish's discharge frequency, the fish will lower its frequency, and vice versa. The sign of the frequency difference is determined by analyzing the "beat" pattern of the incoming interference which consists of the combination of the two fish's discharge patterns.

Neuroethologists performed several experiments under *Eigenmannia*'s natural conditions to study how it determined the sign of the frequency difference. They manipulated the fish's discharge by injecting it with curare which prevented its natural electric organ from discharging. Then, an electrode was placed in its mouth and another was placed at the tip of its tail. Likewise, the neighboring fish's electric field was mimicked using another set of electrodes. This experiment allowed neuroethologists to manipulate different discharge frequencies and observe the fish's behavior. From the results, they were able to conclude that the electric field frequency, rather than an internal frequency measure, was used as a reference. This experiment is significant in that not only does it reveal a crucial neural mechanism underlying the behavior but also demonstrates the value neuroethologists place on studying animals in their natural habitats.

Feature Analysis in Toad Vision

The recognition of prey and predators in the toad was first studied in depth by Jörg-Peter Ewert (Ewert 1974). He began by observing the natural prey-catching behavior of the common toad (*Bufo bufo*) and concluded that the animal followed a sequence that consisted of stalking, binocular fixation, snapping, swallowing and mouth-wiping. However, initially, the toad's actions were dependent on specific features of the sensory stimulus: whether it demonstrated worm or antiworm configurations. It was observed that the worm configuration, which signaled prey, was initiated by movement along the object's long axis, whereas antiworm configuration, which signaled predator, was due to movement along the short axis. (Zupanc 2004).

Ewert and coworkers adopted a variety of methods to study the predator versus prey behavior response. They conducted recording experiments where they inserted electrodes into the brain, while the toad was presented with worm or anti-worm stimuli. This technique was repeated at different levels of the visual system and also allowed feature detectors to be identified. In focus was the discovery of prey-selective neurons in the optic tectum, whose axons could be traced towards the snapping pattern generating cells in the hypoglossal nucleus. The discharge patterns of prey-selective tectal neurons in response to prey objects – in freely moving toads – "predicted" prey-catching reactions such as snapping. Another approach, called stimulation experiment, was carried out in freely moving toads. Focal electrical stimuli were applied to different regions of the brain, and the toad's response was observed. When the thalamic-pretectal region was stimulated, the toad exhibited escape responses, but when the tectum was stimulated in an area close to prey-selective neurons, the toad engaged in prey catching behavior (Carew 2000). Furthermore, neuroanatomical experiments were carried out where the toad's thalamic-pretectal/tectal connection was lesioned and the resulting deficit noted: the prey-selective properties were abolished both in the responses of prey-selective neurons and in the prey catching behavior. These and other experiments suggest that prey selectivity results from pretecto-tectal influences.

Ewert and coworkers showed in toads that there are stimulus-response mediating pathways that translate perception (of visual sign stimuli) into action (adequate behavioral responses). In addition there are modulatory loops that initiate, modify or specify this mediation (Ewert 2004). Regarding the latter, for example, the telencephalic caudal ventral striatum is involved in a loop gating the stimulus-response mediation in a manner of directed attention. The telencephalic ventral medial pallium („primordium hippocampi"), however, is involved in loops that either modify prey-selection due to associative learning or specify prey-selection due to non-associative learning, respectively.

Computational Neuroethology

Computational neuroethology (CN or CNE) is concerned with the computer modelling of the neural mechanisms underlying animal behaviors. Computational neuroethology was first argued for in depth by Randall Beer and by Dave Cliff both of whom acknowledged the strong influence of Michael Arbib's *Rana Computatrix* computational model of neural mechanisms for visual guidance in frogs and toads.

CNE systems work within a closed-loop environment; that is, they perceive their (perhaps artificial) environment directly, rather than through human input, as is typical in AI systems. For example, Barlow et al. developed a time-dependent model for the retina of the horseshoe crab *Limulus polyphemus* on a Connection Machine (Model CM-2). Instead of feeding the model retina with idealized input signals, they exposed the simulation to digitized video sequences made underwater, and compared its response with those of real animals.

Fixed Action Pattern

The term fixed action pattern (FAP), or modal action pattern, is sometimes used in ethology to denote an instinctive behavioral sequence that is relatively invariant within the species and almost inevitably runs to completion.

Fixed action patterns, or similar behaviour sequences, are produced by a neural network known as the innate releasing mechanism in response to an external sensory stimulus known as a sign stimulus or releaser. A fixed action pattern is one of the few types of behaviors which was thought to be "hard-wired" and instinctive.

Terminology

The term "sign stimulus", or "releaser", is used to denote a simple feature of a complex stimulus that can elicit a FAP. For example, the red belly of a male stickleback elicits a head-down, attack behaviour in other male sticklebacks. This same response can be elicited by artificial models or objects that contain the sign stimulus of red, for example, a red coloured card.

The terms "sign stimulus" and "releaser" are sometimes used interchangeably; however, they have different meanings. The term "sign stimulus" is used to denote a feature of an animal's environment that elicits a particular response. The term "releaser" is used for a stimulus that has evolved to facilitate communication between conspecifics (animals of the same species).

Examples

Kelp gull chicks peck at red spot on mother's beak to stimulate regurgitating reflex.

Male Stickleback Behaviour

Another example of fixed action patterns is the courtship and aggression behaviour of the male three-spined stickleback during the mating season, described in a series of studies by Niko Tinbergen. During spring, male sticklebacks change colour, establish a territory and build a nest. They attack male sticklebacks that enter their territory, but court females and entice them to enter the nest to lay eggs. Tinbergen used crude models of sticklebacks to investigate which features of male and female sticklebacks elicited attack and courtship behaviour from male sticklebacks. Tinbergen's main findings were that male sticklebacks responded in a relatively invariant way and attacked a model with a red belly, but in contrast, courted a model with a swollen belly.

Egg-retrieval Behaviour

Another example of a behaviour that has been described as a FAP is the egg-retrieval behavior of the graylag goose, reported in classic studies by Niko Tinbergen and Konrad Lorenz. Like many ground-nesting birds, if an egg becomes displaced from the nest, the greylag rolls it back to the nest with its beak. The sight of the displaced egg is the sign stimulus and elicits the egg-retrieval behaviour. If the egg is removed from the goose during the performance of egg-rolling, the bird often continues with the behavior, pulling its head back as if an imaginary egg is still being maneuvered by the underside of its beak. The greylag will also attempt to retrieve other egg-shaped objects, such as a golf ball, door knob, or even a model egg too large to have possibly been laid by the goose itself (i.e. a supernormal stimulus).

Others

Other examples include:

- Some mating dances, commonly carried out by birds, are examples of fixed action patterns. In these cases, the sign stimulus is typically the presence of the opposite sex.
- Kelp gull chicks are stimulated by a red spot on the mother's beak to peck at the spot, which induces regurgitation.
- Some moths instantly fold their wings and drop to the ground if they encounter ultrasonic signals such as those produced by bats; see ultrasound avoidance.
- Mayflies drop their eggs when they encounter a certain pattern of light polarization which indicates they are over water.
- Some lizards and snakes exhibit strike-induced chemosensory searching behavior as a means of relocating bitten and envenomated prey by scent-trailing.

Significance

FAPs, or inflexible behaviour patterns, are significant in animal behavior because they represent the simplest type of behavior in which a particular stimulus nearly always results in an invariable behavioral response. FAPs have been said to be "hard-wired". They are unusual in that they are relatively un-influenced by the environment, once the behaviour has been elicited.

The egg-rolling retrieval behavior of a greylag goose is a widely cited example of a fixed action pattern.

The existence of FAPs is rather unusual in that a fixed response can lead to maladaptive results, whereas flexible behaviors are generally more likely to be adaptive by increasing fitness. Because of this, most behaviors which are both FAPs and occur in more complex animals are usually essential to the animal's fitness, or in which speed (i.e. an absence of learning) is a factor. For instance, the greylag goose's egg rolling behavior is so essential to the survival of its chicks that the fitness of the parent bird is increased by the behavior being relatively invariant. A stickleback will attack any male fish who enters his territory while the female is sexually receptive, reacting to their red colour, while the female stickleback triggers behavior in the male resulting in the fertilization of her eggs. Relatively invariant behaviors are also predictable, which can lead to their exploitation by humans or other animals.

Exploitation

Some species have evolved to exploit the FAPs of other species by mimicry of their sign stimuli. Replicating the releaser required to trigger a FAP is known as *code-breaking*. A well-known example of this is brood parasitism, where one species will lay its eggs in the nest of another species, which will then parent its young. A young North American cowbird, for example, provides a supernormal stimulus to its foster parent, which will cause it to forage rapidly to satisfy the larger bird's demands. A nestling will provide higher levels of stimulus with noisier, more energetic behavior, communicating its urgent need for food. Parents in this situation have to work harder to provide food, otherwise their own offspring are likely to die of starvation.

Brood parasites, such as the cuckoo, provide a supernormal stimulus to the parenting species.

References

- Zahavi, Amotz (1997). The handicap principle: a missing piece of Darwin's puzzle. Oxford: Oxford University Press. ISBN 0-19-510035-2.
- Maynard Smith, J. (1989). "Evolution in structured populations". Evolutionary Genetics. Oxford: Oxford University Press. pp. 168–169, 181–183. ISBN 0198542151.
- Dawkins, R. (1989). "You scratch my back, I'll ride on yours.". The Selfish Gene. Oxford: Oxford University Press. pp. 183–185. ISBN 0192860925.
- Maynard Smith, J. (1989). "Evolution in structured populations.". Evolutionary Genetics. Oxford: Oxford University Press. pp. 173–175. ISBN 0198542151.
- Williams, G.C. (1972) Adaptation and Natural Selection: A Critique of Some Current Evolutionary Thought. Princeton University Press.ISBN 0-691-02357-3
- Nicholas B. Davies; John R. Krebs; Stuart A. West (April 9, 2012). "11". An Introduction to Behavioural Ecology. John Wiley & Sons. pp. 307–333. ISBN 978-1-4443-3949-9.
- Teske, Nathan (2009). Political Activists in America: The Identity Construction Model of Political Participation. University Park, Pa.: Pennsylvania State University Press. p. 101. ISBN 9780271035468.

- Bell, Graham (2008). Selection: The Mechanism of Evolution. Oxford: Oxford University Press. pp. 367–368. ISBN 0-19-856972-6.
- Davies, Nicholas B.; Krebs, John R.; West, Stuart A. (2012). An Introduction to Behavioral Ecology. West Sussex, UK: Wiley-Blackwell. pp. 307–333. ISBN 978-1-4051-1416-5.
- Davies, N. B., Krebs, J. R and West, S. A., (2012). An Introduction to Behavioral Ecology. West Sussex, UK: Wiley-Blackwell. pp. 266. ISBN 978-1-4051-1416-5.
- Nicholas B. Davies; John R. Krebs; Stuart A. West (2012). An Introduction to Behavioral Ecology. West Sussex, UK: Wiley-Blackwell. pp. 193–202. ISBN 978-1-4051-1416-5.
- Hohmann, Ulf; Bartussek, Ingo; Böer, Bernhard (2001). Der Waschbär (in German). Reutlingen, Germany: Oertel+Spörer. ISBN 978-3-88627-301-0.
- Cheney, D. L.; Seyfarth, R. M. (1990). How monkeys see the world: Inside the mind of another species. University of Chicago Press. ISBN 978-0-226-10246-7.
- Randall D. Beer (1990). Intelligence as Adaptive Behavior: An experiment in computational neuroethology. Academic Press. ISBN 0-12-084730-2.

4

Animal Communication and Migration

Animal communication is the communication between one or a group of animals. This subject is rapidly growing as a field of study. Bee learning and communication, deception in animals, ultrasound avoidance, waggle dance and bird migration are some of the topics discussed in this section. The chapter strategically encompasses and incorporates the major components and key concepts of animal communication and animal migration, providing a complete understanding.

Animal Communication

Animal communication is the transfer of information from one or a group of animals (sender or senders) to one or more other animals (receiver or receivers) that affects the current or future behaviour of the receivers. Information may be sent intentionally, as in a courtship display, or unintentionally, as in the transfer of scent from predator to prey. Information may be transferred to an "audience" of several receivers. Animal communication is a rapidly growing area of study in disciplines including animal behaviour, sociobiology, neurobiology and animal cognition. Many aspects of animal behaviour, such as symbolic name use, emotional expression, learning and sexual behaviour, are being understood in new ways.

Great egret (*Ardea alba*) in a courtship display communicating the desire to find a mate

When the information from the sender changes the behaviour of a receiver, the information is referred to as a "signal". Signalling theory predicts that for a signal to be maintained in the population, both the sender and receiver should usually receive some benefit from the interaction. Signal production by senders and the perception and subsequent response of receivers are thought to coevolve. Signals often involve multiple mechanisms, e.g. both visual and auditory, and for a signal to be understood the coordinated behaviour of both sender and receiver require careful study.

Modes

A lamb investigates a rabbit, an example of interspecific communication using body posture and olfaction.

Visual

- Gestures: The best known form of communication involves the display of distinctive body parts, or distinctive bodily movements; often these occur in combination, so a movement acts to reveal or emphasize a body part. A notable example is the presentation of a parent herring gull's bill to its chick as a signal for feeding. Like many gulls, the herring gull has a brightly coloured bill, yellow with a red spot on the lower mandible near the tip. When the parent returns to the nest with food, it stands over its chick and taps the bill on the ground; this elicits a begging response from a hungry chick (pecking at the red spot), which stimulates the parent to regurgitate food. The complete signal therefore involves a distinctive morphological feature (body part), the red-spotted bill, and a distinctive movement (tapping towards the ground) which makes the red spot highly visible to the chick. While all primates use some form of gesture, Frans de Waal concluded that apes and humans are unique in that only they use intentional gestures to communicate. He tested the hypothesis that gestures evolve into language by studying the gestures of bonobos and chimps.

- Facial expression: Facial gestures play an important role in animal communication. Often a facial gesture is a signal of emotion. Dogs, for example, express anger through snarling and showing their teeth. In alarm their ears perk up, in fear the ears flatten while the dogs expose their teeth slightly and squint their eyes. Jeffrey Mogil studied the facial expressions of mice during increments of increasing pain; there were five recognizable facial expressions; orbital tightening, nose and cheek bulge, and changes in ear and whisker carriage.

- Gaze following: Social animals coordinate their communication by monitoring of each other's head and eye orientation. Such behavior has long been recognized as an important component of communication during human development, and gaze-following has recently received much attention in animals. Studies have been conducted on apes, monkeys, dogs, birds, and tortoises, and have focused on two different tasks: "follow[ing] another's gaze into distant space" and "follow[ing] another's gaze geometrically around a visual barrier e.g. by repositioning themselves to follow a gaze cue when faced with a barrier blocking their view". The first ability has been found among a broad range of animals, while the second has been demonstrated only for apes, dogs, wolves, and corvids (ravens); attempts to demonstrate this "geometric gaze following" in marmoset and ibis gave negative results. Researchers do not yet have a clear picture of the cognitive basis of gaze following, but developmental evidence indicates that "simple" gaze following and "geometric" gaze following probably rely on different cognitive mechanisms.

- Color change: Color change can be separated into changes that occur during growth and development, and those triggered by mood, social context, or abiotic factors such as temperature. The latter are seen in many taxa. Some cephalopods, such as the octopus and the cuttlefish, have specialized skin cells (chromatophores) that can change the apparent colour, opacity, and reflectiveness of their skin. In addition to their use for camouflage, rapid changes in skin color are used while hunting and in courtship rituals. Cuttlefish may display two entirely different signals simultaneously from opposite sides of their body. When a male cuttlefish courts a female in the presence of other males, he displays a male pattern facing the female and a female pattern facing away, to deceive other males. Some color signals occur in cycles. For example, in, when a female olive baboon begins to ovulate, her anogenital area swells and turns a bright red/pink. This signals to males that she is ready to mate.

- Bioluminescent communication: Communication by the production of light occurs commonly in vertebrates and invertebrates in the oceans, particularly at depths (e.g. angler fish). Two well known forms of land bioluminescence occur in fireflies and glow worms. Other insects, insect larvae, annelids, arachnids and even species of fungi possess bioluminescent abilities. Some bioluminescent animals produce the light themselves whereas others have a symbiotic relationship with bioluminescent bacteria.

Auditory

Many animals communicate through vocalization. Vocal communication serves many purposes, including mating rituals, warning calls, conveying location of food sources, and social learning. In a number of species, males perform calls during mating rituals as a form of competition against other males and to signal females. Examples include hammer-headed bats, red deer, humpback whales, elephant seals, and songbirds. Other instances of vocal communication include the alarm calls of the Campbell monkey, the territorial calls of gibbons, and the use of frequency in greater spear-nosed bats to distinguish between groups. The vervet monkey gives a distinct alarm call for each of its four different predators, and the reactions of other monkeys vary appropriately according to the call. For example, if an alarm call signals a python, the monkeys climb into the trees, whereas the "eagle" alarm causes monkeys to seek a hiding place on the ground. Prairie dogs also use complex calls that signal predator differences. According to Con Slobodchikoff and others, prairie dog calls communicate the type, size, and speed of an approaching predator. Whale vocalizations have been found to have different dialects based on region.

Bird calls can serve as alarms or keep members of a flock in contact, while the longer and more complex bird songs are associated with courtship and mating.

Not all animals use vocalization as a means of auditory communication. Many arthropods rub specialized body parts together to produce sound. This is known as stridulation. Crickets and grasshoppers are well known for this, but many others use stridulation as well, including crustaceans, spiders, scorpions, wasps, ants, beetles, butterflies, moths, millipedes, and centipedes. Another means of auditory communication is the vibration of swim bladders in bony fish. The structure of swim bladders and the attached sonic muscles varies greatly across bony fish families, resulting in a wide variety of sounds. Striking body parts together can also produce auditory signals. A well-known example of this is the tail tip vibration of rattlesnakes as a warning signal. Other examples include bill clacking in birds, wing clapping in manakin courtship displays, and chest beating in gorillas.

Olfactory

Despite being the oldest method of communication, chemical communication is one of the least understood forms due in part to the sheer abundance of chemicals in our environment and the difficulty of detecting and measuring all the chemicals in a sample. The ability to detect chemicals in the environment serves many functions, a crucial one being the detection of food, a function that first arose in single-celled organisms (bacteria) living in the oceans during the early days of life on Earth. As this function evolved, organisms began to differentiate between chemicals compounds emanating from resources, conspecifics (same species; i.e., mates and kin), and heterospecifics (different species; i.e., competitors and predators). For instance, a small minnow species may do well to avoid habitat with a detectable concentration of chemical cue associated with a predator species such as northern pike. Minnows with the ability to perceive the presence of predators before they are close enough to be seen and then respond with adaptive behaviour (such as hiding) are more likely to survive and reproduce.

Electro

Electrocommunication is a rare form of communication in animals. It is seen primarily in aquatic animals, though some land mammals, notably the platypus and echidnas, sense electric fields that might be used for communication.

Weakly electric fishes provide an example of electrocommunication, together with electrolocation. These fish use an electric organ to generate an electric field, which is detected by electroreceptors. Differences in the waveform and frequency of changes in the field convey information on species, sex, and identity. These electric signals can be generated in response to hormones, circadian rhythms, and interactions with other fish. Some predators, such as sharks and rays, are able to eavesdrop on these electrogenic fish through passive electroreception.

Touch

Touch is a key factor in many social interactions. Here are some examples:

- Fighting: In a fight, touch may be used to challenge an opponent and to coordinate movements during the fight. It may also be used by the loser to indicate submission. *Mating: Mammals often initiate mating by grooming, stroking or rubbing against each other. This provides the opportunity to apply chemical signals and to assess those excreted by the potential mate. Touch may also announce the intention of the male to mount the female, as when a male kangaroo grabs the tail of a female. During mating, touch stimuli are important for pair positioning, coordination and genital stimulation.
- Social integration: Touch is widely used for social integration, a use that is typified by the social grooming of one animal by another. Social grooming has several functions; it removes parasites and debris from the groomed animal, it reaffirms the social bond or hierarchical relationship between the animals, and it gives the groomer an opportunity to examine olfactory cues on the groomed individual, perhaps adding additional ones. This behaviour has been observed in social insects, birds and mammals.
- Foraging: Some ant species recruit fellow workers to new food finds by first tapping them with their antennae and forelegs, then leading them to the food source while keeping physical contact. Another example of this is the waggle dance of honey bees.
- Huddling: Prolonged physical contact or huddling also serves social integration. Huddling promotes heat exchange, together with the transfer of olfactory or tactile information. Some organisms live in constant contact in a colony, for example colonial corals. When individuals are linked tightly in this way an entire colony can react on the aversive or alarm movements made by only a few individuals. In several herbivorous insect nymphs and larvae, aggregations where there is prolonged contact play a major role in group coordination. These aggregations may take the form of a procession or a rosette.

Seismic

Seismic communication is the exchange of information using self-generated vibrational signals transmitted via a substrate such as the soil, water, spider webs, plant stems, or a blade of grass.

This form of communication has several advantages, for example it can be sent regardless of light and noise levels, and it usually has a short range and short persistence, which may reduce the danger of detection by predators. The use of seismic communication is found in many taxa, including frogs, kangaroo rats, mole rats, bees, nematode worms, and others. Tetrapods usually make seismic waves by drumming on the ground with a body part, a signal that is sensed by the sacculus of the receiver. The sacculus is an organ in the inner ear containing a membranous sac that is used for balance, but can also detect seismic waves in animals that use this form of communication. Vibrations may be combined with other sorts of communication.

Thermal

A number of different snakes have the ability to sense infrared (IR) thermal radiation, which allows these reptiles to derive thermal images from the radiant heat emitted by predators or prey at wavelengths between 5 and 30 μm. The accuracy of this sense is such that a blind rattlesnake can target its strike to the vulnerable body parts of a prey animal. It was previously thought that the pit organs evolved primarily as prey detectors, but it is now believed that they may also be used to control body temperature.

A python (top) and rattlesnake illustrating the positions of the pit organs. Red arrows indicate the pit organs whereas black arrows indicate the nostril.

The facial pits enabling thermoregulation underwent parallel evolution in pitvipers and some boas and pythons, having evolved once in pitvipers and multiple times in boas and pythons. The electrophysiology of the structure is similar between lineages, but it differs in gross structureanatomy. Most superficially, pitvipers possess one large pit organ on either side of the head, between the eye and the nostril (loreal pit), while boas and pythons have three or more comparatively smaller pits lining the upper and sometimes the lower lip, in or between the scales. Those of the pitvipers are the more advanced, having a suspended sensory membrane as opposed to a simple pit structure. Within the family Viperidae, the pit organ is seen only in the subfamily Crotalinae: the pitvipers. Despite the detection of IR radiation, the pits' IR mechanism is dissimilar to photoreceptors; while photoreceptors detect light via photochemical reactions, the protein in the facial pits of snakes is a temperature sensitive ion channel. It senses infrared signals through a mechanism involving warming of the pit organ, rather than chemical reaction to light. This is consistent with the thin pit membrane, which allows incoming IR radiation to quickly and precisely warm a given ion channel and trigger a nerve impulse, as well as vascularize the pit membrane to rapidly cool the ion channel back to its original "resting" or "inactive" temperature.

Common vampire bats (*Desmodus rotundus*) have specialized IR sensors in their nose-leaf. Vampire bats are the only mammals that feed exclusively on blood. The IR sense enables Desmodus to localize homeothermic animals such as cattle and horses within a range of about 10 to 15 cm. This infrared perception may be used in detecting regions of maximal blood flow on targeted prey.

Autocommunication

Autocommunication is a type of communication in which the sender and receiver are the same individual. The sender emits a signal that is altered by the environment and eventually is received by the same individual. The altered signal provides information that can indicate food, predators or conspecifics. Because the sender and receiver are the same animal, selection pressure maximizes signal efficacy, i.e. the degree to which an emitted signal is correctly identified by a receiver despite propagation distortion and noise. There are two types of autocommunication. The first is active electrolocation found in the electric fish Gymnotiformes (knifefishes) and Mormyridae (elephantfish) and also in the platypus (*Ornithorhynchus anatinus*). The second type of autocommunication is echolocation, found in bats and Odontoceti.

Functions

There are many functions of animal communication. However, some have been studied in more detail than others. This includes:

- Communication during contests: Animal communication plays a vital role in determining the winner of contest over a resource. Many species have distinct signals that signal aggression or willingness to attack or signals to convey retreat during competitions over food, territories, or mates.
- Mating rituals: Animals produce signals to attract the attention of a possible mate or to solidify pair bonds. These signals frequently involve the display of body parts or postures. For example, a gazelle will assume characteristic poses to initiate mating. Mating signals can also include the use of olfactory signals or calls unique to a species. Animals that form lasting pair bonds often have symmetrical displays that they make to each other. Famous examples are the mutual presentation of reeds by great crested grebes studied by Julian Huxley, the triumph displays shown by many species of geese and penguins on their nest sites, and the spectacular courtship displays by birds of paradise.
- Ownership/territorial: Signals used to claim or defend a territory, food, or a mate.

- Food-related signals: Many animals make "food calls" to attract a mate, offspring, or other members of a social group to a food source. Perhaps the most elaborate food-related signal is the Waggle dance of honeybees studied by Karl von Frisch. One well-known example of begging of offspring in a clutch or litter is altricial songbirds. Young ravens signal will signal to older ravens when they encounter new or untested food. Rhesus macaques will send food calls to inform other monkeys of a food source to avoid punishment. Pheromones are released by many social insects to lead the other members of the society to the food source. For example, ants leave a pheromone trail on the ground that can be followed by other ants to lead them to the food source.
- Alarm calls: Alarm calls communicate the threat of a predator. This allows all members of a social group (and sometimes other species) to respond accordingly. This may include running for cover, becoming immobile, or gathering into a group to reduce the risk of attack. Alarm signals are not always vocalizations. Crushed ants will release an alarm pheromone to attract more ants and send them into an attack state.
- Meta-communication: Signals that will modify the meaning of subsequent signals. One example is the 'play face' in dogs which signals that a subsequent aggressive signal is part of a play fight rather than a serious aggressive episode.

Interpretation of Animal Behaviour

As described above, many animal gestures, postures, and sounds, convey meaning to nearby animals. These signals are often easier to describe than to interpret. It is tempting, especially with domesticated animals and apes, to anthropomorphize, that is, to interpret animal actions in human terms, but this can be quite misleading; for example, an ape's "smile" is often a sign of aggression. Also, the same gesture may have different meanings depending on context within which it occurs. For example, a domestic dog's tail wag and posture may be used in different ways to convey many meanings as illustrated in Charles Darwin's The Expression of the Emotions in Man and Animals published in 1872. Some of Darwin's illustrations are reproduced here.

- Examples of tail position indicating different emotions in dogs

"Small dog watching a cat on a table"

"Dog approaching another dog with hostile intentions"

"Dog in a humble and affectionate frame of mind"

"Half-bred shepherd dog"

"Dog caressing his master"

Interspecific Communication

Much animal communication is intraspecific, that is, it occurs between members of the same species. As for interspecific communication, that between predator and prey is of particular interest.

Prey to Predator

If a prey animal moves, makes a noise or vibrations, or emits a smell in such a way that a predator can detect it, this is consistent with the definition of "communication" given above. This type of communication is known as interceptive eavesdropping if a predator intercepts a message intended for conspecifics.

There are however, some actions of prey species are clearly directed to actual or potential predators. A good example is warning colouration: species such as wasps that are capable of harming potential predators are often brightly coloured, and this modifies the behaviour of the predator, who either instinctively or as the result of experience will avoid attacking such an animal. Some forms of mimicry fall in the same category: for example hoverflies are coloured in the same way as wasps, and although they are unable to sting, the strong avoidance of wasps by predators gives the hoverfly some protection. There are also behavioural changes that act in a similar way to warning colouration. For example, canines such as wolves and coyotes may adopt an aggressive posture, such as growling with their teeth bared, to indicate they will fight if necessary, and rattlesnakes use their well-known rattle to warn potential predators of their venomous bite. Sometimes, a behavioural change and warning colouration will be combined, as in certain species of amphibians which have most of their body coloured to blend with their surroundings, except for a brightly coloured belly. When confronted with a potential threat, they show their belly, indicating that they are poisonous in some way.

This Chihuahua is baring its teeth to signify an attack is imminent if the photographer comes closer to take his bone

Another example of prey to predator communication is the pursuit-deterrent signal. Pursuit-deterrent signals occur when prey indicates to a predator that pursuit would be unprofitable because the signaler is prepared to escape. Pursuit-deterrent signals provide a benefit to both the signaler and receiver; they prevent the sender from wasting time and energy fleeing, and they prevent the receiver from investing in a costly pursuit that is unlikely to result in capture. Such signals can advertise prey's ability to escape, and reflect phenotypic condition (quality advertisement), or can advertise that the prey has detected the predator (perception advertisement). Pursuit-deterrent signals have been reported for a wide variety of taxa, including fish (Godin and Davis, 1995), lizards (Cooper et al., 2004), ungulates (Caro, 1995), rabbits (Holley 1993), primates (Zuberbuhler et al. 1997), rodents (Shelley and Blumstein 2005, Clark, 2005), and birds (Alvarez, 1993, Murphy, 2006, 2007). A familiar example of quality advertisement pursuit-deterrent signal is *stotting* (sometimes called *pronking*), a pronounced combination of stiff-legged running while simultaneously jumping shown by some antelopes such as Thomson's gazelle in the presence of a predator. At least 11 hypotheses for stotting have been proposed. A leading theory today is that it alerts predators that the element of surprise has been lost. Predators like cheetahs rely on surprise attacks, proven by the fact that chases are rarely successful when antelope stot. Predators do not waste energy on a chase that will likely be unsuccessful (optimal foraging behaviour). Quality advertisement can be communicated by modes other than visual. The banner-tailed kangaroo rat produces several complex foot-drumming patterns in a number of different contexts, one of which is when it encounters a snake. The foot-drumming may alert nearby offspring but most likely conveys vibrations through the ground that the rat is too alert for a successful attack, thus preventing the snake's predatory pursuit.

Predator to Prey

Typically, predators attempt to reduce communication to prey as this will generally reduce the effectiveness of their hunting. However, some forms of predator to prey communication occur in ways that change the behaviour of the prey and make their capture easier, i.e. deception by the predator. A well-known example is the angler fish, an ambush predator which waits for its prey to come to it. It has a fleshy bioluminescent growth protruding from its forehead which it dangles in front of its jaws. Smaller fish attempt to take the lure, placing themselves in a better position for the angler fish to catch them. Another example of deceptive communication is observed in the genus of jumping spiders (Myrmarachne). These spiders are commonly referred to as "antmimicking spiders" because of the way they wave their front legs in the air to simulate antennae.

The humpback anglerfish angles for small fish by deceptively dangling a bioluminescent lure in front of its jaws.

Human/Animal

Various ways in which humans interpret the behaviour of animals, or give commands to them, are consistent with the definition of interspecies communication. Skillful interpretation of animal communications may be critical to the welfare of animals that are being cared for or trained by humans. For example, behaviors indicating pain need to be recognized. Indeed, the survival of both the animal and its human caretaker may be at stake if, for example, a human fails to recognize a signal for imminent attack

Since the late 1990s, one scientist, Sean Senechal, has been developing, studying, and using the learned visible, expressive language in dogs and horses. By teaching these animals a gestural (human made) American Sign Language-like language, the animals have been found to use the new signs on their own to get what they need. The recent experiments on animal language are perhaps the most sophisticated attempt yet to establish human/animal communication, though their relation to natural animal communication is uncertain.

Other Aspects

Evolution

The importance of communication is evident from the highly elaborate morphology, behaviour and physiology that some animals have evolved to facilitate this. These include some of the most striking structures in the animal kingdom, such as the peacock's tail, the antlers of a stag and the frill of the frill-necked lizard, but also include even the modest red spot on a European herring gull's bill. Highly elaborate behaviours have evolved for communication such as the dancing of cranes, the pattern changes of cuttlefish, and the gathering and arranging of materials by bowerbirds. Other evidence for the importance of communication in animals is the prioritisation of physiological features to this function, for example, birdsong appears to have brain structures entirely devoted to its production. All these adaptations require evolutionary explanation.

There are two aspects to the required explanation:

- identifying a route by which an animal that lacked the relevant feature or behaviour could acquire it;
- identifying the selective pressure that makes it adaptive for animals to develop structures that facilitate communication, emit communications, and respond to them.

Significant contributions to the first of these problems were made by Konrad Lorenz and other early ethologists. By comparing related species within groups, they showed that movements and body parts that in the primitive forms had no communicative function could be "captured" in a context where communication would be functional for one or both partners, and could evolve into a more elaborate, specialised form. For example, Desmond Morris showed in a study of grass finches that a beak-wiping response occurred in a range of species, serving a preening function, but that in some species this had been elaborated into a courtship signal.

The second problem has been more controversial. The early ethologists assumed that communication occurred for the good of the species as a whole, but this would require a process of group selection which is believed to be mathematically impossible in the evolution of sexually reproducing animals. Altruism towards an unrelated group is not widely accepted in the scientific community, but rather can be seen as reciprocal altruism, expecting the same behaviour from others, a benefit of living in a group. Sociobiologists argued that behaviours that benefited a whole group of animals might emerge as a result of selection pressures acting solely on the individual. A gene-centered view of evolution proposes that behaviours that enabled a gene to become wider established within a population would become positively selected for, even if their effect on individuals or the species as a whole was detrimental;

In the case of communication, an important discussion by John Krebs and Richard Dawkins established hypotheses for the evolution of such apparently altruistic or mutualistic communications as alarm calls and courtship signals to emerge under individual selection. This led to the realization that communication might not always be "honest" (indeed, there are some obvious examples where it is not, as in mimicry). The possibility of evolutionarily stable dishonest communication has been the subject of much controversy, with Amotz Zahavi in particular arguing that it cannot exist in the long term. Sociobiologists have also been concerned with the evolution of apparently excessive signaling structures such as the peacock's tail; it is widely thought that these can only emerge as a result of sexual selection, which can create a positive feedback process that leads to the rapid exaggeration of a characteristic that confers an advantage in a competitive mate-selection situation.

The apparently excessive eye-spot signalling by the male peacock tail may be runaway selection

One theory to explain the evolution of traits like a peacock's tail is 'runaway selection'. This requires two traits—a trait that exists, like the bright tail, and a prexisting bias in the female to select for that trait. Females prefer the more elaborate tails, and thus those males are able to mate successfully. Exploiting the psychology of the female, a positive feedback loop is enacted and the tail becomes bigger and brighter. Eventually, the evolution will level off because the survival costs to the male do not allow for the trait to be elaborated any further. Two theories exist to explain runaway selection. The first is the good genes hypothesis. This theory states that an elaborate display is an honest signal of fitness and truly is a better mate. The second is the handicap hypothesis. This explains that the peacock's tail is a handicap, requiring energy to keep and makes it more visible to predators. Thus, the signal is costly to maintain, and remains an honest indicator of the signaler's condition. Another assumption is that the signal is more costly for low quality males to produce than for higher quality males to produce. This is simply because the higher quality males have more energy reserves available to allocate to costly signaling.

Cognitive Aspects

Ethologists and sociobiologists have characteristically analysed animal communication in terms of more or less automatic responses to stimuli, without raising the question of whether the animals concerned understand the meaning of the signals they emit and receive. That is a key question in animal cognition. There are some signalling systems that seem to demand a more advanced understanding. A much discussed example is the use of alarm calls by vervet monkeys. Robert Seyfarth and Dorothy Cheney showed that these animals emit different alarm calls in the presence of different predators (leopards, eagles, and snakes), and the monkeys that hear the calls respond appropriately - but that this ability develops over time, and also takes into account the experience of the individual emitting the call. Metacommunication, discussed above, also seems to require a more sophisticated cognitive process.

It has been reported that bottlenose dolphins can recognize identity information from whistles even when otherwise stripped of the characteristics of the whistle; making dolphins the only animals other than humans that have been shown to transmit identity information independent of the caller's voice or location. The paper concludes that:

The fact that signature whistle shape carries identity information independent from voice features presents the possibility to use these whistles as referential signals, either addressing individuals or referring to them, similar to the use of names in humans. Given the cognitive abilities of bottlenose dolphins, their vocal learning and copying skills, and their fission–fusion social structure, this possibility is an intriguing one that demands further investigation.

— *V. M. Janik, et al.*

Animal Communication and Human Behaviour

Another controversial issue is the extent to which human behaviours resemble animal communication, or whether all such communication has disappeared as a result of our linguistic capacity. Some of our bodily features - eyebrows, beards and moustaches, deep adult male voices, perhaps female breasts - strongly resemble adaptations to producing signals. Ethologists such as Irenäus Eibl-Eibesfeldt have argued that facial gestures such as smiling, grimacing, and the *eyebrow flash* on greeting are universal human communicative signals that can be related to corresponding signals in other primates. Given how recently spoken language has emerged, it is very likely that human body language does include some more or less involuntary responses that have a similar origin to the communication we have.

Humans also often seek to mimic animals' communicative signals in order to interact with them. For example, cats have a mild affiliative response of slowly closing their eyes; humans often mimic this signal towards a pet cat to establish a tolerant relationship. Stroking, petting and rubbing pet animals are all actions that probably work through their natural patterns of interspecific communication.

Dogs have shown an ability to understand human communication. In object choice tasks, dogs utilize human communicative gestures such as pointing and direction of gaze in order to locate hidden food and toys. It has also been shown that dogs exhibit a left gaze bias when looking at human faces, indicating that they are capable of reading human emotions. It is interesting to note that dogs do not make use of direction of gaze or exhibit left gaze bias with other dogs.

A new approach in the 21st century in the field of animal communication uses applied behavioural analysis (ABA), specifically Functional Communication Training (FCT). This FCT previously has been used in schools and clinics with humans with special needs, such as children with autism, to help them develop language. Sean Senechal, at the AnimalSign Center has been using an approach similar to this FCT with domesticated animals, such as dogs (since 2004) and horses (since 2000) with encouraging results and benefits to the animals and people. Functional communication training for animals, Senechal calls "AnimalSign Language". This includes teaching communication through gestures (like simplified American sign language), Picture Exchange Communication System, tapping, and vocalisation. The process for animals includes simplified and modified techniques.

Animal Communication and Linguistics

For linguistics, the interest of animal communication systems lies in their similarities to and differences from human language:

1. Human languages are characterized for having a double articulation (in the characterization of French linguist André Martinet). It means that complex linguistic expressions can be broken down in meaningful elements (such as morphemes and words), which in turn are composed of smallest phonetic elements that affect meaning, called phonemes. Animal signals, however, do not exhibit this dual structure.

2. In general, animal utterances are responses to external stimuli, and do not refer to matters removed in time and space. Matters of relevance at a distance, such as distant food sources, tend to be indicated to other individuals by body language instead, for example wolf activity before a hunt, or the information conveyed in honeybee dance language.It is therefore unclear to what extent utterances are automatic responses and to what extent deliberate intent plays a part.

3. In contrast to human language, animal communication systems are usually not able to express conceptual generalizations. (Cetaceans and some primates may be notable exceptions).

4. Human languages combine elements to produce new messages (a property known as creativity). One factor in this is that much human language growth is based upon conceptual ideas and hypothetical structures, both being far greater capabilities in humans than animals. This appears far less common in animal communication systems, although current research into animal culture is still an ongoing process with many new discoveries.

A recent and interesting area of development is the discovery that the use of syntax in language, and the ability to produce "sentences", is not limited to humans either. The first good evidence of syntax in non-humans, reported in 2006, is from the greater spot-nosed monkey (*Cercopithecus nictitans*) of Nigeria. This is the first evidence that some animals can take discrete units of communication, and build them up into a sequence which then carries a different meaning from the individual "words":

> The greater spot-nosed monkeys have two main alarm sounds. A sound known onomatopoeiacally as the "pyow" warns against a lurking leopard, and a coughing sound that scientists call a "hack" is used when an eagle is flying nearby.
>
> "Observationally and experimentally we have demonstrated that this sequence [of up to three 'pyows' followed by up to four 'hacks'] serves to elicit group movement... the 'pyow-hack' sequence means something like 'let's go!' [a command telling others to move]... The implications are that primates at least may be able to ignore the usual relationship between an individual alarm call, and the meaning it might convey under certain circumstances... To our knowledge this is the first good evidence of a syntax-like natural communication system in a non-human species."

Similar results have also recently been reported in the Campbell's mona monkey.

Bee Learning and Communication

Honey bees are sensitive to odors (including pheromones), tastes, and colors, including ultraviolet. They can demonstrate capabilities such as color discrimination through classical and operant conditioning and retain this information for several days at least; they communicate the location and nature of sources of food; they adjust their foraging to the times at which food is available; they may even form cognitive maps of their surroundings.

Bees learn and communicate in a variety of ways.

Learning

Swarming bees require good communication to all congregate in the same place

Learning is essential for efficient foraging. Honey bees are unlikely to make many repeat visits if a plant provides little in the way of reward. A single forager will visit different flowers in the morning and, if there is sufficient attraction and reward in a particular kind of flower, she will make visits to that type of flower for most of the day, unless the plants stop producing reward or weather conditions change. Honey bees are quite adept at associative learning, and many of the standard phenomena of classical conditioning take the same form in honey bees as they do in the vertebrates that are the more usual subjects of such experiments.

Foragers were trained to enter a simple Y-shaped maze that had been marked at the entrance with a particular color. Inside the maze was a branching point where the bee was required to choose between two paths. One path, which led to the food reward, was marked with the same color that had been used at the entrance to the maze, while the other was marked with a different color. Foragers learned to choose the correct path, and continued to do so when a different kind of marker (black and white stripes oriented in various directions) was substituted for the colored markers. When the experimental conditions were reversed, rewarding bees for choosing the inner passage marked with a symbol that was different from the entrance symbol, the bees again learned to choose the correct path. Extending the length of the tunnel to increase the time between seeing the one marker indicating the correct path and a second marker identifying the correct path show that the bees can retain the information in their visual working memory for about 5 seconds, equivalent to the short-term memory of birds.

Color Learning in Honeybees

One of the most common ways that honey bees, *Apis mellifera*, demonstrate associative learning is in the context of color recognition and discrimination tasks. Just as vertebrate species such as mice or pigeons that can be trained to perform associative learning tasks, honey bees make excellent subjects for tasks involving discrimination and color memory. Beginning in the early 1900s, scientists Karl von Frisch and later Randolf Menzel began asking questions about the existence, learning rates, memory, and timing of color vision in bees.

Color Discrimination

The Austrian zoologist Karl von Frisch began the exploration of color vision in honey bees when he asked the first question in 1919: does color vision in bees exist? By making use of bees associative learning abilities he performed an elegant experiment to show that honey bees were in fact capable of color discrimination.

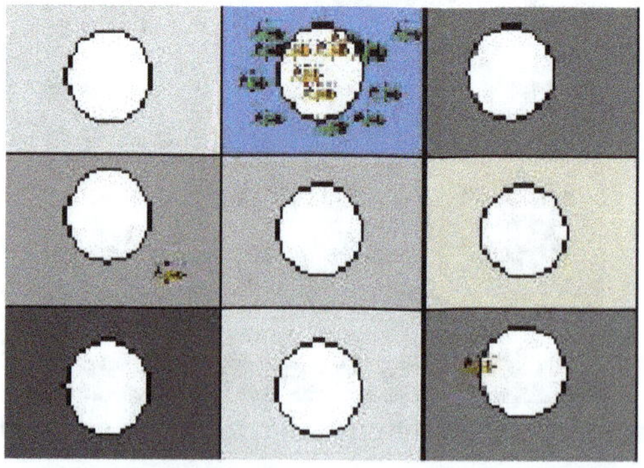

Figure 1. Testing for color vision in honey bees. The majority of bees flew directly to the dish with the blue background as they had been trained to do. Thus, they were able to discriminate between gray and blue backgrounds, showing their capability for color vision.

To test color vision, von Frisch first trained his honey bees to feed from a small dish filled with a nectar-like sugar water. This dish was placed over a small piece of blue colored cardboard so that the color was visible to the bees as they fed. Once the bees had become accustomed to the blue cardboard, von Frisch surrounded the blue piece of cardboard with other identically sized pieces in varying shades of gray and placed small dishes over each piece. If bees could not discriminate between colors, they would be unable to distinguish the blue piece from the many gray toned pieces. In the case that bees did not have color vision then, von Frisch predicted that the bees would visit the gray and blue pieces with equal frequency, as they would not be able to tell a difference between them.

When he allowed bees access to the dishes, however, he found that the vast majority of the bees flew directly to the blue piece of cardboard on which they had previously obtained their sugar-water reward. The bees largely ignored the gray pieces which had not been rewarded. This directed exploration and targeting of the blue cardboard showed the bees could indeed discriminate between the gray and blue shades, showing that bees do possess color vision. Von Frisch repeated this same basic experiment to show that bees produced the same results with other colors like violet and yellow. Later other researchers were able to apply this excellent experimental design to other vertebrates as well, making it an invaluable insight into testing color vision in many organisms.

Color Learning Rates and Preferences

After von Frisch's initial studies, the German scientist Randolf Menzel continued the study of color vision in honey bees and performed more detailed tests. He was curious about which colors honey bees would be able to learn fastest and whether or not bees had a greater aptitude for learning certain colors.

He used lights of varying color and intensity to illuminate circles of light on a solid surface. This set up was similar to the pieces of colored cardboard employed by von Frisch, but by using light instead of cardboard, Menzel was able to change the intensity and color of light easily. He could simply adjust the projection of the light to create a wide variety of different experimental set-ups.

Figure 2. Honey bee collecting pollen

To test the intricacies of the bee color vision von Frisch had established, Menzel performed an experiment that aimed to test bees ability to distinguish between two different colors. To do this, Menzel used a projected circle of colored light surrounding a small dish that could hold a sugar-water reward. Menzel then projected a second circle of differently colored light surrounding a second dish some distance away from the first. Next, a single bee was placed equidistant between these two different lights and allowed to choose which dish to search for a sugar-water treat. Only one of the colored light circles surrounded a dish that contained sugar-water; the other was empty. Menzel was then able to measure how quickly the bees learned to preferentially search only the rewarded light and to ignore the dish surrounded by unrewarded light.

The results of the experiment showed that bees did not learn to discriminate between all color pairs equally well. The fastest rate of learning was when violet light was rewarded. The color that the bees had the most difficulty learning was green, and all other colors fell somewhere in between. This evidence of inherent bias is evolutionarily reasonable given that color vision in bees allows them to distinguish between different nectar-bearing flowers, much like the rewarded dishes. As more flowers are purple than green it makes sense that bees would be more sensitive to colors likely to result in nourishment.

Color Memory

After this work on color preferences, Menzel extended his experiments to study memory in honey bee color vision. He wanted to know how many trials were necessary for honey bees to reliably choose a previously rewarded color when presented with several choices for potential rewards and how long honey bees could retain information about which color would be rewarded.

To test these questions, Menzel performed a variety of experiments. First, he presented individual bees with a sugar reward on a colored background for just a single trial. He then kept these bees in small cages for several days without any further trials. After a few days, he presented each bee with several dishes each on a different colored background at once. One of the colors was the same as that used during the initial trial. The others were novel, unrewarded colors. Amazingly, after just one trial and several days without any exposure to the rewarded color, bees correctly chose to explore the color used in the first trial more than fifty-percent of the time.

Menzel then repeated this experiment with another group of bees, keeping all factors the same except that in the second round of testing he gave the bees three initial trials with the rewarded color instead of just one. After several days in confinement when the bees were presented with a choice of colors just as in the first experiment, they virtually always chose the color that had been used during the first three trials.

This ability to retain information about color-linked rewards over a period of several days and after only minimal exposure to the colored background indicates the great strength of honeybees memory with respect to color vision.

Timing in Color Learning

One of von Frisch's students, Elizabeth Opfinger, observed that bees would learn color when approaching a feeder. Menzel took this question further: when do bees register and learn color? He wanted to know if bees registered color before, during, or after receiving their sugar-water reward. In order to explore this intriguing question, Menzel displayed the color beneath a rewarded dish at different stages of the honey bee feeding process: during approach, feeding and departure.

The outcome of this experiment revealed that bees register color during both the approach and feeding stages of the exposure process. In order for a bee to accurately remember a given color, it must be present for approximately five seconds in total. Although it varies slightly, Menzel and his colleagues found that bees usually remember best when the stimulus is present for about three seconds during the approach and two seconds after landing and beginning to feed.

Neurobiology of Color Vision

Color vision in honey bees can also be examined from a neurobiological perspective in terms of the structure and organization of their compound eyes.

Figure 3. Western honey bee

In 1975 Menzel published a seminal paper describing the morphology and spectral sensitivity of the honey bee eye. He examined color-coding the honey bee retina by using a technique to mark individual cells with a fluorescent dye and record from these cells as single units. Such fine structure analysis allowed him to determine that there are three types of receptors in the honey bee eye: 1) UV receptors, 2) blue receptors, and 3) green receptors. The three receptors are dominated by three rhodopsin-like pigments. These pigments have maximal absorbance at wavelengths corresponding to 350 nm, 440 nm, and 540 nm.

As the cells were examined in detail, certain features were distinguishable for each type of receptor cell. UV cells were found to form the longest visual fibres. These long visual fibers penetrated the lamina with arborizations, a tree-like branching of the fibers and spines. Blue and green receptor cells have more shallow fibers.

Interestingly, Menzel found that most of the cells he studied had secondary sensitivities that corresponded to wavelength regions at which the other two receptor types were maximally active. He used spectral efficiency experiments to show that such corresponding wavelength receptivity is the result of electric coupling.

Communication

Foragers communicate their floral findings in order to recruit other worker bees of the hive to forage in the same area. The factors that determine recruiting success are not completely known but probably include evaluations of the quality of nectar and/or pollen brought in.

There are two main hypotheses to explain how foragers recruit other workers—the "waggle dance" or "dance language" theory and the "odor plume" theory. The dance language theory is far more widely accepted, and has far more empirical support. The theories also differ in that the former allows for an important role of odor in recruitment (i.e., effective recruitment relies on dance *plus* odor), while the latter claims that the dance is essentially irrelevant (recruitment relies on odor alone). The academic debate between these two theories is extremely polarized and often hostile.

Dance Communication

It has long been known that successfully foraging Western honey bees perform a dance on their return to the hive, known as *waggle dance,* indicating that food is farther away, while the *round dance* is a short version of the *waggle dance,* indicating that food is nearby. The laden forager dances on the comb in a circular pattern, occasionally crossing the circle in a zig-zag or waggle pattern. Aristotle described this behaviour in his *Historia Animalium.* It was thought to attract the attention of other bees.

Figure-Eight-Shaped waggle dance of the honeybee (Apis mellifera). A waggle run oriented 45° to the right of 'up' on the vertical comb indicates a food source 45° to the right of the direction of the sun outside the hive. The abdomen of the dancer appears blurred because of the rapid motion from side to side.

In 1947, Karl von Frisch correlated the runs and turns of the dance to the distance and direction of the food source from the hive. The orientation of the dance correlates to the relative position of the sun to the food source, and the length of the waggle portion of the run is correlated to the distance from the hive. Also, the more vigorous the display is, the better the food. There is no evidence that this form of communication depends on individual learning. Honeybees detect the dances of conspecifics by sensing near field sound and electric fields using the Johnston's organ.

Von Frisch performed a series of experiments to validate his theory. He was awarded the Nobel Prize in Physiology or Medicine in 1973 for his discoveries.

One of the most important lines of evidence on the origin and utility of the dance is that all of the known species and races of honey bees exhibit the behavior, but details of its execution vary among the different species. For example, in *Apis florea* and *Apis andreniformis* (the "dwarf honeybees") the dance is performed on the dorsal, horizontal portion of the nest, which is exposed. The runs and dances point directly toward the resource in these species. Each honey bee species has a characteristically different correlation of "waggling" to distance, as well. Such species-specific behavior suggests that this form of communication does not depend on learning but is rather determined genetically. It also suggests how the dance may have evolved.

Various experiments document that changes in the conditions under which the dance is performed lead to characteristic changes in recruitment to external resources, in a manner consistent with von Frisch's original conclusions. Experiments with robotic dummies were indeed to induce some recruitment, which should not have been possible if the dance contains no information. Various experimental results demonstrate that the dance does convey information, but the use of this information may be context-dependent, and this may explain why the results of earlier studies were inconsistent. Researchers have also discovered other forms of honeybee dance communication, such as the tremble dance.

Odor Plume

While many researchers believe that bee dances give enough information to locate resources, proponents of the odor plume theory argue that the dance gives little, or no actual guidance to a nectar source. They argue that bees instead are primarily recruited by odor. The purpose of the dance is simply to gain attention to the returning worker bee so she can share the odor of the nectar with other workers who will then follow the odor trail to the source. Most scientists agree that odor is used in recruitment to resources, but they differ strongly in opinion as to the information content of the dance.

The primary lines of evidence used by the odor plume advocates are

1. experiments with odorless sugar sources which show that worker bees are unable to recruit to those sources and

2. logical difficulties of a small-scale dance (a few centimeters across) giving directions precise enough to hold the other bees on course during a flight that could be several kilometers long. Misreading by even a few degrees would lead the bee off course by hundreds of meters at the far end.

Neither of these points invalidate the dance theory, but simply suggest that odor might be involved, which is indeed conceded by all proponents of dance theory. Critics of the odor plume theory counter that most natural nectar sources are relatively large—orchards or entire fields— so, precision may not be necessary or even desirable. They have also challenged the reproducibility of the odorless source experiment.

Odor learning in bees is usually tested by the proboscis extension reflex. Significant to the argument are the experiments of William F. Towne, of the Kutztown University in Pennsylvania, in which hives are moved to "mirror image" terrain settings, and the bees are thereby fooled into dancing about the wrong location for a nectar source. Foragers were successfully recruited to the wrong location, but only when the sun was obscured by clouds, forcing them to rely on terrain-based navigation rather than "solar ephemeris"-based navigation. As the cloud cover broke up, more and more bees corrected their dances to indicate the actual location of nectar, and forager visits shifted to the correct location.

Odor is an essential and even necessary at various stages of the recruitment process, including once a recruited forager reaches the vicinity of the resource while some scientists think that dancing may be a simple idiothetic movement that conveys no information. Others see the dance as conveying information, but doing it poorly compared to other means and potentially used backup approach.

Note: much of the research on the two competing hypotheses of communication has been restricted to Western honey bees (see the work of F.C. Dyer though). Other species of *Apis* use variants on the same theme, and other types of bees use other methods altogether.

Trophallaxis

The exchange of food, trophallaxis, can be used to communicate the quality of a food source, temperature, a need for water, and the condition of the queen (Sebeok, 1990).

Primer Pheromones

Research that was published in November 2004, by scientists under the leadership of Zachary Huang, Michigan State University indicates that so called primer pheromones play an important part in how a honey bee colony adjusts its distribution of labor most beneficially. In order to survive as a bee colony of sometimes 50,000–100,000 individual bees, the communal structure has to be adaptable to seasonal changes and the availability of food. The division of labor has to adjust itself to the resources available from foraging. While the division of labor in a bee colony is quite complex, the work can be roughly seen as work inside the hive and outside the hive. Younger bees play a role inside the hive while older bees play a role outside the hive mostly as foragers. Huang's team found that forager bees gather and carry a chemical called ethyl oleate in the stomach. The forager bees feed this primer pheromone to the worker bees, and the chemical keeps them in a nurse bee state. The pheromone prevents the nurse bees from maturing too early to become forager bees. As forager bees die off, less of the ethyl oleate is available and nurse bees more quickly mature to become foragers. It appears that this control system is an example of decentralized decision making in the bee colony.

Other bees like *Trigona corvina* rely on pheromones for much of their communication with nest mates and rivals. They produce pheromones from their labial glands. The function of signaling depends on the profitability, but they commonly will scent mark a food source either for self-orientation, to deter rivals or to direct a nest mate to the resource. Once an individual finds a good food source, they will return to the same source for many days. If an individual detects the scent of a rival bee, they will avoid the plant in order to avoid conflict and to save time. It has also been shown that pheromones are a method of sexual selection between male drones and queens.

Cognition

Experiments by James Gould suggest that honey bees may have a cognitive map for information they have learned, and utilize it when communicating.

In one test reported in a 1983 issue of *Science News*, he moved a supply of sugar water 25% further away from a hive each day. The bees communicated to each other as usual on its location. Then he placed the sugar water on a boat anchored in the middle of a small lake. When scouts returned to the hive to communicate their find, other bees refused to go with them, even though they frequently flew over the lake to reach pollen sources on the opposite shore.

In another test related in the August 1986 issue of *Discover* ("A Honey of a Question: Are Bees Intelligent?"), Gould lured some bees to a dish of artificial nectar, then gradually moved it farther from the hive after they became accustomed to it. He marked the trained bees, placed them in a darkened jar, and relocated them to a spot where the hive was still visible, but not the dish. When released one by one, the bees would appear disoriented for a few seconds, then fly directly for the covert dish. Seventy-three of 75 bees reached it in about 28 seconds. They apparently accomplished this feat by devising a new flight path based on a cognitive map of visible landmarks.

Deception in Animals

Deception in animals is the transmission of misinformation by one animal to another, of the same or different species, in a way that propagates beliefs that are not true. Deception in animals does not automatically imply a conscious act, but can occur at different levels of cognitive ability.

Mimicry and camouflage enable animals to appear to be other than they are. Prey animals may appear as predators, or *vice versa*; both predators and prey may be hard to see (crypsis), or may be mistaken for other objects (mimesis). In Batesian mimicry, harmless animals may appear to be distasteful or poisonous. In automimicry, animals may have eyespots in less important parts of the body than the head, helping to distract attack and increase the chance of survival.

In more active forms of anti-predator adaptation, animals may feign death when they detect a predator, or may quickly conceal themselves or take action to distract a predator, such as when a cephalopod releases ink. In deimatic behaviour, a harmless animal adopts a threatening pose or displays startling, brightly coloured parts of its body to startle a predator or rival.

Some animals may use tactical deception, with behaviour that is deployed in a way that other animals misinterpret what is happening to the advantage of the agent. Some of the evidence for this is anecdotal, but in the great apes in particular, experimental studies in ethology suggest that deception is actively practised by some animals.

Overview

Some types of deception in animals are completely involuntary (e.g. disruptive colouration), but others are under voluntary control and may involve an element of learning. Most instances of voluntary deception in animals involve a simple behaviour, such as a cat arching its back and raising its hackles, to make itself appear larger than normal when attacked. There are relatively few examples of animal behaviour which might be attributed to the manipulative type of deception which we know occurs in humans, i.e. "tactical deception". It has been argued that true deception assumes the deceiver knows that (1) other animals have minds, (2) different animals' minds can believe different things are true (when only one of these is actually true), and (3) it can make another mind believe that something false is actually true. True deception requires the deceiver to have the mental capacity to assess different representations of reality. Animal behaviour scientists are therefore wary of interpreting a single instance of behaviour to true deception, and explain it with simpler mental processes such as learned associations. In contrast, human activities such as military deception are certainly intentional, even when they involve methods such as camouflage which physically parallel camouflage methods used by animals.

Levels of Deception in Animals

Mitchell and Thompson list four levels of deception in animals:

1. false markings on animals, such as butterfly markings that indicate their heads are at the back end of their bodies as an aid to escape, or markings to make predators appear safe
2. false behaviour, such as a predator acting in a way to hide its predatory nature around prey
3. Feigned injury to get or divert attention; for example, a parent bird feigning a broken wing to attract a predator away from its defenceless offspring
4. Verbal deception such as a chimp misleading other chimps to hide a food source, or a human lying in order to deceive another

Mimicry

Mimicry is the similarity of one species to another which protects one or both species. This similarity can be in appearance, behaviour, sound, scent, and location, with the mimics found in similar places to their models. There are many forms of mimicry, and an individual example may fall into more than one of the recognised categories.

Defensive Mimicry

Defensive or protective mimicry takes place when organisms are able to avoid encounters that would be harmful to them by deceiving enemies by appearing to be something that they are not. For example, mantis shrimp typically spread their front limbs (known as “smashers”) to threaten rivals in a behaviour called the “meral spread”. Newly moulted mantis shrimps frequently deceive potential competitors by spreading their front limbs, even though their still-soft exoskeletons meant that they could not use their smashers without damaging themselves.

Batesian Mimicry

An Example of Batesian Mimicry

Batesian mimicry is a form of mimicry typified by a situation where a harmless species has evolved to imitate the warning signals of a harmful species directed at a common predator. The harmful species (the model) might have spines, stingers, or toxic chemistry, while its apparent double has no defence other than resembling the unpalatable species. Protection of the mimic from predators is afforded by its resemblance to the unpalatable species, which the predator associates with a certain appearance and a bad experience.

A venomous coral snake

A harmless red milk snake mimics the bright colours of the venomous coral snake

Examples of Batesian mimicry are the several species of butterflies that mimic the toxic Heliconid butterflies. Another butterfly mimic is the non-toxic great Mormon of Indonesia. Each female butterfly (regardless of her colouration) can produce one or more different female forms which mimic any of five other species of foul-tasting butterflies. Batesian mimicry is also found in the harmless milk snake, which mimics venomous coral snakes. Both snakes are marked with alternating yellow, red, and black bands, causing potential predators to avoid both. The snakes can often be distinguished by using an old saying: "Red against yellow: kill a fellow. red against black: friend to Jack." The deadly coral snake has bands in the order of red, yellow, black, while the innocuous species have the pattern of red, black, yellow (although there are exceptions).

Deception by Batesian mimicry need not be visual, as it may involve any of the senses. For example, some moths use a highly effective defence against bats. In response to hearing ultrasound emitted by hunting bats, they produce loud ultrasonic clicks to mimic the unpalatable tiger moth – a case of auditory Batesian mimicry.

Müllerian Mimicry

An example of Müllerian Mimicry

Müllerian mimicry occurs when two or more unpalatable species have come to mimic each other's warning signals. Typically the species share one or more common predators, though they may or may not be closely related.

Viceroy butterfly

monarch butterfly

For example, the viceroy butterfly appears very similar to the noxious-tasting monarch butterfly. Although it was for a long time purported to be an example of Batesian mimicry, the viceroy has recently been discovered to be just as unpalatable as the monarch, making this a case of Müllerian mimicry. Poison dart frogs of South America and mantella frogs of Madagascar are examples of Müllerian mimicry with their conspicuous colouration (bright colours against black markings) and toxic composition.

Müllerian mimicry may also use any of the senses. For example, many snakes share the same auditory warning signals.

Aggressive Mimicry

Aggressive mimicry refers to characteristics or behaviours of predators (or parasites) that mimic those of harmless species, allowing the predator to approach and/or attract its prey. Several examples follow.

Anglerfish are named for their characteristic method of predation. Anglerfish typically have at least one long filament (the *illicium*) sprouting from the middle of the head, protruding above the fish's eyes and terminating in an irregular growth of flesh (the *esca*) at the tip of the filament. The filament can move in all directions and the esca can be wiggled so as to resemble a prey animal, thus acting as bait to lure other predators close enough for the anglerfish to devour them. Some deep-sea anglerfishes of the bathypelagic zone emit light from their escas to attract prey. This bioluminescence is a result of symbiosis with bacteria.

Several turtle species and the frogmouth catfish have tongue extensions that lure prey to a position where they become an easy catch. In caudal luring the predator uses tail movements to attract prey. This form of mimicry is used by several snake and lizard species, as well by as the tasselled wobbegong shark.

In another example of aggressive mimicry, males are lured toward what seems to be a sexually receptive female, only to be eaten. For example, fireflies of the genus *Photuris* emit the same light signals that females of the genus *Photinus* emit as a mating signal. Male fireflies from several different genera are attracted to these "femmes fatales" because the predatory females can identify the male's species and emit the signal used by the female of the male's species.

Aggressive mimicry need not involve the sense of vision. For example, the assassin bug preys on spiders, entering their web and plucking its silk threads. This produces vibrations that match the pattern of vibrations made by typical prey caught in the web, causing the spider to approach.

Automimicry

Automimicry refers to instances in which one body part of an animal mimics another. This may help and animal survive an attack, or help a predators to appear innocuous. Examples include many moth, butterfly, and fish species that have "eye-spots". These are large dark markings that help prey escape by causing predators to attack a false target. For example, the gray hairstreak (*Strymon melinus*) shows the false head at its rear; it has a better chance of surviving an attack to that part than an attack to the head. Another example is the "two-headed" snake of Central Africa which has a tail that resembles a head. The snake even moves its tail in the way most snakes move their heads. This adaptation functions to deceive prey as to the source of an attack.

Four-eye butterflyfish showing its concealed eye and false eyespot near the tail

Camouflage

Camouflage is the use of any combination of materials, colouration, or behaviour that helps to conceal an animal by making it hard to see (crypsis) or by disguising it as something else (mimesis).

Crypsis

There are several methods of achieving crypsis. These include, resemblance to the surroundings, disruptive colouration, eliminating shadow, self-decoration, cryptic behaviour, motion camouflage, changeable skin appearance, countershading, counter-illumination, transparency, and silvering to reflect the environment.

An example of cryptic camouflage by a Uroplatus gecko resembling the surroundings

Many species are cryptically colored to resemble their surroundings. For example, *Uroplatus* geckos can be almost completely invisible, even to a nearby observer. Similarly, the katydids, a group of grasshopper-like insects found worldwide, are nocturnal and use their cryptic colouration to remain unnoticed during the day. They remain perfectly still, often in a position that increases the effectiveness of their camouflage.

An example of transparency in the transparent goby

Some animals have colouration which makes them highly conspicuous when outside their normal environment but highly cryptic when in it. For example, the blue morpho, a forest butterfly, has iridescent blue upper wings and a 17 cm wingspan. However, because the underwings are dark, when the morpho flies through the flickering light of the forest or even out in daylight, it seems to disappear. Other forest species, especially mammals, use disruptive colouration and have spotted or striped pelage which helps break up the animal's outline. In the shade created by trees or other foliage, even large mammals such as leopards, jaguars, ocelots, and okapi are difficult to see because of such disruptive colouration.

Underwater animals adopt a wide range of methods of camouflage including transparency, reflection, counter-illumination, countershading, self-decoration, and others.

Most forms of camouflage are less effective when the camouflaged animal moves because motion is easily seen by the observing predator or prey. However, insects such as hoverflies and dragonflies use motion camouflage: the hoverflies to approach possible mates, and the dragonflies to approach rivals when defending territories. Motion camouflage is achieved by moving so as to stay on a straight line between the target and a fixed point in the landscape; the pursuer thus deceives the target animal by appearing not to move but only to loom larger in the target's field of vision.

Mimesis

Katydids have evolved a wide range of camouflage adaptations so their body colouring and shape match entire leaves, half-eaten leaves, dying leaves, leaves with bird droppings, sticks, twigs, and tree bark. Other well-known mimetic animals include beetles, mantids, caterpillars, moths, snakes, lizards, frogs, and fish.

An example of mimesis in the leafy sea dragon with seaweed-like colouration, protuberances and behaviour

A well known response of cephalopods when threatened is to release large volumes of ink. Some cephalopods also release pseudomorphs ("false bodies"); smaller clouds of ink with a greater mucus content, which allows them to hold their shape for longer. These are expelled slightly away from the cephalopod; they are roughly the same volume and look like the cephalopod that released them. Predators have often been seen attacking a pseudomorph, allowing the cephalopod to escape.

Active Camouflage

There are two mechanisms of active camouflage in animals: counter-illumination camouflage, and colour change (sometimes called "metachrosis").

An example of active camouflage. Four frames of a peacock flounder taken a few minutes apart.

In counter-illumination camouflage an animal produces light that causes it to blend in against a lit background. In water, light comes down from the surface, so when animals are seen from below, they appear darker than the background. Some species of cephalopod, such as the midwater squid and the sparkling enope squid, produce light in photophores on their undersides to match the background. Bioluminescence is common among marine animals, so counter-illumination camouflage may be a widespread mode of deception.

Colour change permits camouflage against different backgrounds. In the context of deception, this can be used as a defence or predatory strategy, or during courtship and mating. Colour change is made possible by chromatophores; pigment-containing and light-reflecting organelles in cells found in amphibians, fish, reptiles, crustaceans and cephalopods. Inside the chromatophore cell of cephalopods, pigment granules are enclosed in an elastic sac. To change colour, the animal distorts the sac by muscular contraction, changing its translucency, reflectivity or opacity. This differs from the mechanism used in fish, amphibians and reptiles, in that the shape of the sac is being changed rather than a translocation of pigment vesicles within the cell.

Some chameleon and anole species are able to voluntarily change their skin colours. Different chameleon species are able to change different colours which can include pink, blue, red, orange, green, black, brown, light blue, yellow, turquoise and purple. Some species, such as the Smith's dwarf chameleon, adjust their colours for camouflage in accordance with the vision of the specific predator species (bird or snake) that they are being threatened by.

Some octopuses can use muscles in the skin to change both the colour and texture of their mantle to achieve a greater camouflage. In some species, the mantle can take on the spiky appearance of seaweed, or the scraggly, bumpy texture of a rock, among other disguises. A few species, such as the mimic octopus, have another defence mechanism. They can combine their highly flexible bodies with their colour-changing ability to accurately mimic other, more dangerous animals, such as lionfish, sea snakes, and eels.

Feigning Death

A well-researched form of deception is feigning death, often referred to by non-specialists as "playing dead" or "playing possum", although specialists use the terms "tonic immobility" or "thanatosis". A wide range of animals, e.g. lizards, birds, rodents, and sharks, will behave as if dead, usually as a defensive method of avoiding predation, as predators will usually take only live prey.

A grass snake feigning death

In beetles, artificial selection experiments have shown that there is heritable variation for length of death-feigning. Those selected for longer death-feigning durations are at a selective advantage to those at shorter durations when a predator is introduced. Birds often feign death to escape predation; for example tonic immobility in quail reduces the probability of attacks by cats.

Death feigning may also play a role in reproduction, For example, in the nursery web spider, the male sometimes feigns death to avoid getting eaten by females during mating) In some cases, death feigning is used by a predator. Fore example, the predatory cichlid *Haplochromis livingstoni* will lie on its side on the bottom sediments until approached by scavengers attracted to what appears to be a dead fish, whereupon *H. livingstoni* abandons the pretence, rights itself and attacks the scavenger).

Death feigning behaviour can be deliberately induced by humans, a prominent example being the "hypnosis" of chickens or pigeons. For example, if a pigeon is grasped firmly, quickly inverted and held briefly on its back on a table, it often remains immobile for a minute or two. According to Gilman et al. the investigation of 'animal hypnosis' dates back to the year 1646 in a report by Kircher. It has been shown that the intensity and duration of death feigning is related to the intensity of fear prior to the feigning state being induced. This has been used to show that hens in cages are more fearful than those in pens, hens on the top tier of battery cages are more fearful than those on the lower levels, hens carried by hand are more fearful than hens carried on a mechanical conveyor, and hens undergoing longer transportation times are more fearful than those undergoing transport of a shorter duration.

Concealment

Many animals hide from predators behind rocks, in holes, in brush, and in many other ways. Some actually carry around parts of the environment to use for that purpose. For example, at least four individual veined octopuses were seen retrieving discarded coconut shells, carrying them up to 20 meters, manipulating them, and then assembling them to use as shelter. . This may be the first known example of tool use in cephalopods.

Cephalopods also conceal themselves by releasing large amounts of dark ink when they are threatened. The ink obsucures the vision of the threatening animal and allows the cephalopod to escape.

Distraction Displays

Distraction displays, also known as deflection displays, diversionary displays, or paratrepsis, are behaviours that draw the attention of a predator away from an object, typically the nest or young. They are particularly well known in birds but also occur in fish. A familiar example is the *broken-wing* display seen in nesting waders, plovers and doves such as the mourning dove. In this display, a bird walks away from its nest with one wing dragging on the ground. It seems to be an easy target, thus distracting the predator's attention away from the nest. Once the bird is far enough away it "recovers" and quickly flies off.

Deimatic Displays

A deimatic displays is a pattern of threatening or startling behaviour, such as suddenly displaying conspicuous eyespots, used to scare off or momentarily distract a predator, thus giving the prey animal an opportunity to escape. For example, some moths look threatening while at rest by displaying a sinister lurking face, such as those of genus *Speiredonia*, or the more aggressive face of a snake poised to attack, such as many species of genus *Spirama*. Adult Atlas moths of the genera *Attacus* and *Rothschildia* also display snake heads. The eyed hawkmoth displays its large eyespots on its wings and moves them slowly as if it were a vertebrate predator such as an owl.

Spirama retorta female

Tactical Deception

Tactical deception (also referred to as functional deception) is the use of signals or displays from an animal's normal repertoire mislead or deceive another individual. Some researchers limit the use of this term to an intraspecific behaviour, meaning that it occurs between members of the same species. Most other kinds of deception are meant to fool members of a different species. Tactical deception can also be achieved when the deceiver withholds information by failing to perform an expected action, such as giving a warning call when danger is observed.

Tactical deception can be costly to the user, because it occurs mostly in social animals that may lose trust in a group-member when that member's deceit is discovered. Indeed, a notable view in ethology is that animal displays are usually accurate signals, and that widespread deception cannot become a stable feature of a communication system. In this view, widespread use of deception would cause communication to break down, as the recipients of false signals would become "skeptical" about signal validity in general and fail to respond appropriately. However, as exemplified below, limited use of deception does seem to be a stable feature in various communities.

It is unclear what cognitive abilities are necessary for an animal to exhibit tactical deception. It might require the ability to understand another animal's point of view, or it might require only the use of specialized learned behaviours. These alternatives are hotly debated, and the answer tends to vary depending on the species being observed. In the first scenario, the deceiver would need a Theory of Mind which is the ability to attribute mental states (beliefs, knowledge, intents, desires, etc.) to another individual, The alternative view denies the need for such a capacity, but it does suggest the evolution of specialized brain functions responsible for social learning.

Tactical deception has been used as a measure of advanced social cognition as it relates to brain function. Primates have larger brains, relative to body size, than in any other mammal except for dolphins, and this size difference is mainly due to an enlarged neocortex. Research has suggested that the evolution of the primate brain is selected-for in highly social species. One study used 18 species with varying brain volumes (three strepsirrhines, four New World monkeys, seven Old World monkeys, and four ape species). The study used the frequency of tactical deception as a measure of social cognition, and it found a strong correlation between the use of social deception and size of the neocortex.

Cephalopods: Some colour changes in cuttlefish might be called tactical deception, as these fish sometimes present entirely different displays to two different observers. When a male cuttlefish courts a female in the presence of other males, he displays a male pattern facing the female (courtship), and a female pattern facing away, to deceive other males.

Domestic pigs: Research with pigs showed that in a setting where the behavior of a trained animal could reveal the source of food to another animal, the trained animal spent longer at the food source before other pigs arrived.

Birds: In an anecdotal account, Simmons reported that a female marsh harrier courted a male to obtain access to food he had stored. She then took this food and fed it to chicks that had been fathered by another male. More extensive studies focused on possibly deceitful behaviour in the pied flycatcher, a species in which males may possess more than one territory. Females gain from mating with a male that has no other mates and males may try to deceive females about their mating status (mated or unmated). Females frequently visit the male, and if he is always alone on his territory he is probably unmated. Thus, by repeated sampling of male behaviour, females are usually able to avoid mating with previously mated males.

Group-foraging common ravens hoard their food in a number of places, and also raid the caches made by others. Cachers withdraw from conspecifics when hiding their food and usually place their caches behind structures, out of sight of potential observers. Raiders remain inconspicuous, keeping at a distance from cachers near their cache sites, but within sight. In response, cachers often interrupt caching, change cache sites, or empty their caches. These behaviours suggest that ravens can withhold information about their intentions, which may qualify as tactical deception. Similarly, if a Eurasian jay (*Garrulus glandarius*) is being watched by another jay, it tends to cache food behind an opaque barrier rather than a transparent barrier, apparently to reduce the likelihood of other jays pilfering their caches.

Great apes: Observations on great apes have been widely reported as evidence of tactical deception. Following are a few examples.

Several great apes have been trained to use sign language and in some instances these animals seem to have used language in an attempt to deceive human observers. Koko, a female gorilla, was trained to use a form of American Sign Language. It has been claimed that she once tore a steel sink out of its moorings and when her handlers confronted her, Koko signed "cat did it" and pointed at her innocent pet kitten. Nim Chimpsky was a chimpanzee also trained in American Sign Language. In a documentary about the chimp ("Project Nim") trainers claimed that when Nim got bored of learning to sign words she would sign 'dirty' indicating she wanted to go to the toilet, which caused the trainer to stop the lesson.

Another example involves a chimpanzee that was approached from behind by a loud aggressive rival. Here, the chimpanzee moved his lips until he lost his fear grin thereby concealing his fear. Only then did he turn around to face the challenger.

Studies of deceit in great apes have also been performed under experimental conditions, one of which is summarised by Kirkpatrick.

> "...food was hidden and only one individual, named Belle, in a group of chimpanzees was informed of the location. Belle was eager to lead the group to the food but when one chimpanzee, named Rock, began to refuse to share the food, Belle changed her behaviour. She began to sit on the food until Rock was far away, then she would uncover it quickly and eat it. Rock figured this out though and began to push her out of the way and take the food from under her. Belle then sat farther and farther away waiting for Rock to look away before she moved towards the food. In an attempt to speed the process up, Rock looked away until Belle began to run for the food. On several occasions he would even walk away, acting disinterested, and then suddenly spin around and run towards Belle just as she uncovered the food."

Old World monkeys: In addition to the great apes, deceptive behaviour has been observed in baboons (*Papio ursinus*) which are Old World monkeys. In one of their articles, Byrne and Whiten recorded observations of "intimate tactical deception" within a group of baboons, and documented examples that they classified as follows: A juvenile using warning screams to gain access to underground food storages which otherwise would have been inaccessible; an exaggerated "looking" gesture (which in an honest context would mean detection of a predator) produced by a juvenile to avoid attack by an adult male; recruitment of a "fall-guy" (a third party used by the deceiver to draw attention or aggression); and using one's own movement pattern to draw group-mates away from food caches. Byrne and Whiten also broke these categories into subcategories denoting the modality of the action (e.g. vocalization) and what the action would have signified if observed in an honest context. They noted whether the individual that had been manipulated was in turn used to manipulate others, what the costs had been to the manipulated individual, and whether or not there were additional costs to third parties. Byrne and Whiten expressed concern that these observations might be exceptions, and that such deceptive behaviors might not be common to the species.

New World monkeys: Tufted capuchin (Cebus apella) monkey subordinates have been found to employ a vocal form of tactical deception when competing with dominant monkeys over valuable food resources. They will use alarm calls normally reserved for predator sightings— either barks (used specifically for aerial stimuli), peeps, or hiccups— to elicit a response in fellow group members and then take advantage of the distraction to pilfer food. In a series of experiments directed by Brandon Wheeler a group of tufted capuchin monkeys was provided with bananas on feeding platforms. Here, subordinate monkeys made nearly all of the alarm calls that could be classified as "false", and in many of the false alarms, the caller was on or within two meters of the feeding platform. The calls made dominant monkeys leave the platform while the subordinate caller stayed behind to eat.

Costs of Tactical Deception

Withholding information, a form of tactical deception, can be costly to the deceiver. For example, rhesus monkeys discovering food announce their discoveries by calling on 45% of occasions. Discoverers who fail to call, but are detected with food by other group members, receive significantly more aggression than vocal discoverers. Moreover, silent female discoverers eat significantly less food than vocal females. Presumably because of such costs to deceivers, tactical deception occurs rather rarely. It is thought to be more common in forms and species where the cost of ignoring the possibly deceptive act is even higher than the cost of believing. For example, tufted capuchin monkeys sometimes emit false alarm calls. The cost of ignoring one of these calls could be death, which may lead to a "better safe than sorry" philosophy even when the caller is a known deceiver.

Ultrasound Avoidance

Ultrasound avoidance is an escape or avoidance reflex displayed by certain animal species that are preyed upon by echolocating predators. Ultrasound avoidance is known for several groups of insects that have independently evolved mechanisms for ultrasonic hearing. Insects have evolved a variety of ultrasound-sensitive ears based upon a vibrating tympanic membrane tuned to sense the bat's echolocating calls. The ultrasonic hearing is coupled to a motor response that causes evasion of the bat during flight.

Although ultrasonic signals are used for echolocation by toothed whales, no known examples of ultrasonic avoidance in their prey have been found to date.

Ultrasonic hearing has evolved multiple times in insects: a total of 19 times. Bats appeared in the Eocene era, (about 50 million years ago); antibat tactics should have evolved then. Antibat tactics are known in four orders of Insecta : moths (Lepidoptera), crickets (Orthoptera), mantids (Dictyoptera), and green lacewings (Neuroptera). There are hypotheses of ultrasound avoidance being present in Diptera (flies) and Coleoptera (beetles).

Ultrasound Avoidance in Moths

The idea that moths were able to hear the cries of echolocating bats dates back to the late 19th century. White, in an 1877 letter to *Nature* made the association between the moth's high-pitched sounds and the high-pitched bat calls and wondered whether the moths would be able to hear it. However, it was not until the early 1960s that Kenneth Roeder *et al.* made the first electrophysiological recordings of a noctuid moth's auditory nerve and were able to confirm this suspicion.

Later research showed that moths responded to ultrasound with evasive movements. Moths, as do crickets and most insects that display bat avoidance behaviors, have tympanic organs that display phonotactic and directional hearing; they fly away from the source of the sound and will only have the diving behavior considered above when the sound is too loud—or when, in a natural setting, the bat would be presumably too close to simply fly away.

It was found that the moths' responses vary according to ultrasound intensity, diving towards the ground if the pulse was of a high amplitude, or flying directly away from the sound source if the sound amplitude was low (if the sound was softer). Acoustic sensory receptors in noctuid moths are mechanoreceptors located in a chamber formed by the wall of the abdomen and the tympanic membrane, are most sensitive to lower frequencies of ultrasound (between 20–30 kHz.).

The moth's body axis allows it to be more sensitive to sounds coming from particular directions. Their ears, on either side of the metathorax, have two sensory cells within the membranes. Though the tuning curves of these cells are identical, the sensitivity thresholds differ, allowing for sound localization and a wider range of sensitivity to sound. The movement of the wings during flight also plays a role, since sound thresholds change with wing position. The neural mechanisms for triggering the acoustic startle response are partially understood. However, there is little known about the motor control of flight that ultrasound initiates.

Further research has shown that many species of moths are sensitive to ultrasound. Sensitivities for ultrasound change according to the environment the moth thrives in, and the moth can even change its own sensitivity if it is preyed upon by bats with different echolocating calls. Such is the case of the Australian noctuid moth, Speiredonia spectans, which adapts its acoustic sensitivity according to the characteristics of the call of the bat inside the cave with them.

Ultrasound Avoidance in Crickets

Crickets are preyed on by bats during the night while they fly from one place to another. Avoidance behaviors by crickets were first reported in 1977 by Popov and Shuvalov. They also demonstrated that crickets, like moths, fly away from bats once they've heard their echolocating calls, an example of negative phonotaxis. The cricket will steer itself away from the source of the sound within a very short time frame (40–80 ms) The response is evoked by brief ultrasonic pulses in the 20 to 100 kHz range, pulses which fall within the range of bat ultrasonic echolocating calls (20–100 kHz).

A adult male and a juvenile male of the species *Gryllus bimaculatus*.

As opposed to moths, the cricket ear, located in the foreleg, is complex- having 70 receptors that are arranged in a tonotopic fashion. This is understandable since crickets don't only need to listen to bats, but also to each other. Crickets have broad frequency sensitivity to different types of echolocating calls. One specific auditory interneuron, the AN2 interneuron, exhibits remarkably rapid responses to echolocating call stimuli.

All these receptors synapse on a far lower number of interneurons that relay the receptors' information to the cricket's central nervous system. In the "Teleogryllus" cricket, two ascending interneurons carry information to the brain- one carries information about cricket song (around 5 kHz) while the other gets excited at ultrasound and other high frequencies (15–100 kHz). The ultrasound-sensitive interneuron- labeled INT-1- has been demonstrated as both necessary and sufficient for negative phonotaxis by Nolen and Hoy in 1984:

Stimulating int-1 by current injection is sufficient to initiate negative phonotaxis, while hyperpolarizing int-1 effectively cancels the turning response to ultrasound. Due to this, int-1 has been proposed to be a command neuron of sorts; in the cricket, int-1 is a bat detector when the cricket is in flight and the interneuron's activity reaches a specific threshold. If these conditions are met, the magnitude of the sound is linearly proportional to the magnitude of the avoidance response. This research also demonstrated that the brain is necessary for the response, since decapitated crickets will fly, but show no avoidance response behaviors.

Bats may have potentially found ways to get around this system. In the Teleogryllus oceanicus cricket, its broad sensitivity can be circumvented by the use of frequency-mismatched calls by part of bats like the gleaning bat, N. Geoffroyi. Furthermore, it has been found that the ultrasound avoidance response is restricted to when the crickets are in flight: that is, the response is extinguished when the crickets are on the ground.

It has also been shown that short-winged crickets are less sensitive to ultrasound, but not to low frequencies, than their long-winged counterparts in a wing-dimorphic cricket, "Grillus texensis". A hormone, named 'juvenile hormone' (JH), is believed to play a role in whether the individual develops shorter or longer wings: if the individual has a higher level of JH, its wings will be shorter.

Ultrasound Avoidance in Other Insects

In praying mantises, ultrasound avoidance behaviors are non-directional turns or power dives that are very effective in preventing capture by bats. The mantis ear, located in the midline between the metathoracic (third) legs, comprises two tympana within an auditory chamber that enhances sensitivity. A bilaterally symmetrical pair of auditory interneurons, 501-T3, accurately track the ultrasonic calls during the early stages of a bat attack. Because 501-T3 stops firing just before the evasive response starts, it may be involved in triggering the behavior. The praying mantis ear first appeared ca. 120 million years ago, predating the appearance of echolocating bats by ca. 50 million years, so its original function must be different from its current one.

Arctiid moths use a very different, but highly effective defense against bats. They produce loud ultrasonic clicks in response to ultrasound. Depending on the species of moth and its ecology, the clicks may work by startling the bat, by jamming its echolocation system, or by warning of distastefulness (aposematic).

Green lacewings (Chrysopidae) have sensitive ears on their wings. Ultrasound causes flying lacewings to fold their wings and drop, an effective maneuver for evading capture by bats. Some tettigoniids use a similar strategy, although other species respond much like crickets.

Several other insects have sensitive ultrasonic hearing that probably is used in bat evasion, but direct evidence is not yet available. These include scarab beetles, tiger beetles and a parasitoid fly (Ormia sp.)

Waggle Dance

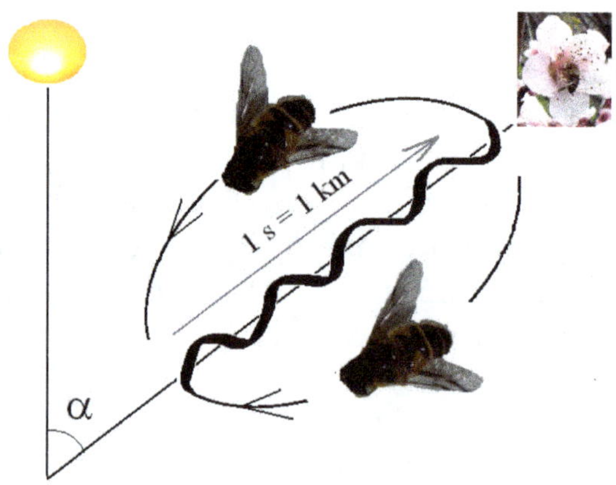

The waggle dance - the direction the bee moves in relation to the hive indicates direction; if it moves vertically the direction to the source is directly towards the Sun. The duration of the waggle part of the dance signifies the distance.

Waggle dance is a term used in beekeeping and ethology for a particular figure-eight dance of the honey bee. By performing this dance, successful foragers can share, with other members of the colony, information about the direction and distance to patches of flowers yielding nectar and pollen, to water sources, or to new nest-site locations. A waggle dance with a very short waggle run used to be characterized as a distinct (round) recruitment dance (see below). Austrian ethologist and Nobel laureate Karl von Frisch was one of the first who translated the meaning of the waggle dance.

Description

A waggle dance consists of one to 100 or more circuits, each of which consists of two phases: the waggle phase and the return phase. A worker bee's waggle dance involves running through a small figure-eight pattern: a waggle run (aka waggle phase) followed by a turn to the right to circle back to the starting point (aka return phase), another waggle run, followed by a turn and circle to the left, and so on in a regular alternation between right and left turns after waggle runs. Waggle-dancing bees produce and release two alkanes, tricosane and pentacosane, and two alkenes, (*Z*)-9-tricosene and (*Z*)-9-pentacosene, onto their abdomens and into the air.

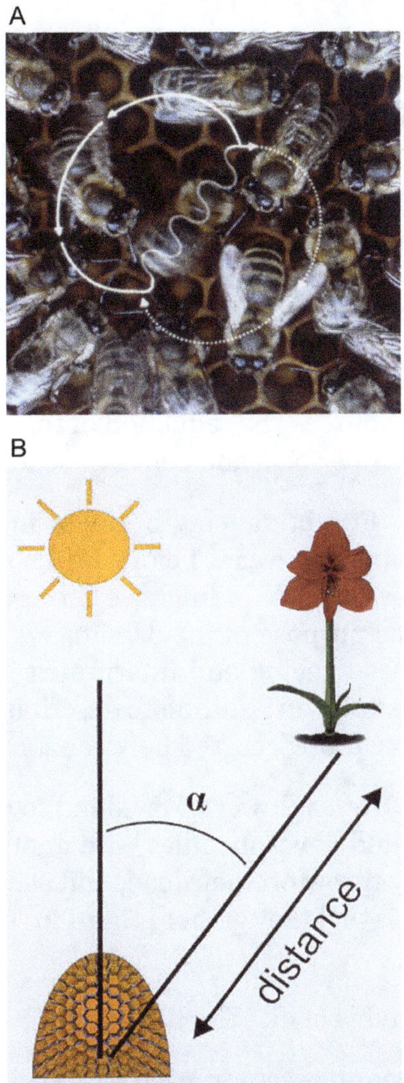

Figure-eight-shaped waggle dance of the honeybee (*Apis mellifera*). A waggle run oriented 45° to the right of 'up' on the vertical comb (A) indicates a food source 45° to the right of the direction of the sun outside the hive (B). The abdomen of the dancer appears blurred because of the rapid motion from side to side.

The direction and duration of waggle runs are closely correlated with the direction and distance of the resource being advertised by the dancing bee. The resource can include the location of a food source or a potential nesting site. For cavity-nesting honey bees, like *Apis mellifera* or *Apis nigrocincta*, flowers that are located directly in line with the sun are represented by waggle runs in an upward direction on the vertical combs, and any angle to the right or left of the sun is coded by a corresponding angle to the right or left of the upward direction. The distance between hive and recruitment target is encoded in the duration of the waggle runs. The farther the target, the longer the waggle phase. The more excited the bee is about the location, the more rapidly it will waggle, so it will grab the attention of the observing bees, and try to convince them. If multiple bees are doing the waggle dance, it's a competition to convince the observing bees to follow their lead, and competing bees may even disrupt other bees' dances or fight each other off. In addition, some open-air nesting honeybees, like *Apis andreniformis*, whose nests hang from twigs or branches, will perform a horizontal dance on a stage above their nest in order to signal to resources.

Waggle dancing bees that have been in the nest for an extended time adjust the angles of their dances to accommodate the changing direction of the sun. Therefore, bees that follow the waggle run of the dance are still correctly led to the food source even though its angle relative to the sun has changed.

The consumption of ethanol by foraging bees has been shown to reduce waggle dance activity and increase occurrence of the tremble dance.

Kevin Abbott and Reuven Dukas of McMaster University in Hamilton, Ontario, Canada discovered that if a dead *Apis mellifera* bee is placed on a flower, bees performed far fewer waggle dances upon returning to the hive. The scientists explain that the bees associate the dead bee with the presence of a predator at the food source. The reduction of the dance repetition frequency, therefore, indicates that the dancing bees perform and communicate a form of risk/benefit analysis.

Though first decoded by Karl von Frisch, dancing behavior in bees had been observed and described multiple times prior. Around 100 years before Frisch's discovery, Nicholas Unhoch described dancing behavior of bees as being an indulgence "in certain pleasures and jollity". He did, however, admit ignorance as to the purpose of the dancing. 35 years prior to Unhoch's observations, Ernst Spitzner observed bees dancing and interpreted it as transmitting forage resource odors to other nestmates. Even Aristotle, in addition to describing flower constancy behavior, suspected that some form of communication occurred between foragers within a nest:

"On each trip the bee does not fly from a flower of one kind to a flower of another, but flies from one violet, say, to another violet, and never meddles with another flower until it has got back to the hive; on reaching the hive they throw off their load, and each bee on her return is followed by three or four companions. What it is that they gather is hard to see, and how they do it has not been observed".

Jürgen Tautz also writes about it in his book "The Buzz about Bees":

Page 112: Many elements of the communication used to recruit miniswarms to feeding sites are also observed in "true" swarming behavior. Miniswarms of foragers are not placed under the same selection pressure as are true swarms, because the fate of the entire colony is not at stake. A truly swarming colony has to be quickly led to a new home, or it will perish. The behavior used to recruit to food sources possibly developed from the "true" swarming behavior.

Tautz,J.: The Buzz about Bees - Biology of a Superorganism (photos by H. R. Heilmann) Springer Heidelberg & Berlin, 2008

Mechanism

Honeybees accumulate an electric charge during flying and when their body parts are moved or rubbed together. Bees emit constant and modulated electric fields during the waggle dance. Both low- and high-frequency components emitted by dancing bees induce passive antennal movements in stationary bees according to Coulomb's Law. The electrically charged flagellum of mechanoreceptor cells are moved by electric fields and more strongly so if sound and electric fields interact. Recordings from axons of the Johnston's organ indicate its sensitivity to electric fields. Therefore, it has been suggested that electric fields emanating from the surface charge of bees stimulate mechanoreceptors and may play a role in social communication during the waggle dance.

Controversy

Workers of *Apis mellifera carnica* on honeycomb.

The Dance Language Vs. the Waggle Dance

As defined by von Frisch, *Tanzsprache* (German for "dance language") is the information about direction, distance, and quality of a resource (such as food or nesting sites) contained within the waggle dance. There is supporting evidence of the waggle dance and "Tanzsprache" in Apis dorsata. Similar to other bees, they utilize the dance language to indicate the critical information regarding food resources. The dancer's body points in the direction of the food source and the sound produced during the dance indicates the profitability of the food. Although there is some evidence for a direct connection between the *Tanzsprache* and the performance of the waggle dance, recent criticism holds that potential foragers need not correctly translate the dance language from the waggle dance to successfully forage. In an experiment on the honeybee *Apis mellifera*, most individuals who thoroughly followed a waggle dance ignored the resource direction and location information. Instead, 93% of the foragers returned to foraging areas they had previous knowledge of.

Bees that follow a waggle dance can successfully forage without decoding the dance language information in several ways:

- Dance follower may use olfactory information from the dancer and find either the same resource or a different one with a similar scent.
- Following a dance may simply trigger foraging behavior. A forager may then search randomly for resources.
- Following a dance may reactivate private knowledge of a resource. After reactivation, the forager may return to the known resource.
- Using information communicated in the waggle dance is more useful to foragers when private information about resources is lacking.

Dance Language as a language

The use of the word "language" may lead to misrepresentations of the waggle dance. The Swiss linguist Ferdinand de Saussure proposed a system of language where a sign is made up two chief components. The signifier is the physical or phonetic representation of a sign. The signified is the conceptual component. If the dance language followed the Saussurian dyadic model of semiotics, the signifier would be the waggle dance and the signified would be the location of the foraging resource. Though the dance language may or may not follow this sort of pattern, it is not considered to be a language with syntactical grammar or a set of symbols.

Efficiency and Adaptation

The waggle dance may be less efficient than once thought. Some bees observe over 50 waggle runs without successfully foraging, while others will forage successfully after observing 5 runs. Likewise, studies have found that honeybees rarely make use of the information communicated in the waggle dance and seem to only do so about ten percent of the time. Evidently there is a conflict between private information, or individual experience, and social information transmitted through dance communication. This sheds light on the fact that following social information is more energetically costly than foraging independently and is not always advantageous. Using olfactory cues and memory of plentiful foraging sites, honeybees are able to successfully forage independently without expending the potentially extensive energy it takes to process and execute the directions communicated by their fellow foragers.

The waggle dance may be adaptive in some environments and not in others, which provides a plausible explanation as to why the information provided by waggle dances are only used sparingly. Depending on weather, other competitors, and food source characteristics, transmitted information may become obsolete quickly. As a result, foragers reported to be attached to their food sites and continue to revisit a single patch many times after it has become unprofitable. For example, the waggle dance plays a significantly larger role in foraging when food sources are not as abundant. In temperate habitats, for instance, honey bee colonies routinely perform the waggle dance, but can still successfully forage when the dance is experimentally obscured. In tropical habitats, honey bee foraging is severely impaired if waggle dancing is prevented. This is thought to be due to the patchiness of resources in tropical environment versus the homogeneity of resources in temperate environments. In the tropics, food resources can come in the form of flowering trees which are rich in nectar but are scattered sparsely and bloom only briefly. Thus, in tropical zones information about forage location might be more valuable than in temperate zones.

Evolution

Ancestors to modern honeybees most likely performed excitatory movements to encourage other nestmates to forage. These excitatory movements include shaking, zig-zagging, buzzing and crashing into nestmates. Similar behavior is observed in other Hymenoptera including stingless bees, wasps, bumblebees and ants.

The waggle dance is thought to have evolved to aid in communicating information about a new nest site, rather than spatial information about foraging sites.

Observations have suggested that different species of honeybees have different "dialects" of the waggle dance, each species or subspecies dance varying by curve or duration. A study from 2008 demonstrated that a mixed colony of Asiatic honeybees (*Apis cerana cerana*) and European honeybees (*Apis mellifera ligustica*) were gradually able to understand one another's "dialects" of waggle dance.

Applications to Operations Research

In line with recent work in swarm intelligence research involving optimization algorithms inspired by the behavior of social insects (including bees, ants and termites), and vertebrates such as fish and birds, there has recently been research on using bee waggle dance behavior for efficient fault-tolerant routing. From the abstract of Wedde, Farooq, and Zhang (2004):

In this paper we present a novel routing algorithm, BeeHive, which has been inspired by the communicative and evaluative methods and procedures of honey bees. In this algorithm, bee agents travel through network regions called foraging zones. On their way their information on the network state is delivered for updating the local routing tables. BeeHive is fault tolerant, scalable, and relies completely on local, or regional, information, respectively. We demonstrate through extensive simulations that BeeHive achieves a similar or better performance compared to state-of-the-art algorithms.

Another bee-inspired stigmergic computational technique called bee colony optimization is employed in Internet Server Optimization.

The Zigbee RF protocol is named after the waggle dance.

Animal Migration

Animal migration is the relatively long-distance movement of individuals, usually on a seasonal basis. It is found in all major animal groups, including birds, mammals, fish, reptiles, amphibians, insects, and crustaceans. The trigger for the migration may be local climate, local availability of food, the season of the year or for mating reasons. To be counted as a true migration, and not just a local dispersal or irruption, the movement of the animals should be an annual or seasonal occurrence, such as Northern hemisphere birds migrating south for the winter; wildebeest migrating annually for seasonal grazing; or a major habitat change as part of their life, such as young Atlantic salmon leaving the river of their birth when they have reached a few inches in size.

Mexican free-tailed bats on their long aerial migration

Overview

Migration can take very different forms in different species, and as such there is no simple accepted definition of migration. One of the most commonly used definitions, proposed by Kennedy is

A Christmas Island red crab on its migration

Migratory behavior is persistent and straightened out movement effected by the animal's own locomotory exertions or by its active embarkation upon a vehicle. It depends on some temporary inhibition of station keeping responses but promotes their eventual disinhibition and recurrence.

Migration encompasses four related concepts: persistent straight movement; relocation of an individual on a greater scale (both spatially and temporally) than its normal daily activities; seasonal 'to-and-fro' movement of a population between two areas; and movement leading to the redistribution of individuals within a population. Migration can be either obligate, meaning individuals must migrate, or facultative, meaning individuals can "choose" to migrate or not. Within a migratory species or even within a single population, often not all individuals migrate. *Complete migration* is when all individuals migrate, *partial migration* is when some individuals migrate while others do not, and *differential migration* is when the difference between migratory and non-migratory individuals is based on age or sex (for example).

The three main types of migration are pilotage, compass orientation, and true navigation.

While most migratory movements occur on an annual cycle, some daily movements are also referred to as migration. Many aquatic animals make a Diel vertical migration, travelling a few hundred metres up and down the water column, while some jellyfish make daily horizontal migrations, traveling a few hundred metres across a lake.

Irregular (non-cyclical) migrations such as irruptions can occur under pressure of famine, overpopulation of a locality, or some more obscure influence.

In Specific Groups

Different kinds of animal migrate in different ways.

Flocks of birds assembling before migration southwards

In Birds

Approximately 1,800 of the world's 10,000 bird species migrate long distances each year in response to the seasons. Many of these migrations are north-south, with species feeding and breeding in high northern latitudes in the summer, and moving some hundreds of kilometres south for the winter. Some species extend this strategy to migrate annually between the Northern and Southern Hemispheres. The Arctic tern is famous for its migration; it flies from its Arctic breeding grounds to the Antarctic and back again each year, a distance of at least 19,000 km (12,000 mi), giving it two summers every year. The average arctic tern will travel the equivalent of going around the Earth 60 times in their lifetime.

In Fish

Most fish species are relatively limited in their movements, remaining in a single geographical area and making short migrations for wintering, to spawn, or to feed. A few hundred species migrate long distances, in some cases of thousands of kilometres. About 120 species of fish, including several species of salmon, migrate between saltwater and freshwater (they are 'diadromous').

Many species of salmon migrate up rivers to spawn

Forage fish such as herring and capelin migrate around substantial parts of the North Atlantic ocean. The capelin for example spawn around the southern and western coasts of Iceland; their larvae drift clockwise around Iceland, while the fish swim northwards towards Jan Mayen island to feed, and return to Iceland parallel with Greenland's east coast.

In the 'sardine run', billions of Southern African pilchard *Sardinops sagax* spawn in the cool waters of the Agulhas Bank and move northward along the east coast of South Africa between May and July.

In Insects

Some winged insects such as locusts and certain butterflies and dragonflies with strong flight migrate long distances. Among the dragonflies, species of *Libellula* and *Sympetrum* are known for mass migration, while *Pantala flavescens*, known as the globe skimmer or wandering glider dragonfly, makes the longest ocean crossing of any insect, between India and Africa. Exceptionally, swarms of the desert locust, *Schistocerca gregaria*, flew westwards across the Atlantic ocean for 4500 km during October 1988, using air currents in the Inter-Tropical Convergence Zone.

An aggregation of migratory *Pantala flavescens* dragonflies, known as globe skimmers, in Coorg, India

In some migratory butterflies, such as the monarch butterfly and the painted lady, no individual completes the whole migration. Instead the butterflies mate and reproduce on the journey, and successive generations travel the next stage of the migration.

In Other Animals

Mass migration occurs in mammals such as the Serengeti 'great migration', an annual circular pattern of movement with some 1.7 million wildebeest and hundreds of thousands of other large game animals including gazelles and zebra.

Wildebeest on the Serengeti 'great migration'

Migration is important in other mammals including Cetaceans, whales, dolphins and porpoises. Long-distance migrations occur in some bats, notably the mass migration of the Mexican free-tailed bat between Oregon and southern Mexico.

Some reptiles and amphibians migrate.

Some crustaceans migrate, most spectacularly the Christmas Island red crab which moves en masse each year by the million.

Tracking Migration

A migratory butterfly, a monarch, tagged for identification

Scientists gather observations of animal migration by tracking their movements. Animals were traditionally tracked with identification tags such as bird rings for later recovery; no information was obtained about the actual route followed between release and recovery, and only a small fraction of tagged individuals were generally recovered. More convenient, therefore, are electronic devices such as radio tracking collars which can be followed by radio, whether handheld, in a vehicle or aircraft, or by satellite. Tags can include a GPS receiver, enabling accurate positions to be broadcast at regular intervals, but these are inevitably heavier and more expensive than devices without GPS. An alternative is the Argos Doppler tag, also called a 'Platform Transmitter Terminal' (PTT) which sends regularly to the polar-orbiting Argos satellites; using Doppler shift, the animal's location can be estimated, relatively roughly compared to GPS, but at lower cost and weight.

Radio tracking tags can be fitted to insects including dragonflies and bees.

In Culture

Before the phenomenon of animal migration was understood, various folklore and erroneous explanations sprang up to account for the disappearance or sudden arrival of birds in an area. In Ancient Greece, Aristotle proposed that robins turned into redstarts when summer arrived. The barnacle goose was explained in European Medieval bestiaries and manuscripts as either growing like fruit on trees, or developing from goose barnacles on pieces of driftwood. Another example is the swallow, which was once thought, even by naturalists such as Gilbert White, to hibernate either underwater, buried in muddy riverbanks, or in hollow trees.

Bird Migration

Bird migration is the regular seasonal movement, often north and south along a flyway, between breeding and wintering grounds. Many species of bird migrate. Migration carries high costs in predation and mortality, including from hunting by humans, and is driven primarily by availability of food. It occurs mainly in the northern hemisphere, where birds are funnelled on to specific routes by natural barriers such as the Mediterranean Sea or the Caribbean Sea.

A flock of barnacle geese during autumn migration

Historically, migration has been recorded as much as 3,000 years ago by Ancient Greek authors including Homer and Aristotle, and in the Book of Job, for species such as storks, turtle doves, and swallows. More recently, Johannes Leche began recording dates of arrivals of spring migrants in Finland in 1749, and scientific studies have used techniques including bird ringing and satellite tracking. Threats to migratory birds have grown with habitat destruction especially of stopover and wintering sites, as well as structures such as power lines and wind farms.

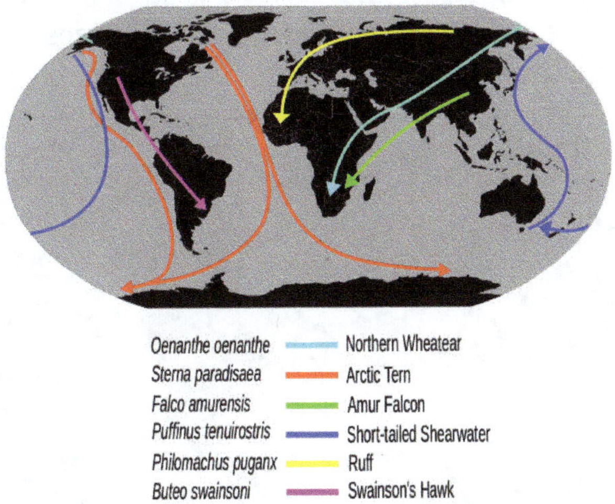

Examples of long distance bird migration routes.

The Arctic tern holds the long-distance migration record for birds, travelling between Arctic breeding grounds and the Antarctic each year. Some species of tubenoses (Procellariiformes) such as albatrosses circle the earth, flying over the southern oceans, while others such as Manx shearwaters migrate 14,000 km (8,700 mi) between their northern breeding grounds and the southern ocean. Shorter migrations are common, including altitudinal migrations on mountains such as the Andes and Himalayas..

The timing of migration seems to be controlled primarily by changes in day length. Migrating birds navigate using celestial cues from the sun and stars, the earth's magnetic field, and probably also mental maps.

Historical Views

Records of bird migration were made as much as 3,000 years ago by the Ancient Greek writers Hesiod, Homer, Herodotus and Aristotle. The Bible also notes migrations, as in the Book of Job (39:26), where the inquiry is made: "Is it by your insight that the hawk hovers, spreads its wings southward?" The author of Jeremiah (8:7) wrote: "Even the stork in the heavens knows its seasons, and the turtle dove, the swift and the crane keep the time of their arrival."

Aristotle noted that cranes traveled from the steppes of Scythia to marshes at the headwaters of the Nile. Pliny the Elder, in his *Historia Naturalis*, repeats Aristotle's observations.

Swallow Migration Versus Hibernation

Aristotle however suggested that swallows and other birds hibernated. This belief persisted as late as 1878, when Elliott Coues listed the titles of no less than 182 papers dealing with the hibernation of swallows. Even the "highly observant" Gilbert White, in his posthumously published 1789 *The Natural History of Selborne*, quoted a man's story about swallows being found in a chalk cliff collapse "while he was a schoolboy at Brighthelmstone", though the man denied being an eyewitness.

However, he also writes that "as to swallows being found in a torpid state during the winter in the Isle of Wight or any part of this country, I never heard any such account worth attending to", and that if early swallows "happen to find frost and snow they immediately withdraw for a time—a circumstance this much more in favour of hiding than migration", since he doubts they would "return for a week or two to warmer latitudes".

Minoan fresco of swallows in springtime at Akrotiri, c. 1500 BC

It was not until the end of the eighteenth century that migration as an explanation for the winter disappearance of birds from northern climes was accepted. Thomas Bewick's *A History of British Birds* (Volume 1, 1797) mentions a report from "a very intelligent master of a vessel" who, "between the islands of Minorca and Majorca, saw great numbers of Swallows flying northward", and states the situation in Britain as follows:

Swallows frequently roost at night, after they begin to congregate, by the sides of rivers and pools, from which circumstance it has been erroneously supposed that they retire into the water.

—*Bewick*

Bewick then describes an experiment which succeeded in keeping swallows alive in Britain for several years, where they remained warm and dry through the winters. He concludes:

These experiments have since been amply confirmed by ... M. Natterer, of Vienna ... and the result clearly proves, what is in fact now admitted on all hands, that Swallows do not in any material instance differ from other birds in their nature and propensities [for life in the air]; but that they leave us when this country can no longer furnish them with a supply of their proper and natural food ...

—*Bewick*

General Patterns

Migration is the regular seasonal movement, often north and south, undertaken by many species of birds. Bird movements include those made in response to changes in food availability, habitat, or weather. Sometimes, journeys are not termed "true migration" because they are irregular (nomadism, invasions, irruptions) or in only one direction (dispersal, movement of young away from natal area). Migration is marked by its annual seasonality. Non-migratory birds are said to be resident or sedentary. Approximately 1800 of the world's 10,000 bird species are long-distance migrants.

Migrating waders in Roebuck Bay, Western Australia

Many bird populations migrate long distances along a flyway. The most common pattern involves flying north in the spring to breed in the temperate or Arctic summer and returning in the autumn to wintering grounds in warmer regions to the south. Of course, in the southern hemisphere the directions are reversed, but there is less land area in the far south to support long-distance migration.

The primary motivation for migration appears to be food; for example, some hummingbirds choose not to migrate if fed through the winter. Also, the longer days of the northern summer provide extended time for breeding birds to feed their young. This helps diurnal birds to produce larger clutches than related non-migratory species that remain in the tropics. As the days shorten in autumn, the birds return to warmer regions where the available food supply varies little with the season.

These advantages offset the high stress, physical exertion costs, and other risks of the migration. Predation can be heightened during migration: Eleonora's falcon *Falco eleonorae,* which breeds on Mediterranean islands, has a very late breeding season, coordinated with the autumn passage of southbound passerine migrants, which it feeds to its young. A similar strategy is adopted by the greater noctule bat, which preys on nocturnal passerine migrants. The higher concentrations of migrating birds at stopover sites make them prone to parasites and pathogens, which require a heightened immune response.

Within a species not all populations may be migratory; this is known as "partial migration". Partial migration is very common in the southern continents; in Australia, 44% of non-passerine birds and 32% of passerine species are partially migratory. In some species, the population at higher latitudes tends to be migratory and will often winter at lower latitude. The migrating birds bypass the latitudes where other populations may be sedentary, where suitable wintering habitats may already be occupied. This is an example of *leap-frog migration*. Many fully migratory species show leap-frog migration (birds that nest at higher latitudes spend the winter at lower latitudes), and many show the alternative, chain migration, where populations 'slide' more evenly north and south without reversing order.

Within a population, it is common for different ages and/or sexes to have different patterns of timing and distance. Female chaffinches *Fringilla coelebs* in Eastern Fennoscandia migrate earlier in the autumn than males do.

Most migrations begin with the birds starting off in a broad front. Often, this front narrows into one or more preferred routes termed flyways. These routes typically follow mountain ranges or coastlines, sometimes rivers, and may take advantage of updrafts and other wind patterns or avoid geographical barriers such as large stretches of open water. The specific routes may be genetically programmed or learned to varying degrees. The routes taken on forward and return migration are often different. A common pattern in North America is clockwise migration, where birds flying North tend to be further West, and flying South tend to shift Eastwards.

Many, if not most, birds migrate in flocks. For larger birds, flying in flocks reduces the energy cost. Geese in a V-formation may conserve 12–20% of the energy they would need to fly alone. Red knots *Calidris canutus* and dunlins *Calidris alpina* were found in radar studies to fly 5 km/h (3.1 mph) faster in flocks than when they were flying alone.

Northern pintail skeletons have been found high in the Himalayas

Birds fly at varying altitudes during migration. An expedition to Mt. Everest found skeletons of northern pintail *Anas acuta* and black-tailed godwit *Limosa limosa* at 5,000 m (16,000 ft) on the Khumbu Glacier. Bar-headed geese *Anser indicus* have been recorded by GPS flying at up to 6,540 metres (21,460 ft) while crossing the Himalayas, at the same time engaging in the highest rates of climb to altitude for any bird. Anecdotal reports of them flying much higher have yet to be corroborated with any direct evidence. Seabirds fly low over water but gain altitude when crossing land, and the reverse pattern is seen in landbirds. However most bird migration is in the range of 150 to 600 m (490 to 1,970 ft). Bird strike aviation records from the United States show most collisions occur below 600 m (2,000 ft) and almost none above 1,800 m (5,900 ft).

Bird migration is not limited to birds that can fly. Most species of penguin (Spheniscidae) migrate by swimming. These routes can cover over 1,000 km (620 mi). Dusky grouse *Dendragapus obscurus* perform altitudinal migration mostly by walking. Emus *Dromaius novaehollandiae* in Australia have been observed to undertake long-distance movements on foot during droughts.

Long-distance Migration

The typical image of migration is of northern landbirds, such as swallows (Hirundinidae) and birds of prey, making long flights to the tropics. However, many Holarctic wildfowl and finch (Fringillidae) species winter in the North Temperate Zone, in regions with milder winters than their summer breeding grounds. For example, the pink-footed goose *Anser brachyrhynchus* migrates from Iceland to Britain and neighbouring countries, whilst the dark-eyed junco *Junco hyemalis* migrates from subarctic and arctic climates to the contiguous United States and the American goldfinch from taiga to wintering grounds extending from the American South northwestward to Western Oregon. Migratory routes and wintering grounds are traditional and learned by young during their first migration with their parents. Some ducks, such as the garganey *Anas querquedula*, move completely or partially into the tropics. The European pied flycatcher *Ficedula hypoleuca* also follows this migratory trend, breeding in Asia and Europe and wintering in Africa.

Often, the migration route of a long-distance migrator bird doesn't follow a straight line between breeding and wintering grounds. Rather, it could follow a hooked or arched line, with detours around geographical barriers. For most land-birds, such barriers could consist in seas, large water bodies or high mountain ranges, because of the lack of stopover or feeding sites, or the lack of thermal columns for broad-winged birds.

The same considerations about barriers and detours that apply to long-distance land-bird migration apply to water birds, but in reverse: a large area of land without bodies of water that offer feeding sites may also be a barrier to a bird that feeds in coastal waters. Detours avoiding such barriers are observed: for example, brent geese *Branta bernicla* migrating from the Taymyr Peninsula to the Wadden Sea travel via the White Sea coast and the Baltic Sea rather than directly across the Arctic Ocean and northern Scandinavia.

In Waders

A similar situation occurs with waders (called *shorebirds* in North America). Many species, such as dunlin *Calidris alpina* and western sandpiper *Calidris mauri*, undertake long movements from their Arctic breeding grounds to warmer locations in the same hemisphere, but others such as semipalmated sandpiper *C. pusilla* travel longer distances to the tropics in the Southern Hemisphere.

Bar-tailed godwit

For some species of waders, migration success depends on the availability of certain key food resources at stopover points along the migration route. This gives the migrants an opportunity to refuel for the next leg of the voyage. Some examples of important stopover locations are the Bay of Fundy and Delaware Bay.

Some bar-tailed godwits *Limosa lapponica* have the longest known non-stop flight of any migrant, flying 11,000 km from Alaska to their New Zealand non-breeding areas. Prior to migration, 55 percent of their bodyweight is stored as fat to fuel this uninterrupted journey.

In Seabirds

Seabird migration is similar in pattern to those of the waders and waterfowl. Some, such as the black guillemot *Cepphus grylle* and some gulls, are quite sedentary; others, such as most terns and auks breeding in the temperate northern hemisphere, move varying distances south in the northern winter. The Arctic tern *Sterna paradisaea* has the longest-distance migration of any bird, and sees more daylight than any other, moving from its Arctic breeding grounds to the Antarctic non-breeding areas. One Arctic tern, ringed (banded) as a chick on the Farne Islands off the British east coast, reached Melbourne, Australia in just three months from fledging, a sea journey of over 22,000 km (14,000 mi). Many tubenosed birds breed in the southern hemisphere and migrate north in the southern winter.

The Arctic tern migrates the longest distance of any bird.

The most pelagic species, mainly in the 'tubenose' order Procellariiformes, are great wanderers, and the albatrosses of the southern oceans may circle the globe as they ride the "roaring forties" outside the breeding season. The tubenoses spread widely over large areas of open ocean, but congregate when food becomes available. Many are also among the longest-distance migrants; sooty shearwaters *Puffinus griseus* nesting on the Falkland Islands migrate 14,000 km (8,700 mi) between the breeding colony and the North Atlantic Ocean off Norway. Some Manx shearwaters *Puffinus puffinus* do this same journey in reverse. As they are long-lived birds, they may cover enormous distances during their lives; one record-breaking Manx shearwater is calculated to have flown 8 million km (5 million miles) during its over-50 year lifespan.

Diurnal Migration in Raptors

Some large broad-winged birds rely on thermal columns of rising hot air to enable them to soar. These include many birds of prey such as vultures, eagles, and buzzards, but also storks. These birds migrate in the daytime. Migratory species in these groups have great difficulty crossing large bodies of water, since thermals only form over land, and these birds cannot maintain active flight for long distances. Mediterranean and other seas present a major obstacle to soaring birds, which must cross at the narrowest points. Massive numbers of large raptors and storks pass through areas such as the Strait of Messina, Gibraltar, Falsterbo, and the Bosphorus at migration times. More common species, such as the European honey buzzard *Pernis apivorus*, can be counted in hundreds of thousands in autumn. Other barriers, such as mountain ranges, can also cause funnelling, particularly of large diurnal migrants. This is a notable factor in the Central American migratory bottleneck. Batumi bottleneck in the Caucasus is one of the heaviest migratory funnels on earth. Avoiding flying over the Black Sea surface and across high mountains, hundreds of thousands of soaring birds funnel through an area around the city of Batumi, Georgia. Birds of prey such as honey buzzards which migrate using thermals lose only 10 to 20% of their weight during migration, which may explain why they forage less during migration than do smaller birds of prey with more active flight such as falcons, hawks and harriers.

Griffon vulture soaring

The bottleneck of the Strait of Messina, point of transit of the migrations, seen from the Peloritani mountains, Sicily

Nocturnal Migration in Smaller Insectivorous Birds

Many of the smaller insectivorous birds including the warblers, hummingbirds and flycatchers migrate large distances, usually at night. They land in the morning and may feed for a few days before resuming their migration. The birds are referred to as *passage migrants* in the regions where they occur for short durations between the origin and destination.

Ruby-throated hummingbird

Nocturnal migrants minimize predation, avoid overheating, and can feed during the day. One cost of nocturnal migration is the loss of sleep. Migrants may be able to alter their quality of sleep to compensate for the loss.

Short-distance and Altitudinal Migration

Cedar waxwing

Many long-distance migrants appear to be genetically programmed to respond to changing day length. Species that move short distances, however, may not need such a timing mechanism, instead moving in response to local weather conditions. Thus mountain and moorland breeders, such as wallcreeper *Tichodroma muraria* and white-throated dipper *Cinclus cinclus*, may move only altitudinally to escape the cold higher ground. Other species such as merlin *Falco columbarius* and Eurasian skylark *Alauda arvensis* move further, to the coast or towards the south. Species like the chaffinch are much less migratory in Britain than those of continental Europe, mostly not moving more than 5 km in their lives.

Short-distance passerine migrants have two evolutionary origins. Those that have long-distance migrants in the same family, such as the common chiffchaff *Phylloscopus collybita*, are species of southern hemisphere origins that have progressively shortened their return migration to stay in the northern hemisphere.

Species that have no long-distance migratory relatives, such as the waxwings *Bombycilla*, are effectively moving in response to winter weather and the loss of their usual winter food, rather than enhanced breeding opportunities.

In the tropics there is little variation in the length of day throughout the year, and it is always warm enough for a food supply, but altitudinal migration occurs in some tropical birds. There is evidence that this enables the migrants to obtain more of their preferred foods such as fruits.

Altitudinal migration is common on mountains worldwide, such as in the Himalayas and the Andes.

Irruptions and Dispersal

Sometimes circumstances such as a good breeding season followed by a food source failure the following year lead to irruptions in which large numbers of a species move far beyond the normal range. Bohemian waxwings *Bombycilla garrulus* well show this unpredictable variation in annual numbers, with five major arrivals in Britain during the nineteenth century, but 18 between the years 1937 and 2000. Red crossbills *Loxia curvirostra* too are irruptive, with widespread invasions across England noted in 1251, 1593, 1757, and 1791.

Mediaeval sketch by Matthew Paris in his *Chronica Majora* (1251) recording that year's major irruption of red crossbills into England

Bird migration is primarily, but not entirely, a Northern Hemisphere phenomenon. This is because land birds in high northern latitudes, where food becomes scarce in winter, leave for areas further south (including the Southern Hemisphere) to overwinter, and because the continental landmass is much larger in the Northern Hemisphere. In contrast, among (pelagic) seabirds, species of the Southern Hemisphere are more likely to migrate. This is because there is a large area of ocean in the Southern Hemisphere, and more islands suitable for seabirds to nest.

Physiology and Control

The control of migration, its timing and response are genetically controlled and appear to be a primitive trait that is present even in non-migratory species of birds. The ability to navigate and orient themselves during migration is a much more complex phenomenon that may include both endogenous programs as well as learning.

Timing

The primary physiological cue for migration are the changes in the day length. These changes are also related to hormonal changes in the birds. In the period before migration, many birds display higher activity or Zugunruhe (German: *migratory restlessness*), first described by Johann Friedrich Naumann in 1795, as well as physiological changes such as increased fat deposition. The occurrence of Zugunruhe even in cage-raised birds with no environmental cues (e.g. shortening of day and falling temperature) has pointed to the role of circannual endogenous programs in controlling bird migrations. Caged birds display a preferential flight direction that corresponds with the migratory direction they would take in nature, changing their preferential direction at roughly the same time their wild conspecifics change course.

In polygynous species with considerable sexual dimorphism, males tend to return earlier to the breeding sites than their females. This is termed protandry.

Orientation and Navigation

Navigation is based on a variety of senses. Many birds have been shown to use a sun compass. Using the sun for direction involves the need for making compensation based on the time. Navigation has also been shown to be based on a combination of other abilities including the ability to detect magnetic fields (magnetoception), use visual landmarks as well as olfactory cues.

The routes of satellite tagged bar-tailed godwits migrating north from New Zealand. This species has the longest known non-stop migration of any species, up to 10,200 km (6,300 mi).

Long distance migrants are believed to disperse as young birds and form attachments to potential breeding sites and to favourite wintering sites. Once the site attachment is made they show high site-fidelity, visiting the same wintering sites year after year.

The ability of birds to navigate during migrations cannot be fully explained by endogenous programming, even with the help of responses to environmental cues. The ability to successfully perform long-distance migrations can probably only be fully explained with an accounting for the cognitive ability of the birds to recognize habitats and form mental maps. Satellite tracking of day migrating raptors such as ospreys and honey buzzards has shown that older individuals are better at making corrections for wind drift.

Migratory birds may use two electromagnetic tools to find their destinations: one that is entirely innate and another that relies on experience. A young bird on its first migration flies in the correct direction according to the Earth's magnetic field, but does not know how far the journey will be. It does this through a radical pair mechanism whereby chemical reactions in special photo pigments sensitive to long wavelengths are affected by the field. Although this only works during daylight hours, it does not use the position of the sun in any way. At this stage the bird is in the position of a boy scout with a compass but no map, until it grows accustomed to the journey and can put its other capabilities to use. With experience it learns various landmarks and this "mapping" is done by magnetites in the trigeminal system, which tell the bird how strong the field is. Because birds migrate between northern and southern regions, the magnetic field strengths at different latitudes let it interpret the radical pair mechanism more accurately and let it know when it has reached its destination. There is a neural connection between the eye and "Cluster N", the part of the forebrain that is active during migrational orientation, suggesting that birds may actually be able to *see* the magnetic field of the earth.

Vagrancy

Migrating birds can lose their way and appear outside their normal ranges. This can be due to flying past their destinations as in the "spring overshoot" in which birds returning to their breeding areas overshoot and end up further north than intended. Certain areas, because of their location, have become famous as watchpoints for such birds. Examples are the Point Pelee National Park in Canada, and Spurn in England.

Reverse migration, where the genetic programming of young birds fails to work properly, can lead to rarities turning up as vagrants thousands of kilometres out of range.

Drift migration of birds blown off course by the wind can result in "falls" of large numbers of migrants at coastal sites.

A related phenomenon called "abmigration" involves birds from one region joining similar birds from a different breeding region in the common winter grounds and then migrating back along with the new population. This is especially common in some waterfowl, which shift from one flyway to another.

Migration Conditioning

It has been possible to teach a migration route to a flock of birds, for example in re-introduction schemes. After a trial with Canada geese *Branta canadensis*, microlight aircraft were used in the US to teach safe migration routes to reintroduced whooping cranes *Grus americana*.

Adaptations

Birds need to alter their metabolism to meet the demands of migration. The storage of energy through the accumulation of fat and the control of sleep in nocturnal migrants require special physiological adaptations. In addition, the feathers of a bird suffer from wear-and-tear and require to be moulted. The timing of this moult - usually once a year but sometimes twice - varies with some species moulting prior to moving to their winter grounds and others molting prior to returning to their breeding grounds. Apart from physiological adaptations, migration sometimes requires behavioural changes such as flying in flocks to reduce the energy used in migration or the risk of predation.

Evolutionary and Ecological Factors

Migration in birds is highly labile and is believed to have developed independently in many avian lineages. While it is agreed that the behavioral and physiological adaptations necessary for migration are under genetic control, some authors have argued that no genetic change is necessary for migratory behavior to develop in a sedentary species because the genetic framework for migratory behavior exists in nearly all avian lineages. This explains the rapid appearance of migratory behavior after the most recent glacial maximum.

Theoretical analyses show that detours that increase flight distance by up to 20% will often be adaptive on aerodynamic grounds - a bird that loads itself with food to cross a long barrier flies less efficiently. However some species show circuitous migratory routes that reflect historical range expansions and are far from optimal in ecological terms. An example is the migration of continental populations of Swainson's thrush *Catharus ustulatus*, which fly far east across North America before turning south via Florida to reach northern South America; this route is believed to be the consequence of a range expansion that occurred about 10,000 years ago. Detours may also be caused by differential wind conditions, predation risk, or other factors.

Climate Change

Large scale climatic changes, as have been experienced in the past, are expected to have an effect on the timing of migration. Studies have shown a variety of effects including timing changes in migration, breeding as well as population variations.

Ecological Effects

The migration of birds also aids the movement of other species, including those of ectoparasites such as ticks and lice, which in turn may carry micro-organisms including those of concern to human health. Due to the global spread of avian influenza, bird migration has been studied as a possible mechanism of disease transmission, but it has been found not to present a special risk; import of pet and domestic birds is a greater threat. Some viruses that are maintained in birds without lethal effects, such as the West Nile Virus may however be spread by migrating birds. Birds may also have a role in the dispersal of propagules of plants and plankton.

Some predators take advantage of the concentration of birds during migration. Greater noctule bats feed on nocturnal migrating passerines. Some birds of prey specialize on migrating waders.

Study Techniques

Early studies on the timing of migration began in 1749 in Finland, with Johannes Leche of Turku collecting the dates of arrivals of spring migrants.

Radars for monitoring bird migration. Kihnu, Estonia.

Bird migration routes have been studied by a variety of techniques including the oldest, marking. Swans have been marked with a nick on the beak since about 1560 in England. Scientific ringing was pioneered by Hans Christian Cornelius Mortensen in 1899. Other techniques include radar and satellite tracking.

Stable isotopes of hydrogen, oxygen, carbon, nitrogen, and sulphur can establish avian migratory connectivity between wintering sites and breeding grounds. Stable isotopic methods to establish migratory linkage rely on spatial isotopic differences in bird diet that are incorporated into inert tissues like feathers, or into growing tissues such as claws and muscle or blood.

An approach to identify migration intensity makes use of upward pointing microphones to record the nocturnal contact calls of flocks flying overhead. These are then analyzed in a laboratory to measure time, frequency and species.

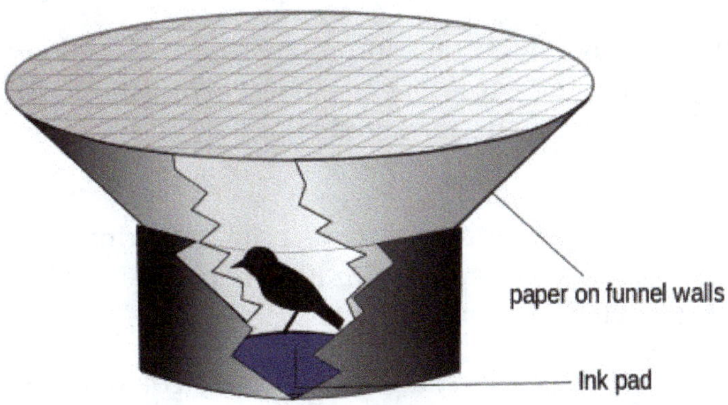

Emlen funnel

An older technique to quantify migration involves observing the face of the moon towards full moon and counting the silhouettes of flocks of birds as they fly at night.

Orientation behaviour studies have been traditionally carried out using variants of a setup known as the Emlen funnel, which consists of a circular cage with the top covered by glass or wire-screen so that either the sky is visible or the setup is placed in a planetarium or with other controls on environmental cues. The orientation behaviour of the bird inside the cage is studied quantitatively using the distribution of marks that the bird leaves on the walls of the cage. Other approaches used in pigeon homing studies make use of the direction in which the bird vanishes on the horizon.

Threats and Conservation

Human activities have threatened many migratory bird species. The distances involved in bird migration mean that they often cross political boundaries of countries and conservation measures require international cooperation. Several international treaties have been signed to protect migratory species including the Migratory Bird Treaty Act of 1918 of the US. and the African-Eurasian Migratory Waterbird Agreement

Migration routes and countries with illegal hunting in Europe

The concentration of birds during migration can put species at risk. Some spectacular migrants have already gone extinct; during the passenger pigeon's (*Ectopistes migratorius*) migration the enormous flocks were a mile (1.6 km) wide, darkening the sky and 300 miles (480 km) long, taking several days to pass.

Other significant areas include stop-over sites between the wintering and breeding territories. A capture-recapture study of passerine migrants with high fidelity for breeding and wintering sites did not show similar strict association with stop-over sites.

Hunting along migration routes threatens some bird species. The populations of Siberian cranes (*Leucogeranus leucogeranus*) that wintered in India declined due to hunting along the route, particularly in Afghanistan and Central Asia. Birds were last seen in their favourite wintering grounds in Keoladeo National Park in 2002. Structures such as power lines, wind farms and offshore oil-rigs have also been known to affect migratory birds. Other migration hazards include pollution, storms, wildfires, and habitat destruction along migration routes, denying migrants food at stopover points. For example, in the East Asian–Australasian Flyway, up to 65% of key intertidal habitat at the Yellow Sea migration bottleneck has been destroyed since the 1950s.

Reverse Migration (birds)

Reverse migration is a phenomenon in bird migration. Although some large birds such as swans learn migration routes from their parents, in most small species, such as passerines, the route is genetically programmed, and young birds can innately navigate to their wintering area. Sometimes this programming goes wrong, and the young bird, in its first autumn, migrates on a route 180° from the correct route. This is shown in the diagram, where the breeding range of a hypothetical Asian species is shown in yellow. The correct migration route to the wintering grounds in south east Asia is indicated in red.

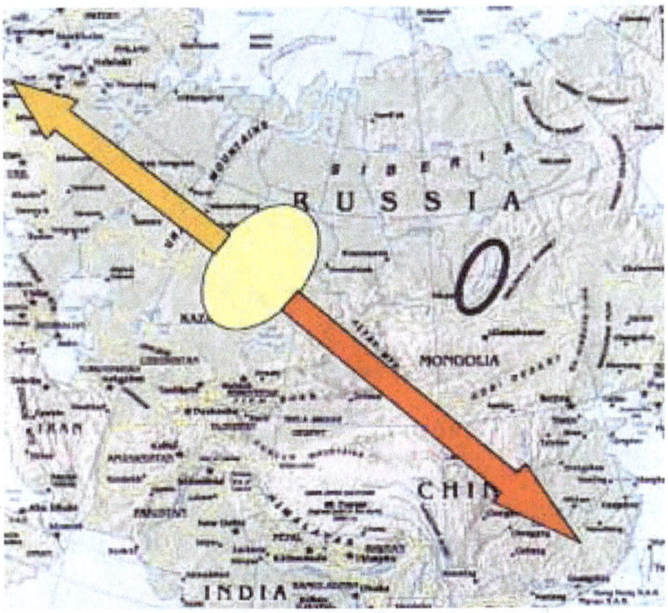

Reverse Migration

If a bird sets off in the opposite direction, shown by the orange arrow, it will end up in Western Europe instead of South East Asia. This is a mechanism that leads to birds such as Pallas's warbler turning up thousands of kilometres from where they should be. Keith Vinicombe suggested that birds from east of Lake Baikal in Siberia (circled) could not occur in western Europe because the migration routes were too north-south. Most of these lost young birds perish in unsuitable wintering grounds, but there is some evidence that a few survive, and either re-orient in successive winters, or even return to the same area.

An article in *British Birds* by James Gilroy and Alexander Lees in September 2003 suggested that misorientation occurs in random directions, but differential survival in different directions combined with asymmetric observer coverage leads to the observed distribution of vagrants. Although "reverse migration" per se certainly does occur, and has been documented well in numerous instances, it does not account for the occurrence in Europe in autumn of Asian vagrants that winter in East Africa, for instance, or the rarity of many southern European species in the UK that winter in West Africa.

Fish Migration

Many types of fish migrate on a regular basis, on time scales ranging from daily to annually or longer, and over distances ranging from a few metres to thousands of kilometres. Fish usually migrate to feed or to reproduce, but in other cases the reasons are unclear.

Many species of salmon are anadromous and migrate long distances up rivers and streams to spawn

Migrations involves the fish moving from one part of a water body to another on a regular basis. Some particular types of migration are anadromous, in which adult fish live in the sea and migrate into fresh water to spawn, and catadromous, in which adult fish live in fresh water and migrate into salt water to spawn.

Marine forage fish often make large migrations between their spawning, feeding and nursery grounds. Movements are associated with ocean currents and with the availability of food in different areas at different times of year. The migratory movements may partly be linked to the fact that the fish cannot identify their own offspring and moving in this way prevents cannibalism. Some species have been described by the United Nations Convention on the Law of the Sea as highly migratory species. These are large pelagic fish that move in and out of the exclusive economic zones of different nations, and these are covered differently in the treaty from other fish.

Salmon and striped bass are well-known anadromous fish, and freshwater eels are catadromous fish that make large migrations. The bull shark is an euryhaline species that moves at will from fresh to salt water, and many marine fish make a diel vertical migration, rising to the surface to feed at night and sinking to lower layers of the ocean by day. Some fish such as tuna move to the north and south at different times of year following temperature gradients. The patterns of migration are of great interest to the fishing industry. Movements of fish in fresh water also occur; often the fish swim upriver to spawn, and these traditional movements are increasingly being disrupted by the building of dams.

Classification

As with various other aspects of fish life, zoologists have developed empirical classifications for fish migrations. Two terms in particular have been in long-standing wide usage:

- *Anadromous* fish migrate from the sea up into fresh water to spawn; examples are salmon and striped bass.
- *Catadromous* fish migrate from fresh water down into the sea to spawn.

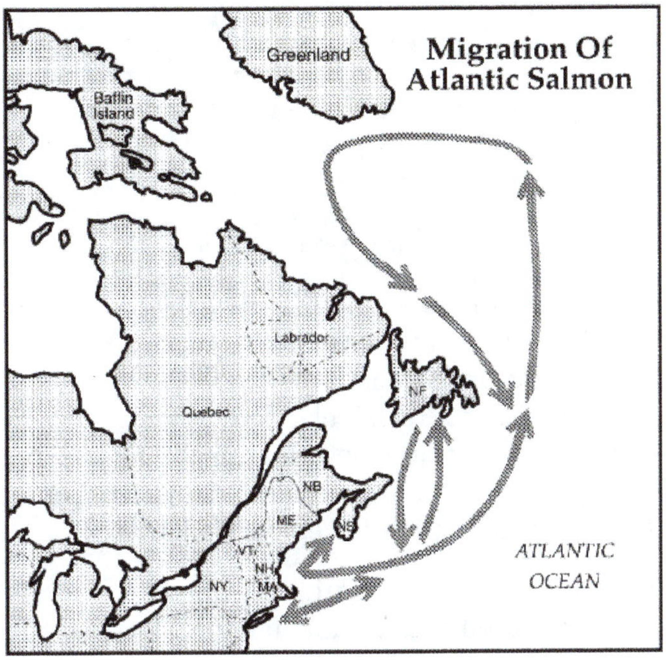

Ocean migration of Atlantic salmon from Connecticut River

In a 1949 journal article, George S. Myers coined the inclusive term *diadromous* to refer to all fishes that migrate between the sea and fresh water. Like the two well known terms, it was formed from classical Greek. *Diadromous* proved a useful word, but terms proposed by Myers for other types of diadromous fishes did not catch on. These included *amphidromous* (fishes that migrate from fresh water to the seas, or vice versa, but not for the purpose of breeding), *potamodromous* (fishes whose migrations occur wholly within fresh water), and *oceanodromous* (fishes that live and migrate wholly in the sea).

Although these classifications were originated for fishes, they are, in principle, applicable to any aquatic organism.

Forage Fish

Forage fish often make great migrations between their spawning, feeding and nursery grounds. Schools of a particular stock usually travel in a triangle between these grounds. For example, one stock of herrings have their spawning ground in southern Norway, their feeding ground in Iceland, and their nursery ground in northern Norway. Wide triangular journeys such as these may be important because forage fish, when feeding, cannot distinguish their own offspring.

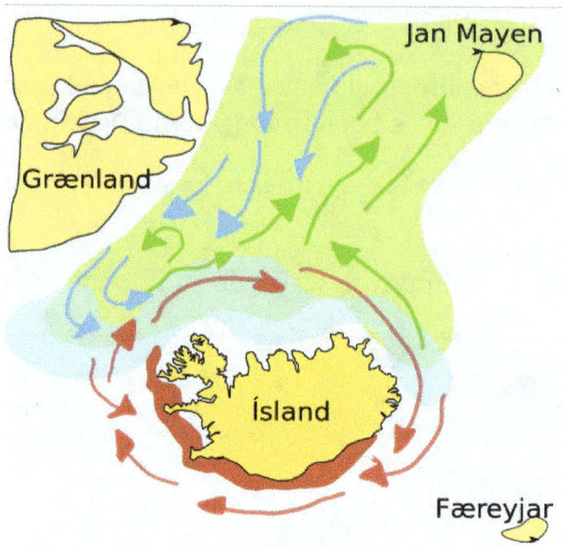

Migration of Icelandic capelin

Capelin are a forage fish of the smelt family found in the Atlantic and Arctic oceans. In summer, they graze on dense swarms of plankton at the edge of the ice shelf. Larger capelin also eat krill and other crustaceans. The capelin move inshore in large schools to spawn and migrate in spring and summer to feed in plankton rich areas between Iceland, Greenland, and Jan Mayen. The migration is affected by ocean currents. Around Iceland maturing capelin make large northward feeding migrations in spring and summer. The return migration takes place in September to November. The spawning migration starts north of Iceland in December or January.

The diagram on the right shows the main spawning grounds and larval drift routes. Capelin on the way to feeding grounds is coloured green, capelin on the way back is blue, and the breeding grounds are red.

In a paper published in 2009, researchers from Iceland recount their application of an interacting particle model to the capelin stock around Iceland, successfully predicting the spawning migration route for 2008.

Highly Migratory Species

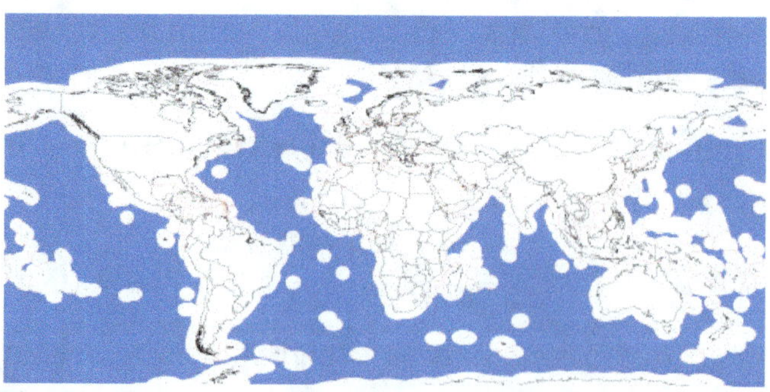

The high seas, highlighted in blue, are the seas which are outside the 200 mile exclusive economic zones

The term highly migratory species (HMS) has its origins in Article 64 of the United Nations Convention on the Law of the Sea (UNCLOS). The Convention does not provide an operational definition of the term, but in an annex (UNCLOS Annex 1) lists the species considered highly migratory by parties to the Convention. The list includes: tuna and tuna-like species (albacore, bluefin, bigeye tuna, skipjack, yellowfin, blackfin, little tunny, southern bluefin and bullet), pomfret, marlin, sailfish, swordfish, saury and oceangoing sharks, dolphins and other cetaceans.

These high trophic level oceanodromous species undertake migrations of significant but variable distances across oceans for feeding, often on forage fish, or reproduction, and also have wide geographic distributions. Thus, these species are found both inside the 200 mile exclusive economic zones and in the high seas outside these zones. They are pelagic species, which means they mostly live in the open ocean and do not live near the sea floor, although they may spend part of their life cycle in nearshore waters.

Highly migratory species can be compared with straddling stock and transboundary stock. Straddling stock range both within an EEZ as well as in the high seas. Transboundary stock range in the EEZs of at least two countries. A stock can be both transboundary and straddling.

Other Examples

Some of the best-known anadromous fishes are the Pacific salmon species, such as Chinook (king), coho (silver), chum (dog), pink (humpback) and sockeye (red) salmon. These salmon hatch in small freshwater streams. From there they migrate to the sea to mature, living there for two to six years. When mature, the salmon return to the same streams where they were hatched to spawn. Salmon are capable of going hundreds of kilometers upriver, and humans must install fish ladders in dams to enable the salmon to get past. Other examples of anadromous fishes are sea trout, three-spined stickleback, and shad.

Several Pacific salmon (Chinook, coho and Steelhead) have been introduced into the US Great Lakes, and have become potamodromous, migrating between their natal waters to feeding grounds entirely within fresh water.

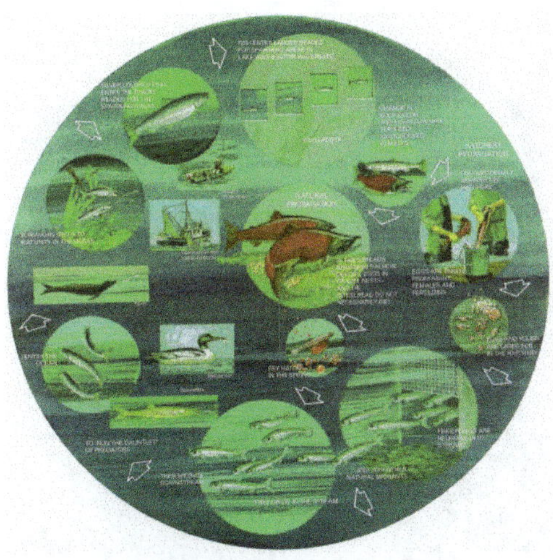

Life cycle of anadromous fish. From a U.S. Government pamphlet. (Click image to enlarge.)

The most remarkable catadromous fishes are freshwater eels of genus *Anguilla*, whose larvae drift from spawning grounds in the Sargasso sea, sometimes for months or years, before entering freshwater river and streams as glass eels or elvers.

An example of a euryhaline species is the bull shark, which lives in Lake Nicaragua of Central America and the Zambezi River of Africa. Both these habitats are fresh water, yet bull sharks will also migrate to and from the ocean. Specifically, Lake Nicaragua bull sharks migrate to the Atlantic Ocean and Zambezi bull sharks migrate to the Indian Ocean.

Diel vertical migration is a common behavior; many marine species move to the surface at night to feed, then return to the depths during daytime.

A number of large marine fishes, such as the tuna, migrate north and south annually, following temperature variations in the ocean. These are of great importance to fisheries.

Freshwater (potamodromous) fish migrations are usually shorter, typically from lake to stream or vice versa, for spawning purposes. However, potamodromous migrations of the endangered Colorado pikeminnow of the Colorado River system can be extensive. Migrations to natal spawning grounds easily be 100 km, with maximum distances of 300 km reported from radiotagging studies. Colorado pikeminnow migrations also display a high degree of homing and the fish may make upstream or downstream migrations to reach very specific spawning locations in whitewater canyons.

Historic Exploitation

Since prehistoric times humans have exploited certain anadromous fishes during their migrations into freshwater streams, when they are more vulnerable to capture. Societies dating to the Millingstone Horizon are known which exploited the anadromous fishery of Morro Creek and other Pacific coast estuaries. In Nevada the Paiute tribe has harvested migrating Lahontan cutthroat trout along the Truckee River since prehistoric times. This fishing practice continues to current times, and the U.S. Environmental Protection Agency has supported research to assure the water quality in the Truckee can support suitable populations of the Lahontan cutthroat trout.

Insect Migration

Monarch butterflies roosting on migration in Texas

Insect migration is the seasonal movement of insects, particularly those by species of dragonflies, beetles, butterflies and moths. The distance can vary with species and in most cases these movements involve large numbers of individuals. In some cases the individuals that migrate in one direction may not return and the next generation may instead migrate in the opposite direction. This is a significant difference from bird migration.

Definition

All insects move to some extent. The range of movement can vary from within a few centimeters for some sucking insects and wingless aphids to thousands of kilometres in the case of other insects such as locusts, butterflies and dragonflies. The definition of migration is therefore particularly difficult in the context of insects. A behaviour oriented definition proposed is

Migratory behaviour is persistent and straightened-out movement effected by the animal's own locomotory exertions or by its active embarkation on a vehicle. It depends upon some temporary inhibition of station-keeping responses but promotes their eventual disinhibition and recurrence.

—*Kennedy, 1985*

This definition disqualifies movements made in the search of resources and which are terminated upon finding of the resource. Migration involves longer distance movement and these movements are not affected by the availability of the resource items. All cases of long distance insect migration concern winged insects.

General Patterns

Migrating butterflies fly within a boundary layer, with a specific upper limit above the ground. The air speeds in this region are typically lower than the flight speed of the insect. These 'boundary-layer' migrants include the larger day-flying insects, and their low-altitude flight is obviously easier to observe than that of most high-altitude windborne migrants.

Many migratory species tend to have polymorphic forms, a migratory one and a resident phase. The migratory phases are marked by their well developed and long wings. Such polymorphism is well known in aphids and grasshoppers. In the migratory locusts, there are distinct long and short-winged forms.

The energetic cost of migration has been studied in the context of life-history strategies. It has been suggested that adaptations for migration would be more valuable for insects that live in habitats where resource availability changes seasonally. Others have suggested that species living in isolated islands of suitable habitats are more likely to evolve migratory strategies. The role of migration in gene flow has also been studied in many species. Parasite loads affect migration. Severely infected individuals are weak and have shortened lifespans. Infection creates an effect known as culling whereby migrating animals are less likely to complete the migration. This results in populations with lower parasite loads.

Orientation

Migration is usually marked by well defined destinations which need navigation and orientation. A flying insect needs to make corrections for crosswinds. It has been demonstrated that many migrating insects sense windspeed and direction and make suitable corrections. Day-flying insects primarily make use of the sun for orientation, however this requires that they compensate for the movement of the sun. Endogenous time-compensation mechanisms have been proposed and tested by releasing migrating butterflies that have been captured and kept in darkness to shift their internal clocks and observing changes in the directions chosen by them. Some species appear to make corrections while it has not been demonstrated in others.

Most insects are capable of sensing polarized light and they are able to use the polarization of the sky when the sun is occluded by clouds. The orientation mechanisms of nocturnal moths and other insects that migrate have not been well studied, however magnetic cues have been suggested in short distance fliers.

Recent studies suggest that migratory butterflies may be sensitive to the Earth's magnetic field on the basis of the presence of magnetite particles. In an experiment on the monarch butterfly, it was shown that a magnet changed the direction of initial flight of migrating monarch butterflies. However this result was not a strong demonstration since the directions of the experimental butterflies and the controls did not differ significantly in the direction of flight.

Lepidoptera

Migration of butterflies and moths is particularly well known. The Bogong moth is a native insect of Australia that is known to migrate to cooler climates. The Madagascan sunset moth (*Chrysiridia rhipheus*) has migrations of up to thousands of individuals, occurring between the eastern and western ranges of their host plant, when they become depleted or unsuitable for consumption.

In southern India, mass migrations of many species occur before monsoons. As many as 250 species of butterflies in India are migratory. These include members of the Pieridae and Nymphalidae. The Australian painted lady periodically migrates down the cost of Australia, and occasionally, in periods of strong migration in Australia, migrate to New Zealand.

The monarch butterfly migrates from southern Canada to wintering sites in central Mexico where they spend the winter. In the late winter or early spring, the adult monarchs leave the Transvolcanic mountain range in Mexico to travel north. Mating occurs and the females seek out milkweed to lay their eggs, usually first in northern Mexico and southern Texas. The caterpillars hatch and develop into adults that move north, where more offspring can go as far as Central Canada until the next migratory cycle. The entire annual migration cycle involves five generations.

The painted lady (*Vanessa cardui*) is a butterfly whose annual 15,000 km round trip from Scandinavia and Great Britain to West Africa involves up to six generations.

Orthoptera

Short-horned grasshoppers sometime form swarms that will make long flights. These are often irregular and may be related to resource availability and thus not fulfilling some definitions of insect migration. There are however some populations of species such as locusts (*Schistocerca gregaria*) that make regular seasonal movements in parts of Africa; exceptionally, the species migrates very long distances, as in 1988 when swarms flew across the Atlantic ocean.

Locusts (*Schistocerca gregaria*) regularly migrate with the seasons.

Odonata

Dragonflies are among the longest distance insect migrants. Many species of *Libellula*, *Sympetrum* and *Pantala* are known for their mass migration. *Pantala flavescens* is thought to make the longest ocean crossings among insects, flying between India and Africa on their migrations. Their movements are often assisted by winds.

Coleoptera

Ladybird beetles such as *Hippodamia convergens*, *Adalia bipunctata* and *Coccinella undecimpunctata* have been noted in large numbers in some places. In some cases, these movements appear to be made in the search for hibernation sites.

Lepidoptera Migration

Monarch butterfly is the best known migrant amongst the Lepidoptera.

Tirumala septentrionis migrate in millions between Eastern Ghats and Western Ghats in India.

Many populations of Lepidoptera (butterflies or moths) migrate, sometimes long distances, to and from areas which are only suitable for part of the year. Lepidopterans migrate on all continents except Antarctica, including from or within subtropical and tropical areas. By migrating, these species can avoid unfavorable circumstances, including weather, food shortage, or over-population. In some lepidopteran species, all individuals migrate; in others, only some migrate.

The best-known Lepidopteran migration is that of the eastern population of the Monarch butterfly which migrates from southern Canada to wintering sites in central Mexico. In late winter/early spring, the adult monarchs leave the Transvolcanic mountain range in Mexico for a more northern climate. Mating occurs and the females begin seeking out milkweed to lay their eggs, usually first in northern Mexico and southern Texas. The caterpillars hatch and develop into adults that move north, where more offspring can go as far as Central Canada until next migratory cycle.

The Danaids in South India are prominent migrants, between Eastern Ghats and Western Ghats. Three species will be involved in this, namely *Tirumala septentrionis*, *Euploea core*, and *Euploea sylvester*. Sometimes they are joined by Lemon Pansy (*Junonia lemonias*), Common Emigrant (*Catopsilia pomona*), Tawny coster (*Acraea terpsicore*) and Blue Tiger (*Tirumala limniace*).

Definition

Migration in Lepidoptera means a regular, predictable movement of a population from one place to another, determined by the seasons. There is no unambiguous definition of *migratory butterfly* or *migratory moth*, and this also applies to proposals to divide them into classes. Migration means different things to behavioral scientists and ecologists. The former emphasize the act of moving whereas the latter discriminate between whether the movement has been ecologically significant or not. Migration may be viewed as "a behavioural process with ecological consequences".

Migration in Lepidoptera takes place in two of the three modes of migration identified by Johnson (Johnson, 1969). In the first mode (also Johnson's first), the Lepidoptera move in one direction in their short life-span and do not return. An example is the pierid butterfly, *Ascia monuste*, which breeds in Florida but sometimes migrates along the coast up to 160 kilometers to breed in more suitable areas.

In the second mode (Johnson's third), migration takes place to a place of hibernation or aestivation where they undergo diapause and the same generation survives to return. The classic example is that of the nymphalid Monarch butterfly (*Danuas plexippus*).

Adventive

Species that are recorded in unexpected areas (adventive species) are not considered to be migratory species, because these did not leave their habitat on their own strength. Examples are species that are imported as egg or caterpillar alongside of their host plants or individuals that were reared by a collector but have escaped. An example of an introduced species is *Galleria mellonella*, which is found all over the world, because it is reared as food for captive birds and reptiles. At times it is difficult to decide if a species is adventive or migratory. Migratory species like *Chrysodeixis chalcites* and *Helicoverpa armigera* would be able to reach Western Europe on their own, but are also common in greenhouses.

Seasonal Migration

Lepidoptera migration is often seasonal. With species of which all individuals migrate, the population moves between areas in the summer and winter season or the dry and wet season.

Australian species *Agrotis infusa* spends the summer as an imago in the Australian Alps, and is sometimes found in buildings in large numbers.

For species of which only part of the population migrates, seasonal migration is hard to determine. They can maintain themselves in part of their habitat but also reach areas where they cannot establish a permanent population. They only live there in the season that is most favorable for the species. Some of the species have the habit of returning to their permanent residence at the end of the season.

Difference with Bird Migration

An important difference with bird migration is that an individual butterfly or moth usually migrates in one direction, while birds migrate back and forth multiple times within their lifespan. This is due to the short lifespan as an imago. Amazingly the Monarch receives no navigation instruction for the migration from their parents, unlike birds. Species that migrate back and forth, usually do so in different generations. There are however, some exceptions:

- The famous migration of the Monarch butterfly in North America. This species migrates back and forth in one generation, though it completes only part of the journey in both directions in that generation. No individual completes the entire journey which is spread over a number of generations. The imago of the last summer generation is born in North America, migrates to Mexico, Florida, or California and stays there for the winter. After the winter it migrates back to the north to reproduce. In a couple of generations, the monarch migrates north to Canada.
- The migration of *Agrotis infusa* in Australia. This species migrates from south-eastern Australia to the Australian Alps in the summer to avoid the heat. After the summer it returns to reproduce.

Flight Behaviour

Migratory Lepidoptera are, in most cases, excellent flyers. Species like the *Vanessa atalanta* are capable of managing a fierce headwind. In case of headwind, they usually fly low and are more goal-oriented. During migration, some species can be found on high altitudes, ranging to up to two kilometers This is especially noteworthy for day-flying species like *Vanessa atalanta*, since the temperatures on these altitudes are low and day-flying species depend on the outside temperature to stay warm. It is thought that *Vanessa atalanta* produces enough body warmth during flight since it has also been recorded migrating at night.

The small Diamondback moth is a migratory species which can be found at altitudes of a 100 meters. It is capable of covering over 3,000 kilometers.

In the case of transcontinental migration where distances are large, the flying speed of the butterfly (of the order of 3 metres per second or less) is inadequate for timely completion of journey. The migration is carried out by relying on heavy winds; a persistent wind speed of 10 metres per second being able to provide a displacement of 300 to 400 kilometers in a single day. For example, the Painted Lady (*Vanessa cardui*) migrates from Africa to Spain aided by tail winds.

That the migratory species are good flyers, is not the same as saying they are robust flyers. The small Diamondback moth is also a migratory species that migrates 3,000 kilometers and can be found up to altitudes of 100 meters or more.

Navigation

To be able to migrate over long distances, species must be able to navigate. There are several ways they do this.

Landscape: Lepidoptera use coastal lines, mountains, but also man-made roads to orient themselves. Above sea it has been observed that the flight direction is much more accurate if the landscape on the coast is still visible.

Celestial navigation: Butterflies are known to be capable of navigation with the help of the sun. They can also navigate by using polarized light. The polarization of the sun's light changes with the angle of the rays, hence they can also navigate with cloudy weather. There are indications that they can even make corrections depending on the time on a day. Diamondback moths are known to fly in a straight trajectory which is not dependent on the angle of the sun's rays. Tests have been performed to interfere with the biological clock of certain species by keeping them in the dark and then observing if they would choose for other flight paths. The conclusion was that some species did, and others did not. Research on monarchs demonstrates that with removal of antennae, the location of the circadian clock, individuals do not localize in any one direction during flight as they do with antennae intact. Night flyers cannot use sun light for navigation. Most of these species rely on the moon and stars instead.

Earth's magnetic field: A number of moths use the Earth's magnetic field to navigate, as a study of the stray Heart and Dart suggests. Another study, this time of the migratory behaviour of the Silver Y, showed that this species, even at high altitudes, can correct its course with changing winds, and prefers flying with favourable winds, which suggests a great sense of direction. *Aphrissa statira* in Panama loses its navigational capacity when exposed to a magnetic field, suggesting it uses the Earth's magnetic field.

Areas Where Migratory Butterflies and Moths Can be Found

Migratory Lepidoptera can be found on all continents, migrating within or from the Tropics and Subtropics. North, they can be found up to Spitsbergen, above the Arctic Circle. Some migratory Lepidoptera have spread over most of the world. Some of these are pest insects, such as the diamondback moth, *Helicoverpa armigera*, and *Trichoplusia ni*.

Blue Tiger (*Tirumala limniace*) butterflies resting while migrating, Western Ghats, South India.

Examples of Migratory Lepidoptera

Some examples of Migratory Lepidoptera are:

- In the Indian subcontinent, migrating Lepidoptera are common just before the monsoon season. Over 250 species are known to migrate during this period.
- On Madagascar, *Chrysiridia rhipheus* migrates between populations of four plant species from the *Omphalea* genus, the host plant of this species. The three western *Omphalea* species live in dry coniferous forests, the eastern species is found in the rainforest and is the only species that is green year round.
- *Rhodometra sacraria* is normally found in Africa, large parts of Asia and Southern Europe. At times, it migrates north and can reach central and northern Europe.
- *Vanessa cardui* is found all over the world, except South America, on altitudes of up to 3,000 meters above sea level. Its home, however, are subtropical steppe areas.
- Several South and Central American species of *Uraniidae* display a great peak in migratory behavior in certain years. In the years with a great number of migrating individuals, there are "population explosions". Individuals migrate to the south and east. There is no real re-migration to speak off. These moths feed on *Omphalea* species. These can be found in scattered populations all across South and Central America, but only part of the areas where they are found are a permanent habitat for these moths. Research suggests that the cause of the migration peak is an increase in toxicity when much of a single plant is eaten and decreasing toxicity when only small amounts of a certain plant are eaten.
- The migration of *Euploea core, Euploea sylvester, Tirumala septentrionis,* and *Tirumala limniace* between Western Ghats and Eastern Ghats in India covering up to 350 to 400 km. This migration happens twice a year. The most probable reasons seem to be the heavy monsoons, reproductive diapause, and a lack of host plants, nectar and alkaloidal resource availability.
- The Australian Painted Lady periodically migrates down the cost of Australia. Occasionally, in periods of strong migration in Australia, these butterflies have been known to migrate to New Zealand.
- The Gatekeeper butterfly (*Pyronia tithonus*) from Southern and Eastern Britain has been expanding to Northern Britain over the past three decades, despite their preference for warmer conditions.

Tirumala septentrionis

Vanessa cardui

Example Species: Macroglossum Stellatarum

Macroglossum stellatarum is a moth that is recorded in the subtropical part of the Palearctic ecozone year round. In summer, the species disperses north up to Scandinavia and Iceland. In winter it migrates further south, deeper into Africa and to the Indian subcontinent.

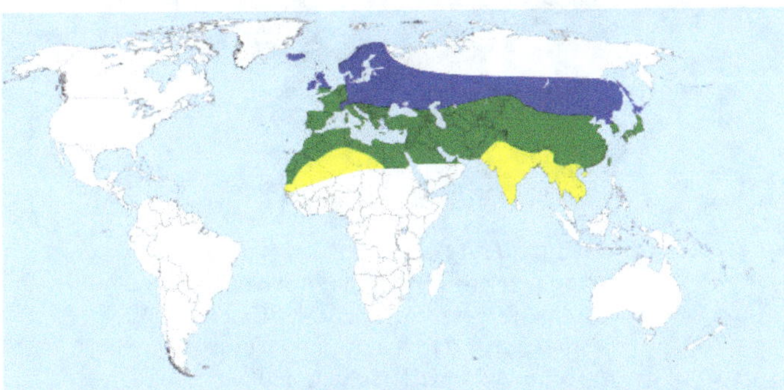

The green areas are the areas where *Macroglossum stellatarum* can survive year round. In the blue areas it can be found in summer. In the yellow area it is found in winter

In the Netherlands and Belgium, there are 100 to 200 records per year in an average year. In warm summers however, like the years 2005 and 2006, several thousand are recorded. In mild winters, small numbers can survive this far north, but these numbers are insufficient to call it a real population.

Causes

Usually, butterflies and moths migrate to escape from potentially harmful circumstances. Examples of this are a shortage of proper food plants, an unfavorable climate, like cold or extreme rain or overpopulation.

Migration and Evolution

A phenomenon like migration is an evolutionary development. By migrating, the species has survived the process of natural selection. There are a number of advantages and disadvantages to migration. An example of this is the protozoan *Ophryocystis elektroscirrha* a parasite of the monarch. Severely infected individuals are weak, unable to expand their wings, or unable to eclose, and have shortened lifespans, but parasite levels vary in populations. This is not the case in laboratory or commercial rearing, where after a few generations, all individuals can be infected. Infection with this parasite creates an effect known as culling whereby migrating monarchs that are infected are less likely to complete the migration. This results in overwintering populations with lower parasite loads.

Recording

Butterflies (and to a lesser extent moths) migrating in large numbers are a noteworthy sight, which is easily to observe and track. There are several historic records about migrating butterflies. There are records dating back to 1100 about migrating butterflies (probably a *Pieris* species) from Bavaria to the Duchy of Saxony and from 1248 about the migration of yellow butterflies in Japan.

Abbildung von einer unzehligen
Menge frembden großen geflügelten
Ungeziefers welches bey Türnau
und angräntzenden Gebürg im
Eingang Monaths Septembris 1719.
gefallen Denen Garten u: Feld
Früchten großen Schaden zugefüget,
doch aber sich bald wieder verlohren
haben. wor bey meist zu consi
deriren daß sie über ihren Maul
einen zusamen gerollten dünnen
Rüssel bey 1½ Finger lang ge-
habt, den sie gleich einer langen
Nadel auß strecken könen, was diese
deuten, ist Gott bekannt.

Record of a massive appearance of (presumably) *Agrius convolvuli* in 1719 near Turnau.

When flying at high altitudes, spotting migrating butterflies or moths can be hard. Low flying species are easily spotted or caught using a light trap. When individuals fly too high for these methods, air balloons equipped with nets are used at times. Alternatively, radar is used to monitor migration.

Another registration technique, is marking the wings with tiny stickers, a technique comparable with Bird ringing. This technique has not proven to be very successful though. Advances in technology might make it possible to equip individuals with micro transmitters in the future.

Migration and Climate Change

Global warming has caused an increase of migratory butterflies and moths that reach north-western European countries like the Netherlands, Belgium and the United Kingdom. Research in the United Kingdom confirms that an increasing number of migrants reach the country. Because one would expect that migratory species can adapt to new circumstances quite well, the researches warn for new species that can have a negative impact on native species and possible damage to both health (species like the Oak Processionary) and agriculture.

References

- Carew, Thomas J. (2000). "9. Associative Learning in Honeybees". Behavioral Neurobiology: The Cellular Organization of Natural Behavior. Sinauer Associates. ISBN 978-0-87893-084-5.
- Mitchell, Robert W.; Thompson, Nicholas S. (1986). Deception, Perspectives on Human and Nonhuman Deceit. SUNY Press. pp. 21–29. ISBN 1438413327.
- King, R. C.; Stansfield, W. D.; Mulligan, P. K. (2006). A Dictionary of Genetics (7th ed.). Oxford: Oxford University Press. p. 278. ISBN 0-19-530762-3.
- Helfman, G.S., Collette, B.B. and Facey, D.E., (1997). The Diversity of fishes. Wiley-blackwell. pp. 324. ISBN 978-0-86542-256-8
- Ruxton, G.D., Sherratt, T.N. and Speed, M.P., (2004). Avoiding attack: the evolutionary ecology of crypsis, warning signals and mimicry. Oxford University Press. ISBN 0-19-852859-0. p. 198
- Baskett, T.S., Sayre, M.W. and Tomlinson, R.E., (1993). Ecology and Management of the mourning dove. Stackpole Books, p. 167, ISBN 0-8117-1940-5.
- green, Malcom (2005). Book of Lies (1st ed.). Kansas City, MO: Andrews McMeel Publishing. p. 61. ISBN 9780740755606.
- Radloff, Sara E.; Hepburn, H. Randall; Engel, Michael S. (2011). Honeybees of Asia. Berlin: Springer Science & Business Media. ISBN 978-3642164217.
- Berthold, Peter; Bauer, Hans-Günther; Westhead, Valerie (2001). Bird Migration: A General Survey. Oxford: Oxford University Press. ISBN 0-19-850787-9.
- Peter Berthold; Hans-Günther Bauer; Valerie Westhead (2001). Bird Migration: A General Survey. Oxford: Oxford University Press. ISBN 0-19-850787-9.
- Newton, Ian (2010). "13. Large-Scale Movement Patterns". The Migration Ecology of Birds. Academic Press. pp. 396, and throughout. ISBN 9780080554839.
- Vinicombe, Keith; David Cottridge (1996). Rare birds in Britain and Ireland a photographic record. London: Collins. p. 192. ISBN 0002199769.

5

Animal Sexual Behavior

Animal sexual behavior is the reproductive process that takes place in animals. The systems involved in animal sexual behavior include monogamy, polyandry, polygamy and promiscuity. The topics discussed in the section are of great importance to broaden the existing knowledge on animal sexual behavior.

Animal Sexual Behaviour

Animal sexual behaviour takes many different forms, including within the same species. Common mating or reproductively motivated systems include monogamy, polyandry, polygamy, and promiscuity. Other sexual behaviour may be reproductively motivated (e.g. sex apparently due to duress or coercion and situational sexual behaviour) or non-reproductively motivated (e.g. interspecific sexuality, sexual arousal from objects or places, sex with dead animals, homosexual sexual behaviour, bisexual sexual behaviour).

Stags fighting – a common sexual behaviour

When animal sexual behaviour is reproductively motivated, it is often termed *mating* or *copulation*; for most non-human mammals, mating and copulation occur at oestrus (the most fertile period in the mammalian female's reproductive cycle), which increases the chances of successful impregnation. Some animal sexual behaviour involves competition, sometimes fighting, between multiple males. Females often select males for mating only if they appear strong and able to protect themselves. The male that wins a fight may also have the chance to mate with a larger number of females and will therefore pass on his genes to their offspring.

Greater sage-grouse at a lek, with multiple males displaying for the less conspicuous females.

Historically, it was believed that only humans and a small number of other species performed sexual acts other than for reproduction, and that animals' sexuality was instinctive and a simple "stimulus-response" behaviour. However, in addition to homosexual behaviours, a range of species masturbate and may use objects as tools to help them do so. Sexual behaviour may be tied more strongly to establishment and maintenance of complex social bonds across a population which support its success in non-reproductive ways. Both reproductive and non-reproductive behaviours can be related to expressions of dominance over another animal or survival within a stressful situation (such as sex due to duress or coercion).

Anatomical structures on the head and throat of a domestic turkey. 1. Caruncles, 2. Snood, 3. Wattle (dewlap), 4. Major caruncle, 5. Beard. During sexual behaviour, these structures enlarge or become brightly coloured.

Mating Systems

In sociobiology and behavioural ecology, the term "mating system" is used to describe the ways in which animal societies are structured in relation to sexual behaviour. The mating system specifies which males mate with which females, and under what circumstances. There are four basic systems:

The four basic mating systems		
	Single female	**Multiple females**
Single male	Monogamy	Polygyny
Multiple males	Polyandry	Polygynandry

Monogamy

Monogamy occurs when one male mates with one female exclusively. A monogamous mating system is one in which individuals form long-lasting pairs and cooperate in raising offspring. These pairs may last for a lifetime, such as in pigeons, or it may occasionally change from one mating season to another, such as in emperor penguins. In contrast with tournament species, these pair-bonding species have lower levels of male aggression, competition and little sexual dimorphism. Zoologists and biologists now have evidence that monogamous pairs of animals are not always sexually exclusive. Many animals that form pairs to mate and raise offspring regularly engage in sexual activities with extra-pair partners. This includes previous examples, such as swans. Sometimes, these extra-pair sexual activities lead to offspring. Genetic tests frequently show that some of the offspring raised by a monogamous pair come from the female mating with an extra-pair male partner. These discoveries have led biologists to adopt new ways of talking about monogamy; According to Ulrich Reichard (2003):

Social monogamy refers to a male and female's social living arrangement (e.g., shared use of a territory, behaviour indicative of a social pair, and/or proximity between a male and female) without inferring any sexual interactions or reproductive patterns. In humans, social monogamy takes the form of monogamous marriage. Sexual monogamy is defined as an exclusive sexual relationship between a female and a male based on observations of sexual interactions. Finally, the term genetic monogamy is used when DNA analyses can confirm that a female-male pair reproduce exclusively with each other. A combination of terms indicates examples where levels of relationships coincide, e.g., sociosexual and sociogenetic monogamy describe corresponding social and sexual, and social and genetic monogamous relationships, respectively.

Whatever makes a pair of animals socially monogamous does not necessarily make them sexually or genetically monogamous. Social monogamy, sexual monogamy, and genetic monogamy can occur in different combinations.

Social monogamy is relatively rare in the animal kingdom. The actual incidence of social monogamy varies greatly across different branches of the evolutionary tree. Over 90% of avian species are socially monogamous. This stands in contrast to mammals. Only 3% of mammalian species are socially monogamous, although up to 15% of primate species are. Social monogamy has also been observed in reptiles, fish, and insects.

Sexual monogamy is also rare among animals. Many socially monogamous species engage in extra-pair copulations, making them sexually non-monogamous. For example, while over 90% of birds are socially monogamous, "on average, 30% or more of the baby birds in any nest [are] sired by someone other than the resident male." Patricia Adair Gowaty has estimated that, out of 180 different species of socially monogamous songbirds, only 10% are sexually monogamous.

The incidence of genetic monogamy, determined by DNA fingerprinting, varies widely across species. For a few rare species, the incidence of genetic monogamy is 100%, with all offspring genetically related to the socially monogamous pair. But genetic monogamy is strikingly low in other species. Barash and Lipton note:

The highest known frequency of extra-pair copulations are found among the fairy-wrens, lovely tropical creatures technically known as *Malurus splendens* and *Malurus cyaneus*. More than 65% of all fairy-wren chicks are fathered by males outside the supposed breeding group.

Such low levels of genetic monogamy have surprised biologists and zoologists, forcing them to rethink the role of social monogamy in evolution. They can no longer assume social monogamy determines how genes are distributed in a species. The lower the rates of genetic monogamy among socially monogamous pairs, the less of a role social monogamy plays in determining how genes are distributed among offspring.

Polygyny

Polygyny occurs when one male gets exclusive mating rights with multiple females. In some species, notably those with harem-like structures, only one of a few males in a group of females will mate. Technically, polygyny in sociobiology and zoology is defined as a system in which a male has a relationship with more than one female, but the females are predominantly bonded to a single male. Should the active male be driven out, killed, or otherwise removed from the group, in a number of species the new male will ensure that breeding resources are not wasted on another male's young. The new male may achieve this in many different ways, including:

- competitive infanticide: in lions, hippopotamuses, and some monkeys, the new male will kill the offspring of the previous alpha male to cause their mothers to become receptive to his sexual advances since they are no longer nursing.
- harassment to miscarriage: amongst wild horses and baboons, the male will "systematically harass" pregnant females until they miscarry.
- Pheromone-based spontaneous abortion
- in some rodents such as mice, a new male with a different scent will cause females who are pregnant to spontaneously fail to implant recently fertilised eggs. This does not require contact; it is mediated by scent alone. It is known as the Bruce effect.

Von Haartman specifically described the mating behaviour of the European pied flycatcher as successive polygyny. Within this system, the males leave their home territory once their primary female lays her first egg. Males then create a second territory, presumably in order to attract a secondary female to breed. Even when they succeed at acquiring a second mate, the males typically return to the first female to exclusively provide for her and her offspring.

Polygynous mating structures are estimated to occur in up to 90% of mammal species. As polygyny is the most common form of polygamy among vertebrates (including humans, to some extent), it has been studied far more extensively than polyandry or polygynandry.

Polyandry

Polyandry occurs when one female gets exclusive mating rights with multiple males. In some species, such as redlip blennies, both polygyny and polyandry are observed.

The anglerfish *Haplophryne mollis* is polyandrous. This female is trailing the atrophied remains of males she has encountered.

The males in some deep sea anglerfishes are much smaller than the females. When they find a female they bite into her skin, releasing an enzyme that digests the skin of their mouth and her body and fusing the pair down to the blood-vessel level. The male then slowly atrophies, losing first his digestive organs, then his brain, heart, and eyes, ending as nothing more than a pair of gonads, which release sperm in response to hormones in the female's bloodstream indicating egg release. This extreme sexual dimorphism ensures that, when the female is ready to spawn, she has a mate immediately available. A single anglerfish female can "mate" with many males in this manner.

Polygynandry

Polygynandry occurs when multiple males mate indiscriminately with multiple females. The numbers of males and females need not be equal, and in vertebrate species studied so far, there are usually fewer males. Two examples of systems in primates are promiscuous mating chimpanzees and bonobos. These species live in social groups consisting of several males and several females. Each female copulates with many males, and vice versa. In bonobos, the amount of promiscuity is particularly striking because bonobos use sex to alleviate social conflict as well as to reproduce. This mutual promiscuity is the approach most commonly used by spawning animals, and is perhaps the "original fish mating system." Common examples are forage fish, such as herrings, which form huge mating shoals in shallow water. The water becomes milky with sperm and the bottom is draped with millions of fertilised eggs.

Polygamy

The term polygamy is an umbrella term used to refer generally to non-monogamous matings. As such, polygamous relationships can be polygynous, polyandrous or polygynandrous. In a small number of species, individuals can display either polygamous or monogamous behaviour depending on environmental conditions. An example is the social wasp *Apoica flavissima*. In some species, polygyny and polyandry is displayed by both sexes in the population. Polygamy in both sexes has been observed in red flour beetle (*Tribolium castaneum*).

A tournament species is one in which "mating tends to be highly polygamous and involves high levels of male-male aggression and competition." Tournament behaviour often correlates with high levels of sexual dimorphism, examples of species including chimpanzees and baboons. Most polygamous species present high levels of tournament behaviour, with a notable exception being bonobos.

Parental Investment and Reproductive Success

Female and male sexual behaviour differ in many species. Often, males are more active in initiating mating, and bear the more conspicuous sexual ornamentation like antlers and colourful plumage. This is a result of anisogamy, where sperm are smaller and much less costly (energetically) to produce than eggs. This difference in physiological cost means that males are more limited by the number of mates they can secure, while females are limited by the quality of genes of her mates, a phenomenon known as Bateman's principle. Many females also have extra reproductive burdens in that parental care often falls mainly, or exclusively, on them. Thus, females are more limited in their potential reproductive success. In species where males take on more of the reproductive costs, such as sea horses and jacanas, the role is reversed, and the females are larger, more aggressive and more brightly coloured than the males.

Mating grey slugs, suspended from a slime thread

In hermaphroditic animals, the costs of parental care can be evenly distributed between the sexes, e.g. earthworms. In some species of planarians, sexual behaviour takes the form of penis fencing. In this form of copulation, the individual that first penetrates the other with the penis, forces the other to be female, thus carrying the majority of the cost of reproduction. Post mating, banana slugs will some times gnaw off their partners penis as an act of sperm competition. In the grey slug, the sharing of cost leads to a spectacular display, where the mates suspend themselves high above the ground from a slime thread, ensuring none of them can refrain from taking on the cost of egg-bearer.

Seasonality

Many animal species have specific mating (or breeding) periods e.g. (seasonal breeding) so that offspring are born or hatch at an optimal time. In marine species with limited mobility and external fertilisation like corals, sea urchins and clams, the timing of the common spawning is the only externally visible form of sexual behaviour. In areas with continuously high primary production, some species have a series of breeding seasons throughout the year. This is the case with most primates (who are primarily tropical and subtropical animals). Some animals (opportunistic breeders) breed dependent upon other conditions in their environment aside from time of year.

Brain corals typically spawning in connection with the full moon every August

Mammals

Mating seasons are often associated with changes to herd or group structure, and behavioural changes, including territorialism amongst individuals. These may be annual (e.g. wolves), biannual (e.g. dogs) or more frequently (e.g. horses). During these periods, females of most mammalian species are more mentally and physically receptive to sexual advances, a period scientifically described as estrous but commonly described as being "in season" or "in heat". Sexual behaviour may occur outside estrus, and such acts as do occur are not necessarily harmful.

Some mammals (e.g. domestic cats, rabbits and camilidae) are termed "induced ovulators". For these species, the female ovulates due to an external stimulus during, or just prior, to mating, rather than ovulating cyclically or spontaneously. Stimuli causing induced ovulation include the sexual behaviour of coitus, sperm and pheromones. Domestic cats have penile spines. Upon withdrawal of a cat's penis, the spines rake the walls of the female's vagina, which may cause ovulation.

Amphibians

For many amphibians, an annual breeding cycle applies, typically regulated by ambient temperature, precipitation, availability of surface water and food supply. This breeding season is accentuated in temperate regions, in boreal climate the breeding season is typically concentrated to a few short days in the spring.

Fish

Like many coral reef dwellers, the clownfish spawn around the time of the full moon in the wild. In a group of clownfish, there is a strict dominance hierarchy. The largest and most aggressive female is found at the top. Only two clownfish, a male and a female, in a group reproduce through external fertilisation. Clownfish are sequential hermaphrodites, meaning that they develop into males first, and when they mature, they become females. If the female clownfish is removed from the group, such as by death, one of the largest and most dominant males will become a female. The remaining males will move up a rank in the hierarchy.

Motivation

Various neurohormones stimulate sexual wanting in animals. In general, studies have suggested that dopamine is involved in sexual incentive motivation, oxytocin and melanocortins in sexual attraction, and noradrenaline in sexual arousal. Vasopressin is also involved in the sexual behaviour of some animals.

Neurohormones in the Mating Systems of Voles

The mating system of prairie voles is monogamous; after mating, they form a lifelong bond. In contrast, montane voles have a polygamous mating style. When montane voles mate, they form no attachments and separate after copulation. Studies on the brains of these two species have found that it is two neurohormones and their respective receptors that are responsible for these differences in mating strategies. Male prairie voles release vasopressin after copulation prairie vole and attachment to the female then develops. Female prairie voles release oxytocin after reproducing and an attachment to her partner also develops. Neither male nor female montane voles have such a high quantity of oxytocin and vasopressin when they mate. Even when injected with these neurohormones, the mating system does not change. In contrast, if prairie voles are injected with the neurohormones, they may form a lifelong attachment, even if they have not mated. The reason for this is that prairie voles have more oxytocin and vasopressin receptors than do montane voles, and are therefore far more receptive to the two neurohormones. It is not the quantity of the hormone that determines the mating system and bond-formation, rather, it is number of neurohormone receptors.

Oxytocin and Rat Sexual Behaviour

Mother rats experience a postparum estrus which makes them highly motivated to mate. However, they also have a strong motivation to protect their newly born pups. As a consequence, the mother rat solicits males to the nest but simultaneously becomes aggressive towards them to protect her young. If the mother rat is given injections of an oxytocin receptor antagonist, they no longer experience these maternal motivations.

Prolactin influences social bonding in rats.

Oxytocin and Primate Sexual Behaviour

Oxytocin plays a similar role in non-human primates as it does in humans.

Grooming, sex, and cuddling frequencies correlate positively with levels of oxytocin. As the level of oxytocin increases so does sexual motivation. While oxytocin plays a major role in parent child relationships, it is also found to play a role in adult sexual relationships. Its secretion affects the nature of the relationship or if there will even be a relationship at all.

Studies have shown that oxytocin is higher in monkeys in lifelong monogamous relationships compared to monkeys which are single. Furthermore, the oxytocin levels of the couples correlate positively; when the oxytocin secretion of one increases the other one also increases. Higher levels of oxytocin are related to monkeys expressing more behaviours such as cuddling, grooming and sex, while lower levels of oxytocin reduce motivation for these activities.

Research on oxytocin's role in the animal brain suggests that it plays less of a role in behaviours of love and affection than previously believed. "When oxytocin was first discovered in 1909, it was thought mostly to influence a mother's labour contractions and milk let-down. Then, in the 1990s, research with prairie voles found that giving them a dose of oxytocin resulted in the formation of a bond with their future mate (Azar, 40)." Oxytocin has since been treated by the media as the sole player in the "love and mating game" in mammals. This view, however, is proving to be false as, "most hormones don't influence behaviour directly. Rather, they affect thinking and emotions in variable ways (Azar, 40)." There is much more involved in sexual behaviour in the mammalian animal than oxytocin and vasopressin can explain.

Pleasure

It is often assumed that animals do not have sex for pleasure, or alternatively that humans, pigs (and perhaps dolphins and one or two species of primate) are the only species that do. This is sometimes stated as "animals mate only for reproduction". This view is considered a misconception by some scholars. Jonathan Balcombe argues that the prevalence of non-reproductive sexual behaviour in certain species suggests that sexual stimulation is pleasurable. He also points to the presence of the clitoris in some female mammals, and evidence for female orgasm in primates. On the other hand, it is impossible to know the subjective feelings of animals, and the notion that non-human animals experience emotions similar to humans is a contentious subject.

A 2006 Danish Animal Ethics Council report, which examined current knowledge of animal sexuality in the context of legal queries concerning sexual acts by humans, has the following comments, primarily related to domestically common animals:

Even though the evolution-related purpose of mating can be said to be reproduction, it is not actually the creating of offspring which originally causes them to mate. It is probable that they mate because they are motivated for the actual copulation, and because this is connected with a positive experience. It is therefore reasonable to assume that there is some form of pleasure or satisfaction connected with the act. This assumption is confirmed by the behaviour of males, who in the case of many species are prepared to work to get access to female animals, especially if the female animal is in oestrus, and males who for breeding purposes are used to having sperm collected become very eager, when the equipment they associate with the collection is taken out.

There is nothing in female mammals' anatomy or physiology that contradicts that stimulation of the sexual organs and mating is able to be a positive experience. For instance, the clitoris acts in the same way as with women, and scientific studies have shown that the success of reproduction is improved by stimulation of clitoris on (among other species) cows and mares in connection with insemination, because it improves the transportation of the sperm due to contractions of the inner genitalia. This probably also concerns female animals of other animal species, and contractions in the inner genitals are seen e.g. also during orgasm for women. It is therefore reasonable to assume that sexual intercourse may be linked with a positive experience for female animals.

Koinophilia

Koinophilia is the love of the "normal" or phenotypically common. The term was introduced to scientific literature in 1990, and refers to the tendency of animals seeking a mate to prefer that mate not to have any unusual, peculiar or deviant features. Similarly, animals preferentially choose mates with low fluctuating asymmetry. However, animal sexual ornaments can evolve through runaway selection, which is driven by (usually female) selection for non-standard traits.

Interpretation Bias

The field of study of sexuality in non-human species was a long-standing taboo. In the past, researchers sometimes failed to observe, mis-categorising and mis-described sexual behaviour which did not meet their preconceptions. In earlier periods, bias tended to support what would now be described as conservative sexual mores. An example of overlooking behaviour relates to descriptions of giraffe mating:

When nine out of ten pairings occur between males, "[e]very male that sniffed a female was reported as sex, while anal intercourse with orgasm between males was only [categorized as] 'revolving around' dominance, competition or greetings."

In the 21st century, liberal social or sexual views are often projected upon animal subjects of research. Popular discussions of bonobos are a frequently cited example. Current research frequently expresses views such as that of the Natural History Museum at the University of Oslo, which in 2006 held an exhibition on animal sexuality:

Many researchers have described homosexuality as something altogether different from sex. They must realise that animals can have sex with who they will, when they will and without consideration to a researcher's ethical principles.

Other animal activities may be misinterpreted due to the frequency and context in which animals perform the behaviour. For example, domestic ruminants display behaviours such as mounting and head-butting. This often occurs when the animals are establishing dominance relationships and are not necessarily sexually motivated. Careful analysis must be made to interpret what animal motivations are being expressed by those behaviours.

Types of Sexual Behaviour

Cuckoldry

Alternate male strategies which allow small males to engage in cuckoldry can develop in species such as fish where spawning is dominated by large and aggressive males. Cuckoldry is a variant of polyandry, and can occur with *sneak spawners*. A sneak spawner is a male that rushes in to join the spawning rush of a spawning pair. A spawning rush occurs when a fish makes a burst of speed, usually on a near vertical incline, releasing gametes at the apex, followed by a rapid return to the lake or sea floor or fish aggregation. Sneaking males do not take part in courtship. In salmon and trout, for example, *jack males* are common. These are small silvery males that migrate upstream along with the standard, large, hook-nosed males and that spawn by sneaking into redds to release sperm simultaneously with a mated pair. This behaviour is an evolutionarily stable strategy for reproduction, because it is favoured by natural selection just like the "standard" strategy of large males.

Small male bluegill sunfishes cuckold large males by adopting *sneaker* strategies.

Hermaphroditism

Hermaphroditism occurs when a given individual in a species possesses both male and female reproductive organs, or can alternate between possessing first one, and then the other. Hermaphroditism is common in invertebrates but rare in vertebrates. It can be contrasted with gonochorism, where each individual in a species is either male or female, and remains that way throughout their lives. Most fish are gonochorists, but hermaphroditism is known to occur in 14 families of teleost fishes.

Female groupers change their sex to male if no male is available.

Usually hermaphrodites are *sequential*, meaning they can switch sex, usually from female to male (protogyny). This can happen if a dominant male is removed from a group of females. The largest female in the harem can switch sex over a few days and replace the dominant male. This is found amongst coral reef fishes such as groupers, parrotfishes and wrasses. It is less common for a male to switch to a female (protandry). As an example, most wrasses are protogynous hermaphrodites within a haremic mating system. Hermaphroditism allows for complex mating systems. Wrasses exhibit three different mating systems: polygynous, lek-like, and promiscuous mating systems.

Sexual Cannibalism

Sexual cannibalism is a behaviour in which a female animal kills and consumes the male before, during, or after copulation. Sexual cannibalism confers fitness advantages to both the male and female. Sexual cannibalism is common among insects, arachnids and amphipods. There is also evidence of sexual cannibalism in gastropods and copepods.

Araneus diadematus – cannibalistic mating behaviour

Sexual Coercion

Sex in a forceful or apparently coercive context has been documented in a variety of species. In some herbivorous herd species, or species where males and females are very different in size, the male dominates sexually by force and size.

During mating, the male muscovy duck typically immobilises the female.

Some species of birds appear to combine sexual intercourse with apparent violent assault; these include ducks, and geese. Female white-fronted bee-eaters are subjected to forced copulations. When females emerge from their nest burrows, males sometimes force them to the ground and mate with them. Such forced copulations are made preferentially on females who are laying and who may therefore lay eggs fertilised by the male.

In the *Acilius* genus of water beetles, an "evolutionary arms race" between the two sexes means that there is no courtship system for these beetles. Their mating behaviours include males holding females underwater till exhausted, and allowing only occasional access to the surface to breathe for up to six hours (to prevent them breeding with other males), and females which have a variety of body shapings (to prevent males from gaining a grip).

It has been reported that young male elephants in South Africa sexually coerced and killed rhinoceroses. This interpretation of the elephants' behaviour was disputed by one of the original study's authors, who said there was "nothing sexual about these attacks".

Parthenogenesis

Parthenogenesis is a form of asexual reproduction in which growth and development of embryos occur without fertilisation. Technically, parthenogenesis it is not a behaviour, however, sexual behaviours may be involved.

Whip-tailed lizard females have the ability to reproduce through parthenogenesis and as such males are rare and sexual breeding non-standard. Females engage in "pseudocopulation" to stimulate ovulation, with their behaviour following their hormonal cycles; during low levels of oestrogen, these (female) lizards engage in "masculine" sexual roles. Those animals with currently high oestrogen levels assume "feminine" sexual roles. Lizards that perform the courtship ritual have greater fecundity than those kept in isolation due to an increase in hormones triggered by the sexual behaviours. So, even though asexual whiptail lizards populations lack males, sexual stimuli still increase reproductive success. From an evolutionary standpoint these females are passing their full genetic code to all of their offspring rather than the 50% of genes that would be passed in sexual reproduction.

It is rare to find true parthenogenesis in fishes, where females produce female offspring with no input from males. All-female species include the Texas silverside, *Menidia clarkhubbsi* as well as a complex of Mexican mollies.

Parthenogenesis has been recorded in 70 vertebrate species including hammerhead sharks, blacktip sharks, amphibians and crayfish.

Unisexuality

Unisexuality occurs when a species is all-male or all-female. Unisexuality occurs in some fish species, and can take complex forms. *Squalius alburnoides*, a minnow found in several river basins in Portugal and Spain, appears to be an all-male species. The existence of this species illustrates the potential complexity of mating systems in fish. The species originated as a hybrid between two species, and is diploid, but not hermaphroditic. It can have triploid and tetraploid forms, including all-female forms that reproduce mainly through hybridogenesis.

Others

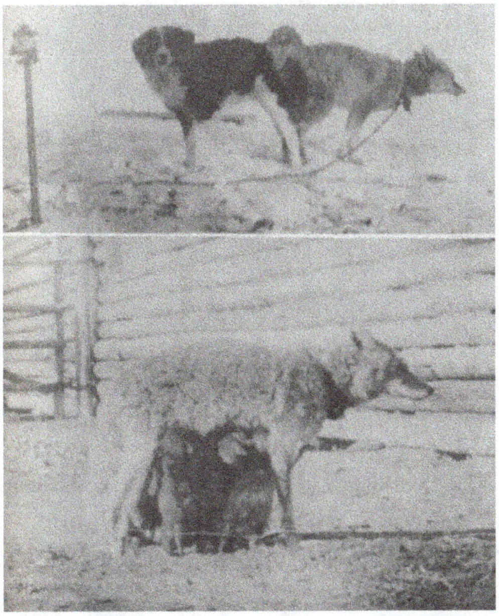

A dog mates with a coyote to produce a dog-coyote hybrid.

- Interbreeding: Hybrid offspring can result from the mating of two organisms of distinct but closely related parent species, although the resulting offspring is not always fertile. According to Alfred Kinsey, genetic studies on wild animal populations have shown a "large number" of inter-species hybrids.
- Prostitution: There are reports that animals occasionally engage in prostitution. A small number of pair-bonded females within a group of penguins took nesting material (stones) after copulating with a non-partner male. The researcher stated "I was watching opportunistically, so I can't give an exact figure of how common it really is." It has been reported that "bartering of meat for sex ... forms part of the social fabric of a troop of wild chimps living in the Tai National Park in the Cote d'Ivoire."
- Pavlovian conditioning: The sexualisation of objects or locations is recognised in the animal breeding world. For example, male animals may become sexually aroused upon visiting a location where they have been allowed to have sex before, or upon seeing a stimulus previously associated with sexual activity such as an artificial vagina. Sexual preferences for certain cues can be artificially induced in rats by pairing scents or objects with their early sexual experiences. The primary motivation of this behaviour is Pavlovian conditioning, and the association is due to a conditioned response (or association) formed with a distinctive "reward".
- Viewing images: A study using four adult male rhesus macaques (*Macaca mulatta*) showed that male rhesus macaques will give up a highly valued item, juice, to see images of the faces or perineum of high-status females. Encouraging captive pandas to mate is problematic. Showing young male pandas "panda pornography" is credited with a recent population boom among pandas in captivity in China. One researcher attributed the success to the sounds on the recordings.

- Copulatory wounding and Traumatic Insemination: Injury to a partner's genital tract during mating occurs in at least 40 taxa, ranging from fruit flies to humans. However, it often goes unnoticed due to its cryptic nature and because of internal wounds not visible outside.

Non-reproductive Sexual Behaviour

There is a range of behaviours that animals perform which appear to be sexually motivated but which can not result in reproduction. These include:

- Masturbation: Some species, both male and female, masturbate, both when partners are available and otherwise.
- Oral sex: Several species engage in both autofellatio and oral sex. This has been documented in brown bears, Tibetan macaques, wolves, goats, primates, hyenas, bats, cape ground squirrels and sheep. In the greater short-nosed fruit bat, copulation by males is dorsoventral and the females lick the shaft or the base of the male's penis, but not the glans which has already penetrated the vagina. While the females do this, the penis is not withdrawn and research has shown a positive relationship between length of the time that the penis is licked and the duration of copulation. Post copulation genital grooming has also been observed.
- Homosexuality: Same-sex sexual behaviour occurs in a range of species, especially in social species, particularly in marine birds and mammals, monkeys, and the great apes. As of 1999, the scientific literature contained reports of homosexual behaviour in at least 471 wild species. Organisers of the Against Nature? exhibit stated that "homosexuality has been observed among 1,500 species, and that in 500 of those it is well documented."

A male black and white tegu mounts a female that has been dead for two days and attempts to mate.

- Genital-genital rubbing: This is sexual activity in which one animal rubs his or her genitals against the genitals of another animal. This is stated to be the "bonobo's most typical sexual pattern, undocumented in any other primate".

- Inter-species mating: Some animals opportunistically mate with individuals of another species.
- Sex involving juveniles: Male stoats (*Mustela erminea*) will sometimes mate with infant females of their species. This is a natural part of their reproductive biology – they have a delayed gestation period, so these females give birth the following year when they are fully grown. Juvenile male chimpanzees have been recorded mounting and copulating with immature chimps. Infants in bonobo societies are often involved in sexual behaviour.
- Necrophilia: This describes when an animal engages in a sexual act with a dead animal. It has been observed in mammals, birds, reptiles and frogs.
- Bisexuality: This describes when an animal shows sexual behaviour towards both males and females.
- Extended female sexuality: This is when females mate with males outside of their conceptive period.

Seahorse

Seahorses, once considered to be monogamous species with pairs mating for life, were described in a 2007 study as "promiscuous, flighty, and more than a little bit gay". Scientists at 15 aquaria studied 90 seahorses of three species. Of 3,168 sexual encounters, 37% were same-sex acts. Flirting was common (up to 25 potential partners a day of both sexes); only one species (the British spiny seahorse) included faithful representatives, and for these 5 of 17 were faithful, 12 were not. Bisexual behaviour was widespread and considered "both a great surprise and a shock", with big-bellied seahorses of both sexes not showing partner preference. 1,986 contacts were male-female, 836 were female-female and 346 were male-male.

Bonobo

The bonobo, which has a matriarchal society, is a fully bisexual species – both males and females engage in sexual behaviour with the same and the opposite sex, with females being particularly noted for engaging in sexual behaviour with each other and at up to 75% of sexual activity being nonreproductive. Primatologist Frans de Waal believes that bonobos use sexual activity to resolve conflict between individuals. Sexual activity occurs between almost all ages and sexes of bonobo societies.

Bonobos mating, Jacksonville Zoo and Gardens.

Dolphin

Male bottlenose dolphins have been observed working in pairs to follow or restrict the movement of a female for weeks at a time, waiting for her to become sexually receptive. The same pairs have also been observed engaging in intense sexual play with each other. Janet Mann, a professor of biology and psychology at Georgetown University, argues that the common same-sex behaviour among male dolphin calves is about bond formation and benefits the species evolutionarily. They cite studies that have shown the dolphins later in life are bisexual and the male bonds forged from homosexuality work for protection as well as locating females with which to reproduce. In 1991, an English man was prosecuted for allegedly having sexual contact with a dolphin. The man was found not guilty after it was revealed at trial that the dolphin was known to tow bathers through the water by hooking his penis around them.

Hyena

The female spotted hyena has a unique urinary-genital system, closely resembling the penis of the male, called a pseudo-penis. The family structure is matriarchal and dominance relationships with strong sexual elements are routinely observed between related females. They are notable for using visible sexual arousal as a sign of submission but not dominance in males as well as females (females have a sizable erectile clitoris). It is speculated that to facilitate this, their sympathetic and parasympathetic nervous systems may be partially reversed in respect to their reproductive organs.

Mating behaviour

Mammals

Mammals mate by vaginal copulation. To achieve this, the male usually mounts the female from behind. The female may exhibit lordosis in which she arches her back ventrally to facilitate entry of the penis. Amongst the land mammals, other than humans, only bonobos mate in a face-to-face position. Some sea mammals copulate in a belly-to-belly position. Some camelids mate in a lying-down position. In most mammals ejaculation occurs after multiple intromissions, but in most ruminant species, a single pelvic thrust occurs during copulation.

Invertebrates

Invertebrates are often hermaphrodites. Some hermaphroditic land snails begin mating with an elaborate tactile courting ritual. The two snails circle around each other for up to six hours, touch-ing with their tentacles, and biting lips and the area of the genital pore, which shows some preliminary signs of the eversion of the penis. As the snails approach mating, hydraulic pressure builds up in the blood sinus surrounding an organ housing a sharpened dart. The dart is made of calcium carbonate or chitin, and is called a love dart. Each snail manoeuvres to get its genital pore in the best position, close to the other snail's body. Then, when the body of one snail touches the other snail's genital pore, it triggers the firing of the love dart. After the snails have fired their darts, they copulate and exchange sperm as a separate part of the mating progression. The love darts are covered with a mucus that contains a hormone-like substance that facilitates the survival of the sperm.

Courting garden snails – the one on the left has fired a love dart into the one on the right.

A male star coral releases sperm into the water.

Penis fencing is a mating behaviour engaged in by certain species of flatworm, such as *Pseudobiceros bedfordi*. Species which engage in the practice are hermaphroditic, possessing both eggs and sperm-producing testes. The species "fence" using two-headed dagger-like penises which are pointed, and white in colour. One organism inseminates the other. The sperm is absorbed through pores in the skin, causing fertilisation.

Corals can be both gonochoristic (unisexual) and hermaphroditic, each of which can reproduce sexually and asexually. Reproduction also allows corals to settle new areas. Corals predominantly reproduce sexually. 25% of hermatypic corals (stony corals) form single sex (gonochoristic) colonies, while the rest are hermaphroditic. About 75% of all hermatypic corals "broadcast spawn" by releasing gametes – eggs and sperm – into the water to spread offspring. The gametes fuse during fertilisation to form a microscopic larva called a planula, typically pink and elliptical in shape. Synchronous spawning is very typical on the coral reef and often, even when multiple species are present, all corals spawn on the same night. This synchrony is essential so that male and female gametes can meet. Corals must rely on environmental cues, varying from species to species, to determine the proper time to release gametes into the water. The cues involve lunar changes, sunset time, and possibly chemical signalling. Synchronous spawning may form hybrids and is perhaps involved in coral speciation.

Butterflies spend much time searching for mates. When the male spots a mate, he will fly closer and release pheromones. He then performs a special courtship dance to attract the female. If the female appreciates the dancing she may join him. Then they join their bodies together end to end at their abdomens. Here, the male passes the sperm to the female's egg-laying tube, which will soon be fertilised by the sperm. The male often dies shortly after mating.

Many animals make plugs of mucus to seal the female's orifice after mating. Normally such plugs are secreted by the male, to block subsequent partners. In spiders the female can assist the process. Spider sex is unusual in that males transfer their sperm to the female on small limbs called pedipalps. They use these to pick their sperm up from their genitals and insert it into the female's sexual orifice, rather than copulating directly. On the 14 occasions a sexual plug was made, the female produced it without assistance from the male. On ten of these occasions the male's pedipalps then seemed to get stuck while he was transferring the sperm (which is rarely the case in other species of spider), and he had great difficulty freeing himself. In two of those ten instances, he was eaten as a result.

Genetic Evidence of Interspecies Sexual Activity

Research into human evolution confirms that, in some cases, interspecies sexual activity may have been responsible for the evolution of new species (speciation). Analysis of animal genes found evidence that after humans had diverged from other apes, interspecies mating nonetheless occurred regularly enough to change certain genes in the new gene pool. Researchers found that the X chromosomes of humans and chimps may have diverged around 1.2 million years after the other chromosomes. One possible explanation is that modern humans emerged from a hybrid of human and chimp populations. A 2012 study questioned this explanation, concluding that "there is no strong reason to involve complicated factors in explaining the autosomal data".

Inbreeding Avoidance

When close relatives mate, progeny may exhibit the detrimental effects of inbreeding depression. Inbreeding depression is predominantly caused by the homozygous expression of recessive deleterious alleles. Over time, inbreeding depression may lead to the evolution of inbreeding avoidance behaviour. Several examples of animal behaviour that reduce mating of close relatives and inbreeding depression are described next.

Reproductively active female naked mole-rats tend to associate with unfamiliar males (usually non-kin), whereas reproductively inactive females do not discriminate. The preference of reproductively active females for unfamiliar males is interpreted as an adaptation for avoiding inbreeding.

When mice inbreed with close relatives in their natural habitat, there is a significant detrimental effect on progeny survival. In the house mouse, the major urinary protein (MUP) gene cluster provides a highly polymorphic scent signal of genetic identity that appears to underlie kin recognition and inbreeding avoidance. Thus there are fewer matings between mice sharing MUP haplotypes than would be expected if there were random mating.

Meerkat females appear to be able to discriminate the odour of their kin from the odour of their non-kin. Kin recognition is a useful ability that facilitates both cooperation among relatives and the avoidance of inbreeding. When mating does occur between meerkat relatives, it often results in inbreeding depression. Inbreeding depression was evident for a variety of traits: pup mass at emergence from the natal burrow, hind-foot length, growth until independence and juvenile survival.

The grey-sided vole (*Myodes rufocanus*) exhibits male-biased dispersal as a means of avoiding incestuous matings. Among those matings that do involve inbreeding the number of weaned juveniles in litters is significantly smaller than that from non-inbred litters indicating inbreeding depression.

In natural populations of the bird *Parus major* (great tit), inbreeding is likely avoided by dispersal of individuals from their birthplace, which reduces the chance of mating with a close relative.

Toads display breeding site fidelity, as do many amphibians. Individuals that return to natal ponds to breed will likely encounter siblings as potential mates. Although incest is possible, *Bufo americanus* siblings rarely mate. These toads likely recognise and actively avoid close kins as mates. Advertisement vocalisations by males appear to serve as cues by which females recognise their kin.

Mating System

A mating system is a way in which a group is structured in relation to sexual behaviour. The precise meaning depends upon the context. With respect to animals, the term describes which males and females mate, under which circumstances; recognised systems include monogamy, polygamy (which includes polygyny, polyandry, and polygynandry), and promiscuity, all of which lead to different mate choice outcomes and thus these systems affect how sexual selection works in the species which practice them. In plants, the term refers to the degree and circumstances of outcrossing. In human sociobiology, the terms have been extended to encompass the formation of relationships such as marriage.

In Plants

The primary mating systems in plants are outcrossing (cross-fertilisation), autogamy (self-fertilisation) and apomixis (asexual reproduction without fertilization, but only when arising by modification of sexual function). Mixed mating systems, in which plants use two or even all three mating systems, are not uncommon.

A number of models have been used to describe the parameters of plant mating systems. The basic model is the mixed mating model, which is based on the assumption that every fertilisation is either self-fertilisation or completely random cross-fertilisation. More complex models relax this assumption; for example, the effective selfing model recognises that mating may be more common between pairs of closely related plants than between pairs of distantly related plants.

In Animals

Chimpanzees have a promiscuous mating system

Male and female gorilla, gorillas have a polygynous mating system

The following are some of the mating systems generally recognized in animals:

- Monogamy: One male and one female have an exclusive mating relationship. The term "pair bonding" often implies this. This is associated with one-male, one-female group compositions.
- Polygamy: Three types are recognized:
    - Polygyny (the most common polygamous mating system in vertebrates so far studied): One male has an exclusive relationship with two or more females. This is associated with one-male, multi-female group compositions. Many perennial *Vespula squamosa* (southern yellowjacket) colonies are polygynous. Different types of polygyny exist, such as lek polygyny and resource defense polygyny. Although most animals opt for only one of these strategies, some exhibit hybrid strategies, such as the bee species, *Xylocopa micans*.
    - Polyandry: One female has an exclusive relationship with two or more males. This is very rare and is associated with multi-male, multi-female group compositions.

- Polygynandry:Polygynandry is a slight variation of this, where two or more males have an exclusive relationship with two or more females; the numbers of males and females don't have to be equal, and in vertebrate species studied so far, the number of males is usually less. This is associated with multi-male, multi-female group compositions.

- Promiscuity: A member of one sex within the social group mates with any member of the opposite sex. This is associated with multi-male, multi-female group compositions.

These mating relationships may or may not be associated with social relationships, in which the sexual partners stay together to become parenting partners. As the alternative term "pair bonding" implies, this is usual in monogamy. In many polyandrous systems, the males and the female stay together to rear the young. In polygynous systems where the number of females paired with each male is low and the male will often stay with one female to help rear the young, while the other females rear their young on their own. In polygynandry, each of the males may assist one female; if all adults help rear all the young, the system is more usually called "communal breeding". In highly polygynous systems, and in promiscuous systems, paternal care of young is rare, or there may be no parental care at all.

These descriptions are idealized, and the social partnerships are often easier to observe than the mating relationships. In particular:

- the relationships are rarely exclusive for all individuals in a species. DNA fingerprinting studies have shown that even in pair-bonding, matings outside the pair (extra-pair copulations) occur with fair frequency, and a significant minority of offspring result from them.

- some species show different mating systems in different circumstances, for example in different parts of their geographical range, or under different conditions of food availability

- mixtures of the simple systems described above may occur.

In Humans

Compared to other vertebrates, where a species usually has a single mating system, human display great variety. Humans also differ by having formal marriages which in many cultures involve negotiation and arrangement between elder relatives. Regarding sexual dimorphism, humans are in the intermediate group with moderate sex differences in body size but with relatively large testes. This indicates relatively frequent sperm competition which is supported by reports of extrapair paternity of 2-22% in socially monogamous and polygynous human societies. One estimate is that 83% of human societies are polygynous, 0.05% are polyandrous, and the rest are monogamous. Even the last group may at least in part be genetically polygynous.

Polygyny is associated with an increased sharing of subsistence provided by women. This is consistent with the theory that if women raise the children alone, men can concentrate on the mating effort. Polygyny is also associated with greater environmental variability in the form of variability of rainfall. This may increase the differences in the resources available to men. An important association is that polygyny is associated with a higher pathogen load in an area which may make having good genes in a male increasingly important. A high pathogen load also decreases the relative importance of sororal polygyny which may be because it becomes increasingly important to have genetic variability in the offspring.

Virtually all the terms used to describe animal mating systems were adopted from social anthropology, where they had been devised to describe systems of marriage. This shows that human sexual behavior is unusually flexible since, in most animal species, one mating system dominates. While there are close analogies between animal mating systems and human marriage institutions, these analogies should not be pressed too far, because in human societies, marriages typically have to be recognized by the entire social group in some way, and there is no equivalent process in animal societies. The temptation to draw conclusions about what is "natural" for human sexual behavior from observations of animal mating systems should be resisted: a socio-biologist observing the kinds of behavior shown by humans in any other species would conclude that all known mating systems were natural for that species, depending on the circumstances or on individual differences.

As culture increasingly affects human mating choices, ascertaining what is the 'natural' mating system of the human animal from a zoological perspective becomes increasingly difficult. Some clues can be taken from human anatomy, which is essentially unchanged from the prehistoric past:

- humans have a large relative size of testes to body mass in comparison to most primates;
- humans have a large ejaculate volume and sperm count in comparison to other primates;
- as compared to most primates, humans spend more time in copulation;
- as compared to most primates, humans copulate with greater frequency;
- the outward signs of estrous in women (i.e. higher body temperature, breast swelling, sugar cravings, etc.), are often perceived to be less obvious in comparison to the outward signs of ovulation in most other mammals;
- for most mammals, the estrous cycle and its outward signs bring on mating activity; the majority of female-initiated matings in humans coincides with estrous, but humans copulate throughout the reproductive cycle;
- after ejaculation/orgasm in males and females, humans release a hormone that has a sedative effect; however human females may remain sexually receptive and may remain in the plateau stage of orgasm if their orgasm has not been completed.

Some have suggested that these anatomical factors signify some degree of sperm competition, though as levels of genetic and societal promiscuity are highly varied across cultures, this evidence is far from conclusive.

In Microorganisms

Bacteria

Mating in bacteria involves transfer of DNA from one cell to another and incorporation of the transferred DNA into the recipient bacteria's genome by homologous recombination. Transfer of DNA between bacterial cells can occur in three main ways. First, a bacterium can take up exogenous DNA released into the intervening medium from another bacterium by a process called transformation. DNA can also be transferred from one bacterium to another by the process of transduction, which is mediated by an infecting virus (bacteriophage). The third method of DNA transfer is *conjugation*, in which a plasmid mediates transfer through direct cell contact between cells.

Transformation, unlike transduction or conjugation, depends on numerous bacterial gene products that specifically interact to perform this complex process, and thus transformation is clearly a bacterial adaptation for DNA transfer. In order for a bacterium to bind, take up and recombine donor DNA into its own chromosome, it must first enter a special physiological state termed natural competence. In *Bacillus subtilis* about 40 genes are required for the development of competence and DNA uptake. The length of DNA transferred during *B. subtilis* transformation can be as much as a third and up to the whole chromosome. Transformation appears to be common among bacterial species, and at least 60 species are known to have the natural ability to become competent for transformation. The development of competence in nature is usually associated with stressful environmental conditions, and seems to be an adaptation for facilitating repair of DNA damage in recipient cells.

Archaea

In several species of archaea, mating is mediated by formation of cellular aggregates. *Halobacterium volcanii*, an extreme halophilic archaeon, forms cytoplasmic bridges between cells that appear to be used for transfer of DNA from one cell to another in either direction.

When the hyperthermophilic archaea *Sulfolobus solfataricus* and *Sulfolobus acidocaldarius* are exposed to the DNA damaging agents UV irradiation, bleomycin or mitomycin C, species-specific cellular aggregation is induced. Aggregation in *S. solfataricus* could not be induced by other physical stressors, such as pH or temperature shift, suggesting that aggregation is induced specifically by DNA damage. Ajon et al. showed that UV-induced cellular aggregation mediates chromosomal marker exchange with high frequency in *S. acidocaldarius*. Recombination rates exceeded those of uninduced cultures by up to three orders of magnitude. Frols et al. and Ajon et al. hypothesized that cellular aggregation enhances species-specific DNA transfer between *Sulfolobus* cells in order to provide increased repair of damaged DNA by means of homologous recombination. This response appears to be a primitive form of sexual interaction similar to the more well-studied bacterial transformation systems that are also associated with species specific DNA transfer between cells leading to homologous recombinational repair of DNA damage.

Protists

Protists are a large group of diverse eukaryotic microorganisms, mainly unicellular animals and plants, that do not form tissues. Eukaryotes emerged in evolution more than 1.5 billion years ago. The earliest eukaryotes were likely protists. Mating and sexual reproduction are widespread among extant eukaryotes. Based on a phylogenetic analysis, Dacks and Roger proposed that facultative sex was present in the common ancestor of all eukaryotes.

However, to many biologists it seemed unlikely until recently, that mating and sex could be a primordial and fundamental characteristic of eukaryotes. A principal reason for this view was that mating and sex appeared to be lacking in certain pathogenic protists whose ancestors branched off early from the eukaryotic family tree. However, several of these protists are now known to be capable of, or to recently have had, the capability for meiosis and hence mating. To cite one example, the common intestinal parasite *Giardia intestinalis* was once considered to be a descendant of a protist lineage that predated the emergence of meiosis and sex. However, *G. intestinalis* was recently found to have a core set of genes that function in meiosis and that are widely present among sexual eukaryotes. These results suggested that *G. intestinalis* is capable of meiosis and thus mating and sexual reproduction. Furthermore, direct evidence for meiotic recombination, indicative of mating and sexual reproduction, was also found in *G. intestinalis*. Other protists for which evidence of mating and sexual reproduction has recently been described are parasitic protozoa of the genus *Leishmania*, *Trichomonas vaginalis*, and acanthamoeba.

Protists generally reproduce asexually under favorable environmental conditions, but tend to reproduce sexually under stressful conditions, such as starvation or heat shock.

Viruses

Both animal viruses and bacterial viruses (bacteriophage) are able to undergo mating. When a cell is mixedly infected by two genetically marked viruses, recombinant virus progeny are often observed indicating that mating interaction had occurred at the DNA level. Another manifestation of mating between viral genomes is multiplicity reactivation (MR). MR is the process by which at least two virus genomes, each containing inactivating genome damage, interact with each other in an infected cell to form viable progeny viruses. The genes required for MR in bacteriophage T4 are largely the same as the genes required for allelic recombination. Examples of MR in animal viruses are described in the articles *Herpes simplex virus*, *Influenza A virus*, *Adenoviridae*, *Simian virus 40*, *Vaccinia virus*, and *Reoviridae*.

Mating Call

A mating call is the auditory signal used by animals to attract mates. It can occur in males or females, but literature is abundantly favored toward researching mating calls in males. In addition, mating calls are often the subject of mate choice, in which the preferences of one gender for a certain type of mating call can drive sexual selection in a species. This can result in sympatric speciation of some animals, where two species diverge from each other while living in the same environment.

There are many different mechanisms to produce mating calls, which can be broadly categorized into vocalizations and mechanical calls. Vocalizations are considered as sounds produced by the larynx and are often seen in species of birds, mammals, amphibians, and insects. Mechanical calls refer to any other type of sound that the animal produces using unique body parts and/or tools for communication with potential mates. Examples include crickets that vibrate their wings, birds that flap their feathers, and frogs that use an air sac instead of lungs.

Vocalizations

Birds

Mating call of Japanese bush warbler, *Horornis diphone*

Song sparrow

The use of vocalizations is widespread in avian species and are often used to attract mates. Different aspects and features of bird song such as structure, amplitude and frequency have evolved as a result of sexual selection.

Large song repertoires are preferred by females of many avian species. One hypothesis for this is that song repertoire is positively correlated with the size of the brain's song control nucleus (HVC). A large HVC would indicate developmental success. In song sparrows, males with large repertoires had larger HVCs, better body condition and lower heterophil-to-lymphocyte ratios indicating better immune health. This supports the idea that song sparrows with large song repertoires have better lifetime fitness and that song repertoires are honest indicators of the males "quality." Possible explanations for this adaptation include direct benefits to the female, such as superior parental care or territory defense, and indirect benefits, such as good genes for their offspring.

Japanese bush warbler songs from island populations have an acoustically simple structure when compared to mainland populations. Song complexity is correlated with higher levels of sexual selection in mainland populations, showing that a more complex song structure is advantageous in an environment with high levels of sexual selection. Another example is in purple-crowned fairy-wrens; larger males of this species sing advertising songs at a lower frequency than smaller rival males. Since body size is a characteristic of good health, lower frequency calls are a form of honest signaling. Negative correlation between body size and call frequency is supported across multiple species within the taxa. In the rock sparrow, song frequency is positively associated with reproductive success. Slower song rate is associated with age and is preferred by females. Reproductive status of the individual is communicated through higher maximum frequency. There was also positive correlation between age and extra pair copulation frequency.

Bird calls are also known to continue after pair formation in several socially monogamous bird species. In one experimental population of zebra finches, there was increased singing activity by the male after breeding. This increase is positively correlated with the partner's reproductive investment. The female finches were bred in cages with two subsequent males that differed with varying amounts of song output. Females produced larger eggs with more orange yolks when paired with a male with a high song output. This suggests that the relative amount of song production in paired zebra finch males might function to stimulate the partner rather than to attract extra-pair females.

Red deer stag during rut

Mammals

During the breeding season, mammals will call out to the opposite sex. Male koalas that are bigger will let out a different sound than smaller koalas. The bigger males which are routinely sought out for are called sires. Females choose sires because of indirect benefits that their offspring could inherit, like larger bodies. Non-sires and females do not vary in their body mass and can reject a male by screaming or hitting him. Male-male competition is rarely exhibited in koalas. Acoustic signaling is a type of call that can be used from a significant distance encoding an organism's location, condition and identity. Sac-winged bats display acoustic signaling, which is often interpreted as songs. When females hear these songs, named a 'whistle', they call onto the males to breed with a screech of their own. This action is termed 'calling of the sexes'. Red deer and spotted hyenas along with other mammals also perform acoustic signaling.

Tungara frog

Amphibians

Most frogs use an air sac located under their mouth to produce mating calls. Air from the lungs channels to the air sac to inflate it, and the air sac resonates to produce a mating call. The larynx is larger and more developed in males, which causes their call to be louder and stronger

In the túngara frog, males use a whining call followed by up to seven clucks. Males who have a whine-cluck call are more successful in attracting females than males whose call is a whine alone. The ability to produce clucks is due to a specialized fibrous mass attached to the frog's vocal folds, creating an unusual vocalization similar to the two-voiced songs found in some birds.

In the common toad, sexual competition is driven in large part by fighting - successful males often physically displaced other males from the backs of a female in order to gain access to mating with that female. Larger males were more successful in such takeovers, and had higher reproductive success as a result. However, the vocalizations of these toads provide a reliable signal of body size and thus fighting ability, allowing contests for possession of females to be settled without risk of injury.

Insects

While mating calls in insects are usually associated with mechanical mating calls, such as in crickets, several species of insects use vocalizations to attract mates. In the Asian corn borer, males emit clicking sounds that mimic the echolocation of bats which prey on the moths. They then take advantage of the female's "freezing" response to mate with the female.

In the Japanese lichen moth, however, the female is able to distinguish between the sounds made by males and those made by bats and other predators.As a result, the males use ultrasonic clicking as a more conventional mating signal, compared to the "deceptive" courtship song used in the Asian Corn Bearer.

Mechanical Calls

Mating calls also take form through mechanical processes. Animals that are unable to vocalize their call may use their body to attract mates.

Crickets

Mating call of field cricket, *Gryllus pennsylvanicus*

In the field cricket, *Gryllus integer*, males rub their wings together to create a rapid trill that produces sound. Males individually vary in the durations of their trilling or, what is more sophisticatedly called, bout length. The bout length of each male is heritable and passed on to his future offspring. Also, females prefer to mate with males that have longer bout lengths. The end result is that males with longer bout lengths produce more offspring than males with shorter bout lengths.

Other factors that influence the formation of these bout lengths include temperature and predation. In field crickets, males prefer warmer sites for mating as shown by an increase in the frequency of their mating calls when they were living in warmer climates. Predation also affects the mating calls of field crickets. When in a potentially dangerous environment, males cease calling for longer periods of time when interrupted by a predator cue. This suggests that there is an interplay between intensity of mating call and risk of predation.

Sonation

As described in Sonation, "the term sonate is described as the deliberate production of sounds, not from the throat, but rather from structures such as the bill, wings, tail, feet and body feathers, or by the use of tools". In several amphibian and fish species, other special structures are used to produce different sounds to attract mates. Birds are common users of sonation, although several amphibian and fish species have been shown to use sonation as a form of mating call as well. In general, sonation is one factor that plays into how a female may choose a mate. There are other features of mating such as territory defense or mate defense, which contribute to the cause of finding suitable mates.

As outlined below, each species uses a distinct method to produce a non-vocal mating call in order to be most successful in attracting mates. The examples below represent the most common examples found in the literature, although many more examples may exist in nature that are still currently unknown.

Feather of male *Pavo cristatus* (Indian peafowl). These feathers are used in sonation to create infrasound with intent of mating.

Birds

The feathers, the beak, the feet, and different tools are all used by different bird species to produce mating calls to attract mates. For example, the snipe uses its feathers to produce a "drumming" sound to attract mates during a special mating dance. Snipes used specialized tail feathers to create a sound described as a "rattle" or "throbbing" noise. Palm cockatoos use sticks to drum on hollow trees, creating a loud noise to attract the attention of mates. Bustards are large, highly terrestrial birds that stamp their feet during mating displays to attract mates. *Mirafra apiata*, commonly known as the Clapper lark, engages in a complex display flight that is characterized by the rattling of the wings.

Many species of birds use sonation for mating calls such as manakins and hummingbirds. However, peacocks exhibit a feature of sonation that reveals intrasexual and intersexual properties of this type of mating call. Males move their feathers to produce low-frequency sonations (infrasound) and sonate more frequently in response to sonations by other males. This is attributable to a male's desire to advertise its presence above other males looking for mates, suggesting that sonation carries an intrasexual function. In addition, females show increased alertness when hearing the infrasound signals produced by males' wing-shaking, which highlights how the two sexes use sonation to interact with each other.

Fish

While most bird species use their feathers, tools, or feet to produce sounds and attract mates, many fish species use specialized internal organs to sonate. In Gadoid fish, special muscles attached to the swimbladder assist in the production of knocking or grunting sounds to attract mates.

Speciation Due to Mating Call Differences

Differences in mating calls can lead to the separation of different populations within a species. These differences can be due to several factors, including body size, temperature, and other ecological factors.These can arise in the form of tonal, temporal, or behavioral variations in mating calls that subsequently lead to the separation of populations. The separation of these populations due to differences in mating call and mating call preferences can lead to the evolution and creation of new, unique species.

This type of speciation is most often sympatric speciation: where two or more species are created from an existing parent species that all live in the same geographic location. Although there is an absence of research on mammals and birds, this phenomenon has been heavily researched in several frog species around the world. The examples below illuminate speciation due to mating call differences in several frog species around the world. These distinct species are included because they are the focus of the majority of current research.

Two *Microhyla olivacea* in a mating position

Microhyla Olivacea *and* Microhyla Carolinensis

These two species of narrow-mouthed frog live in the southern United States and have overlapping ranges in Texas and Oklahoma. Researchers have discovered that these two different species alter the frequencies of their call in the overlap zone of their ranges. For example, the *Microhyla olivacea* mating call has a significantly lower midpoint frequency in the overlap zone than the mating call outside this zone. This leads researchers to suggest that the differences in mating call in the overlap zone of *M. olivacea and M. carolinensis* act as an isolating mechanism between the two species. They also hypothesize that the evolution of these differences in mating call led to the separation of these two different frog species from one common species.

Engystomops petersi

Engystomops Petersi

Female preferences for specific male mating calls can lead to sexual selection in mating calls. Females may prefer a specific type of call that certain males possess, in which only those males will be able to mate with females and pass on their genes and specific mating call. As a result, this female preference may lead to divergence of two species.

In Amazonian frogs, sexual selection for different calls has led to the behavioral isolation and speciation of the túngara frog (*Engystomops petersi).* From genetic and mating call analysis and, researchers were able to identify that two populations of the túngara frog were almost completely reproductively isolated. From their research, scientists believe that differences in female preferences for mating call type have led to the evolution of this speciation process. Specifically, the Yasuní population females prefer the male mating call that includes a whine, while the other population does not prefer this whine. Subsequently, the Yasuní males include the whine in their call, while the other males do not. For this reason, the differences in call have led to the mechanical separation of this species.

Pseudacris Triseriata

Several studies have shown that the species *Pseudacris triseriata* (Chorus Frog) can be divided into two subspecies, *P. t. maculata* and *P. t. triseriata,* due to speciation events from mating call differences. The Chorus Frog has a very large home range, from New Mexico to Southern Canada. These two subspecies have an overlapping range from South Dakota to Oklahoma. In this overlapping range, both the call duration and the calls per second for each species is much different than outside of this range. This means that calls of these two subspecies are more similar outside of this range, and starkly different within the range. For this reason, scientists suggest that these subspecies evolved from differences in mating call type. Additionally, these subspecies are rarely recorded to have hybrid offspring, which further suggests that there is complete speciation due to mating call differences. The differences in mating calls also help to reinforce the speciation process.

A Chorus Frog making a mating call

Monogamy in Animals

Monogamous pairing in animals refers to the natural history of mating systems in which species pair bond to raise offspring. This is associated, usually implicitly, with sexual monogamy.

Animals

The evolution of mating systems in animals has received an enormous amount of attention from biologists. This section briefly reviews three main findings about the evolution of monogamy in animals.

The amount of social monogamy in animals varies across taxa, with over 90% of birds engaging in social monogamy while only 3% of mammals are known to do the same.

This list is not complete. Other factors may also contribute to the evolution of social monogamy. Moreover, different sets of factors may explain the evolution of social monogamy in different species. There is no one-size-fits-all explanation of why different species evolved monogamous mating systems.

Sexual Dimorphism

Sexual dimorphism refers to differences in body characteristics between females and males. A frequently studied type of sexual dimorphism is body size. For example among mammals, males typically have larger bodies than females. In other orders, however, females have larger bodies than males. Sexual dimorphism in body size has been linked to mating behavior. In polygynous species, males compete for control over sexual access to females. Large males have an advantage in the competition for access to females, and they consequently pass their genes along to a greater number of offspring. This eventually leads to large differences in body size between females and males. Polygynous males are often 1.5 to 2.0 times larger in size than females. In monogamous species, on the other hand, females and males have more equal access to mates, so there is little or no sexual dimorphism in body size. From a new biological point of view, monogamy could result from mate guarding and is engaged as a result of sexual conflict.

Some researchers have attempted to infer the evolution of human mating systems from the evolution of sexual dimorphism. Several studies have reported a large amount of sexual dimorphism in *Australopithecus*, an evolutionary ancestor of human beings that lived between 2 and 5 million years ago. These studies raise the possibility that *Australopithecus* had a polygamous mating system. Sexual dimorphism then began to decrease. Studies suggest sexual dimorphism reached modern human levels around the time of *Homo erectus* 0.5 to 2 million years ago. This line of reasoning suggests human ancestors started out polygamous and began the transition to monogamy somewhere between 0.5 million and 2 million years ago.

Attempts to infer the evolution of monogamy based on sexual dimorphism remain controversial for three reasons:

- The skeletal remains of *Australopithecus* are quite fragmentary. This makes it difficult to identify the sex of the fossils. Researchers sometimes identify the sex of the fossils by their size, which, of course, can exaggerate findings of sexual dimorphism.
- Recent studies using new methods of measurement suggest *Australopithecus* had the same amount of sexual dimorphism as modern humans. This raises questions about the amount of sexual dimorphism in *Australopithecus*.
- Humans may have been partially unique in that selection pressures for sexual dimorphism might have been related to the new niches that humans were entering at the time, and how that might have interacted with potential early cultures and tool use. If these early humans had a differentiation of gender roles, with men hunting and women gathering, selection pressures in favor of increased size may have been distributed unequally between the sexes.
- Even if future studies clearly establish sexual dimorphism in *Australopithecus*, other studies have shown the relationship between sexual dimorphism and mating system is unreliable. Some polygamous species show little or no sexual dimorphism. Some monogamous species show a large amount of sexual dimorphism.

Studies of sexual dimorphism raise the possibility that early human ancestors were polygamous rather than monogamous. But this line of research remains highly controversial. It may be that early human ancestors showed little sexual dimorphism, and it may be that sexual dimorphism in early human ancestors had no relationship to their mating systems.

Testis Size

The relative sizes of male testes often reflect mating systems. In species with promiscuous mating systems, where many males mate with many females, the testes tend to be relatively large. This appears to be the result of sperm competition. Males with large testes produce more sperm and thereby gain an advantage impregnating females. In polygynous species, where one male controls sexual access to females, the testes tend to be small. One male defends exclusive sexual access to a group of females and thereby eliminates sperm competition.

Chimpanzee

Male and female gorilla

Studies of primates, including humans, support the relationship between testis size and mating system. Chimpanzees, which have a promiscuous mating system, have large testes compared to other primates. Gorillas, which have a polygynous mating system, have smaller testes than other primates. Humans, which have *a socially monogamous mating system, accompanied by moderate amounts of sexual non-monogamy*, have moderately sized testes. The moderate amounts of sexual non-monogamy in humans may result in a low to moderate amount of sperm competition. Also, notably, in the case of an avowedly sexually monogamous society, the occurrence of sexual nonmonogamy is typically culturally stigmatized, and therefore detecting its prevalence is inherently difficult, if indeed it is at all possible. At best, such statistics can be viewed as general approximations with a wide margin of error.

Monogamy as a Best Response

In species where the young are particularly vulnerable and may benefit from protection by both parents, monogamy may be an optimal strategy. Monogamy tends to also occur when populations are small and dispersed. This is not conductive to polygamous behavior as the male would spend far more time searching for another mate. The monogamous behavior allows the male to have a mate consistently, without having to waste energy searching for other females. Furthermore, there is an apparent connection between the time a male invests in their offspring and their monogamous behavior. A male which is required to care for the offspring to ensure their survival is much more likely to exhibit monogamous behavior over one that does not. The selection factors in favor of different mating strategies for a species of animal, however, may potentially operate on a large number of factors throughout that animal's life cycle. For instance, with many species of bear, the female will often drive a male off soon after mating, and will later guard her cubs from him. It is thought that this may be due to the fact that too many bears close to one another may deplete the food available to the relatively small but growing cubs. Monogamy may be social but rarely genetic. For example, in the cichlid species *Variabilichromis moorii*, a monogamous pair will care for their eggs and young but the eggs are not all fertilized by the same male. Thierry Lodé argued that monogamy should result from conflict of interest between the sexes called sexual conflict.

Monogamous Species

While it is difficult to find monogamous relationships in nature, there are a few species which have adopted monogamy with great success. For instance, the prairie vole will mate exclusively with the first female he ever mates with. The vole is extremely loyal and will go as far as to even attack other females that may approach him. This type of behavior has been linked to the hormone vasopressin. This hormone is released when a male mates and cares for young. Due to this hormone's rewarding effects, the male experiences a positive feeling when they maintain a monogamous relationship. To further test this theory, the receptors that control vasopressin were placed into another species of vole that is promiscuous. After this addition, the originally unfaithful voles became monogamous with their selected partner. These very same receptors can be found in human brain, and have been found to vary at the individual level -- which could explain why some human males tend to be more loyal than others.

Black vultures stay together as it is more beneficial for their young to be taken care of by both parents. They take turns incubating the eggs, and then supplying their fledglings with food. Black vultures will also attack other vultures that are participating in extra pair copulation, this is an attempt to increase monogamy and decrease promiscuous behavior. Similarly, emperor penguins also stay together to care for their young. This is due to the harshness of the arctic weather, predators and the scarcity of food. One parent will protect the chick, while the other finds food. However, these penguins only remain monogamous until the chick is able to go off on their own. After the chick no longer needs their care, approximately 85% of parents will part ways and typically find a new partner every breeding season.

Hornbills are a socially monogamous bird species that usually only have one mate throughout their lives, much like the prairie vole. The females will close herself up in a nest cavity, sealed with a nest plug, for two months. At this time, she will lay eggs and will be cared for by her mate. The males are willing to work to support himself, his mate, and his offspring in order for survival; however, unlike the emperor penguin, the hornbills do not find new partners each season.

It is relatively uncommon to find monogamous relationships in fish, amphibians and reptiles; however, the red-backed salamander as well as the Caribbean cleaner goby practice monogamy as well. However, the male Caribbean cleaner goby fish has been found to separate from the female suddenly, leaving her abandoned. In a study conducted by Oregon State University, it was found that this fish practices not true monogamy, but serial monogamy. This essentially means that the goby will have multiple monogamous relationships throughout its life - but only be in one relationship at a time. The red-backed salamander exhibited signs of social monogamy, which is the idea that animals form pairs to mate and raise offspring, but still will partake in extra pair copulation with various males or females in order to increase their biological fitness. This is relatively new concept in salamanders, and has not been seen frequently - it is also concerning that the act of monogamy may inhibit the salamanders reproductive rates and biological success. However, the study which was conducted in cooperation by the University of Louisiana, Lafayette, and the University of Virginia showed that the salamanders are not inhibited by this monogamy if they show alternative strategies with other mates.

Azara's night monkeys are another species that proved to be monogamous. In an 18 year study conducted by the University of Pennsylvania, these monkeys proved to be entirely monogamous, exhibiting no genetic information or visual information that could lead to the assumption that extra pair copulation was occurring. This explained the question as to why the male owl monkey invested so much time in protecting and raising their own offspring. Because monogamy if often referred to "placing all your eggs in one basket" the male wants to ensure his young survive, and thus pass on his genes.

Other monogamous species include wolves, otters, a few hooved animals, some bats, certain species of fox, and the Eurasian beaver. This beaver is particularly interesting as it is practicing monogamy in its reintroduction to certain parts of Europe; however, its' American counterpart is not monogamous at all and often partakes in promiscuous behavior. The two species are quite similar in ecology but American beavers tend to be less aggressive than European beavers. In this instance, the scarcity of the European beavers' population could drive its monogamous behavior; moreover, it lowers the risk of parasite transmission which is correlated with biological fitness. Monogamy is proving to be very efficient for this beaver, as their population is climbing.

Polygyny Threshold Model

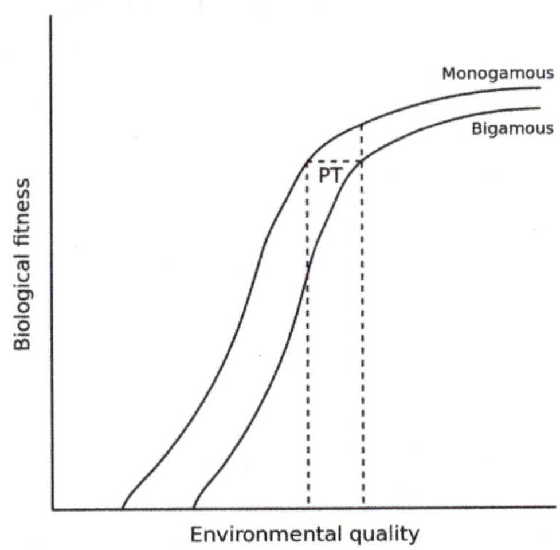

Polygyny threshold model graph

The polygyny threshold model is an explanation of polygyny, the mating of one male of a species with multiple females. The model shows how females may gain a higher level of biological fitness by mating with a male who already has a mate. The female makes this choice despite other surrounding males because the choice male's territory, food supply, or other important characteristics are better than those of his competitors, even with two females on the territory.

Graphical Depiction

The graphical depiction of the model presented in Gordon H. Orians' 1969 paper is often used to explain the concept. The graph shows two curves on a graph of biological fitness versus environmental quality. Environmental quality refers to the quality of the male's territory. The left curve, labeled monogamous, is the perceived biological fitness for a female entering into a monogamous relationship with a given male. The right curve, labeled bigamous, shows the fitness of the same female entering into a relationship with a different male who already has one female mate but who has defended more resources. The second curve is roughly the first curve shifted to the right some amount. The given shapes of the curve will change with other intrinsic factors like genetic quality and male paternal investment. It is important to note that the designation "female" and "male" here are oft accurate; however, in some mating systems the operational sex ratio leans towards females, who then have motivation to engage in resource defence polyandry (provided the requirements of economic defendability are met).

The intersection of the vertical dotted line on the left with the monogamous curve indicates the biological fitness of a female who chooses a monogamous male with a lower environmental quality. The intersection of the vertical dotted line on the right with the bigamous curve indicates the biological fitness of a female who enters into a bigamous relationship with the male of a higher environmental quality. The difference between these two intersection points, labeled PT, is the *polygyny threshold*. It is the gain of environmental quality for the female when she chooses the bigynous relationship and thus the minimum environmental quality difference necessary to make bigyny beneficial for the female. Also important is the vertical line drawn from the intersection of the line with the bigyny curve to the monogyny curve above. This represents the fitness gain of a female who chooses bigyny over monogyny due to, here, resource holding differences.

Orians predicted that animals exhibiting resource defence polygyny, such as the fish Neolamprologus pulcher would fit to this model when living in successive habitats, where territory quality is very variable. Using the territory quality to decide whether to pursue a monogamous or polygynous mating relationship. This is shown in the red-winged blackbird by Pribil and Searcy (2001). Female red-winged blackbirds prefer to mate with males with territories over water and also unmated males. The females were given a choice between unmated males or previously mated males with the superior territories over water. In 12 out of 14 trials (86%) females choice the already mated male with the superior territory.

Costs of Polygyny

According to William A. Searcy and Ken Yasukawa, the term cost of polygyny is defined as the net costs of polygyny after the summation of all of the component costs and benefits. Costs include less parental care and increased competition between females for the male's provision and food among other resources. A benefit could be group defense of the territory and resources. Searcy and Yasukawa graphically defined the distance between curve 1 (monogamous line) and curve 2 (bigamous line) of the polygyny threshold model graph to be the cost of polygyny. N. B. Davies further defined it as the cost of sharing in order to be clear that the term refers to the fitness cost to females that are breeding on the same territory.

Searcy and Yasukawa conducted studies on Pennsylvania red-winged blackbirds that showed that females would mate on territory already settled by another female, which indicated that there was no cost of polygyny. However, in a later study by Pribil and Picman conducted on Ontario populations of red-winged blackbirds the results indicated that there was indeed a cost of polygyny. The females in this study were given a choice between adjacent territories, one in which there was already a settled female (defined by many researchers as the primary female) and the other in which there was no female present. In all 16 situations, the females chose the unsettled land in which she could be monogamous. Davies, Krebs, and West, in their textbook An Introduction to Behavioural Ecology, cited another Pribil study noting the polygyny cost to the red-winged blackbird females. The effect of polygyny made them less effective mothers when they were removed and taken to a more isolated population, proven by the fact that the mothers from monogamous relationships had better adaptation to the new environment. In an earlier text, Davies explores the examples of costs, showing that the cost is not always to the second and subsequently joining females. He asserts that there are situations in which the cost is shared between the primary and secondary female. He also mentions scenarios in which the primary female receives a decrease in her fitness upon addition of the secondary female to the harem.

There are many other studies concerning the polygyny threshold model and costs to polygyny using other species. Staffan Bensch conducted a study on the great reed warbler that showed the only cost of polygyny to these females to be higher mortality of nestlings that were belonging to the primary female. Johnson, Kermott, and Lien conducted a study on the house wren (Trglodytes aedon) showing that there were inherent polygyny costs to these female populations, also. The secondary females lost more of the broods largely because of starvation, and they also experienced lesser reproductive success in other areas. One of the main factors in their decreased fitness was less male aid. Kyle Summers and David Earn studied female poison frogs, genus Dendrobates, to see if the polygyny costs drove the evolution of the parental care system from a female care to biparental or paternal care. They concluded that the costs could not be concluded to be the sole cause of this parental transition. The numerous studies concerning polygyny costs show the different factors that not only cause these costs, but are also affected by these costs.

Parthenogenesis

Parthenogenesis is a natural form of asexual reproduction in which growth and development of embryos occur without fertilization. In animals, parthenogenesis means development of an embryo from an unfertilized egg cell and is a component process of apomixis.

Gynogenesis and pseudogamy are closely related phenomena in which a sperm or pollen triggers the development of the egg cell into an embryo but makes no genetic contribution to the embryo. The rest of the cytology and genetics of these phenomena are mostly identical to that of parthenogenesis.

The asexual, all-female whiptail species *Cnemidophorus neomexicanus* (center), which reproduces via parthenogenesis, is shown flanked by two sexual species having males *C. inornatus* (left) and *C. tigris* (right), which hybridized naturally to form the *C. neomexicanus* species.

The term is sometimes used inaccurately to describe reproduction modes in hermaphroditic species that can reproduce by themselves because they contain reproductive organs of both sexes in a single individual's body. However, these species still use fertilization.

Parthenogenesis occurs naturally in many plants, some invertebrate animal species (including nematodes, water fleas, some scorpions, aphids, some mites, some bees, some Phasmida and parasitic wasps) and a few vertebrates (such as some fish, amphibians, reptiles and very rarely birds). This type of reproduction has been induced artificially in a few species including fish and amphibians.

Normal egg cells form after meiosis and are haploid, with half as many chromosomes as their mother's body cells. Haploid individuals, however, are usually non-viable, and parthenogenetic offspring usually have the diploid chromosome number. Depending on the mechanism involved in restoring the diploid number of chromosomes, parthenogenetic offspring may have anywhere between all and half of the mother's alleles. The offspring having all of the mother's genetic material are called full clones and those having only half are called half clones. Full clones are usually formed without meiosis. If meiosis occurs, the offspring will get only a fraction of the mother's alleles.

Parthenogenetic offspring in species that use either the XY or the XO sex-determination system have two X chromosomes and are female. In species that use the ZW sex-determination system, they have either two Z chromosomes (male) or two W chromosomes (mostly non-viable but rarely a female), or they could have one Z and one W chromosome (female).

Life History Types

Some species reproduce exclusively by parthenogenesis (such as the Bdelloid rotifers), while others can switch between sexual reproduction and parthenogenesis. This is called facultative parthenogenesis (other terms are cyclical parthenogenesis, heterogamy or heterogony). The switch between sexuality and parthenogenesis in such species may be triggered by the season (aphid, some gall wasps), or by a lack of males or by conditions that favour rapid population growth (rotifers and cladocerans like daphnia). In these species asexual reproduction occurs either in summer (aphids) or as long as conditions are favourable. This is because in asexual reproduction a successful genotype can spread quickly without being modified by sex or wasting resources on male offspring who won't give birth. In times of stress, offspring produced by sexual reproduction may be fitter as they have new, possibly beneficial gene combinations. In addition, sexual reproduction provides the benefit of meiotic recombination between non-sister chromosomes, a process associated with repair of DNA double-strand breaks and other DNA damages that may be induced by stressful conditions.

Many taxa with heterogony have within them species that have lost the sexual phase and are now completely asexual. Many other cases of obligate parthenogenesis (or gynogenesis) are found among polyploids and hybrids where the chromosomes cannot pair for meiosis.

The production of female offspring by parthenogenesis is referred to as thelytoky (e.g., aphids) while the production of males by parthenogenesis is referred to as arrhenotoky (e.g., bees). When unfertilized eggs develop into both males and females, the phenomenon is called deuterotoky.

Types and Mechanisms

Parthenogenesis can occur without meiosis through mitotic oogenesis. This is called apomictic parthenogenesis. Mature egg cells are produced by mitotic divisions, and these cells directly develop into embryos. In flowering plants, cells of the gametophyte can undergo this process. The offspring produced by apomictic parthenogenesis are *full clones* of their mother. Examples include aphids.

Parthenogenesis involving meiosis is more complicated. In some cases, the offspring are haploid (e.g., male ants). In other cases, collectively called automictic parthenogenesis, the ploidy is restored to diploidy by various means. This is because haploid individuals are not viable in most species. In automictic parthenogenesis the offspring differ from one another and from their mother. They are called *half clones* of their mother.

Automictic Parthenogenesis

Automixis is a term that covers several reproductive mechanisms, some of which are parthenogenetic.

Diploidy might be restored by the doubling of the chromosomes without cell division before meiosis begins or after meiosis is completed. This is referred to as an *endomitotic* cycle. This may also happen by the fusion of the first two blastomeres. Other species restore their ploidy by the fusion of the meiotic products. The chromosomes may not separate at one of the two anaphases (called restitutional meiosis), or the nuclei produced may fuse or one of the polar bodies may fuse with the egg cell at some stage during its maturation.

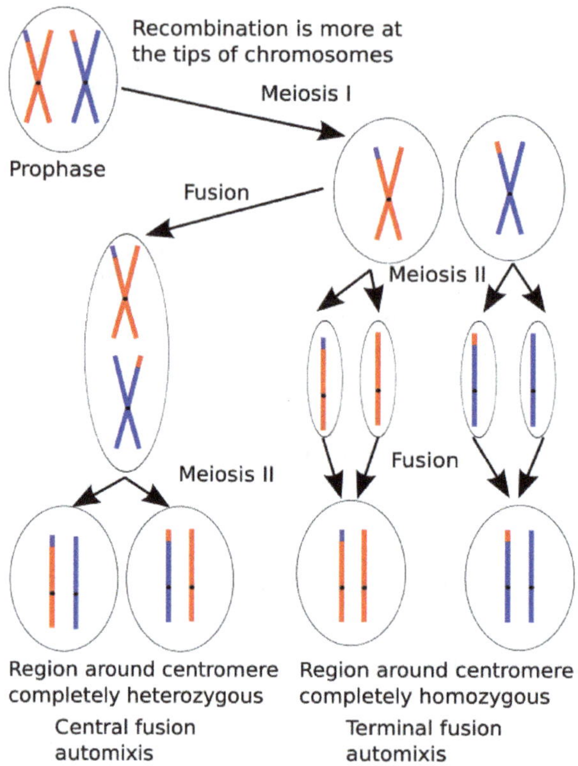

The effects of central fusion and terminal fusion on heterozygosity

Some authors consider all forms of automixis sexual as they involve recombination. Many others classify the endomitotic variants as asexual, and consider the resulting embryos parthenogenetic. Among these authors the threshold for classifying automixis as a sexual process depends on when the products of anaphase I or of anaphase II are joined together. The criterion for "sexuality" varies from all cases of restitutional meiosis, to those where the nuclei fuse or to only those where gametes are mature at the time of fusion. Those cases of automixis that are classified as sexual reproduction are compared to self-fertilization in their mechanism and consequences.

The genetic composition of the offspring depends on what type of apomixis takes place. When endomitosis occurs before meiosis or when *central fusion* occurs (restitutional meiosis of anaphase I or the fusion of its products), the offspring get all to more than half of the mother's genetic material and heterozygosity is mostly preserved (if the mother has two alleles for a locus, it is likely that the offspring will get both). This is because in anaphase I the homologous chromosomes are separated. Heterozygosity is not completely preserved when crossing over occurs in central fusion. In the case of pre-meiotic doubling, recombination -if it happens- occurs between identical sister chromatids.

If *terminal fusion* (restitutional meiosis of anaphase II or the fusion of its products) occurs, a little over half the mother's genetic material is present in the offspring and the offspring are mostly homozygous. This is because at anaphase II the sister chromatids are separated and whatever heterozygosity is present is due to crossing over. In the case of endomitosis after meiosis the offspring is completely homozygous and has only half the mother's genetic material.

This can result in parthenogenetic offspring being unique from each other and from their mother.

Sex of the Offspring

In apomictic parthenogenesis, the offspring are clones of the mother and hence are usually (except for aphids) female. In the case of aphids, parthenogenetically produced males and females are clones of their mother except that the males lack one of the X chromosomes (XO).

When meiosis is involved, the sex of the offspring will depend on the type of sex determination system and the type of apomixis. In species that use the XY sex-determination system, parthenogenetic offspring will have two X chromosomes and are female. In species that use the ZW sex-determination system the offspring genotype may be one of ZW (female), ZZ (male), or WW (non-viable in most species but a fertile, viable female in a few (e.g., boas)). ZW offspring are produced by endoreplication before meiosis or by central fusion. ZZ and WW offspring occur either by terminal fusion or by endomitosis in the egg cell.

In polyploid obligate parthenogens like the whiptail lizard, all the offspring are female.

In many hymenopteran insects such as honeybees, female eggs are produced sexually, using sperm from a drone father, while the production of further drones (males) depends on the queen (and occasionally workers) producing unfertilised eggs. This means that females (workers and queens) are always diploid, while males (drones) are always haploid, and produced parthenogenetically.

Facultative Parthenogenesis

Facultative parthenogenesis is the term for when a female can produce offspring either sexually or via asexual reproduction. Facultative parthenogenesis is extremely rare in nature, with only a few examples of animal taxa capable of facultative parthenogenesis. One of the best known examples of taxa exhibiting facultative parthenogenesis are mayflies; presumably this is the default reproductive mode of all species in this insect order. Facultative parthenogenesis is believed to be a response to a lack of a viable male. A female may undergo facultative parthenogenesis if a male is absent from the habitat or if it is unable to produce viable offspring.

Facultative parthenogenesis is often incorrectly used to describe cases of accidental or spontaneous parthenogenesis in normally sexual animals. For example, many cases of accidental parthenogenesis in sharks, some snakes, Komodo dragons and a variety of domesticated birds were widely perpetuated as facultative parthenogenesis. These cases are, however, examples of accidental parthenogenesis, given the frequency of asexually produced eggs and their hatching rates are extremely low, in contrast to true facultative parthenogenesis where the majority of asexually produced eggs hatch. In addition, asexually produced offspring in vertebrates are virtually always sterile, highlighting that this mode of reproduction is not adaptive, because the ability to reproduce asexually is not inherited to the next generation. The occurrence of such asexually produced eggs in sexual animals can be explained by a meiotic error, leading to eggs produced via automixis.

Obligate Parthenogenesis

Obligate parthenogenesis is the process in which organisms exclusively reproduce through asexual means. Many species have been shown to transition to obligate parthenogenesis over evolutionary time. Among these species, one of the most well documented transitions to obligate parthenogenesis was found in almost all metazoan taxa, albeit through highly diverse mechanisms. These transitions often occur as a result of inbreeding or mutation within large populations. There are a number of documented species, specifically salamanders and geckos, that rely on obligate parthenogenesis as their major method of reproduction. As such, there are over 80 species of unisex reptiles (mostly lizards but including a single snake species), amphibians and fishes in nature for which males are no longer a part of the reproductive process. A female will produce an ovum with a full set (two sets of genes) provided solely by the mother. Thus, a male is not needed to provide sperm to fertilize the egg. This form of asexual reproduction is thought in some cases to be a serious threat to biodiversity for the subsequent lack of gene variation and potentially decreased fitness of the offspring.

Natural Occurrence

Parthenogenesis is seen to occur naturally in aphids, *Daphnia*, rotifers, nematodes and some other invertebrates, as well as in many plants. Among vertebrates, strict parthenogenesis is only known to occur in lizards, snakes, birds and sharks, with fish, amphibians and reptiles exhibiting various forms of gynogenesis and hybridogenesis (an incomplete form of parthenogenesis). The first all-female (unisexual) reproduction in vertebrates was described in the fish *Poecilia formosa* in 1932. Since then at least 50 species of unisexual vertebrate have been described, including at least 20 fish, 25 lizards, a single snake species, frogs, and salamanders. Other, usually sexual, species may occasionally reproduce parthenogenetically and Komodo dragons; the hammerhead and blacktip sharks are recent additions to the known list of spontaneous parthenogenetic vertebrates. As with all types of asexual reproduction, there are both costs (low genetic diversity and therefore susceptibility to adverse mutations that might occur) and benefits (reproduction without the need for a male) associated with parthenogenesis.

Parthenogenesis is distinct from artificial animal cloning, a process where the new organism is necessarily genetically identical to the cell donor. In cloning, the nucleus of a diploid cell from a donor organism is inserted into an enucleated egg cell and the cell is then stimulated to undergo continued mitosis, resulting in an organism that is genetically identical to the donor. Parthenogenesis is different, in that it originates from the genetic material contained within an egg cell and the new organism is not necessarily genetically identical to the parent.

Parthenogenesis may be achieved through an artificial process as described below under the discussion of mammals.

Insects

Parthenogenesis in insects can cover a wide range of mechanisms. The offspring produced by parthenogenesis may be of both sexes, only female (thelytoky, e.g. aphids and some hymenopterans) or only male (arrhenotoky, e.g. most hymenopterans). Both true parthenogenesis and pseudogamy (gynogenesis or sperm-dependent parthenogenesis) are known to occur. The egg cells, depending on the species may be produced without meiosis (apomictically) or by one of the several automictic mechanisms.

A related phenomenon, polyembryony is a process that produces multiple clonal offspring from a single egg cell. This is known in some hymenopteran parasitoids and in Strepsiptera.

In automictic species the offspring can be haploid or diploid. Diploids are produced by doubling or fusion of gametes after meiosis. Fusion is seen in the Phasmatodea, Hemiptera (Aleurodids and Coccidae), Diptera, and some Hymenoptera.

In addition to these forms is hermaphroditism, where both the eggs and sperm are produced by the same individual, but is not a type of parthenogenesis. This is seen in three species of *Icerya* scale insects.

Parasitic bacteria like *Wolbachia* have been noted to induce automictic thelytoky in many insect species with haplodiploid systems. They also cause gamete duplication in unfertilized eggs causing them to develop into female offspring.

Honey Bee on a plum blossom

Among species with the haplo-diploid sex-determination system, such as hymenopterans (ants, bees and wasps) and thysanopterans (thrips), haploid males are produced from unfertilized eggs. Usually eggs are laid only by the queen, but the unmated workers may also lay haploid, male eggs either regularly (e.g. stingless bees) or under special circumstances. An example of non-viable parthenogenesis is common among domesticated honey bees. The queen bee is the only fertile female in the hive; if she dies without the possibility for a viable replacement queen, it is not uncommon for the worker bees to lay eggs. This is a result of the lack of the queen's pheromones and the pheromones secreted by uncapped brood, which normally suppress ovarian development in workers. Worker bees are unable to mate, and the unfertilized eggs produce only drones (males), which can mate only with a queen. Thus, in a relatively short period, all the worker bees die off, and the new drones follow if they have not been able to mate before the collapse of the colony. This behaviour is believed to have evolved to allow a doomed colony to produce drones which may mate with a virgin queen and thus preserve the colony's genetic progeny.

A few ants and bees are capable of producing diploid female offspring parthenogenetically. These include a honey bee subspecies from South Africa, *Apis mellifera capensis*, where workers are capable of producing diploid eggs parthenogenetically, and replacing the queen if she dies; other examples include some species of small carpenter bee, (genus *Ceratina*). Many parasitic wasps are known to be parthenogenetic, sometimes due to infections by *Wolbachia*.

The workers in five ant species and the queens in some ants are known to reproduce by parthenogenesis. In *Cataglyphis cursor*, a European formicine ant, the queens and workers can produce new queens by parthenogenesis. The workers are produced sexually.

In Central and South American electric ants, *Wasmannia auropunctata*, queens produce more queens through automictic parthenogenesis with central fusion. Sterile workers usually are produced from eggs fertilized by males. In some of the eggs fertilized by males, however, the fertilization can cause the female genetic material to be ablated from the zygote. In this way, males pass on only their genes to become fertile male offspring. This is the first recognized example of an animal species where both females and males can reproduce clonally resulting in a complete separation of male and female gene pools. As a consequence, the males will only have fathers and the queens only mothers, while the sterile workers are the only ones with both parents of both genders.

These ants get both the benefits of both asexual and sexual reproduction—the daughters who can reproduce (the queens) have all of the mother's genes, while the sterile workers whose physical strength and disease resistance are important are produced sexually.

Other examples of insect parthenogenesis can be found in gall-forming aphids (e.g., *Pemphigus betae*), where females reproduce parthenogenetically during the gall-forming phase of their life cycle and in grass thrips. In the grass thrips genus *Aptinothrips* there have been, despite the very limited number of species in the genus, several transitions to asexuality.

Crustaceans

Crustacean reproduction varies both across and within species. The water flea *Daphnia pulex* alternates between sexual and parthenogenetic reproduction. Among the better-known large decapod crustaceans, some crayfish reproduce by parthenogensis. "Marmorkrebs" are parthenogenetic crayfish that were discovered in the pet trade in the 1990s. Offspring are genetically identical to the parent, indicating it reproduces by apomixis, i.e. parthenogenesis in which the eggs did not undergo meiosis. Spinycheek crayfish (*Orconectes limosus*) can reproduce both sexually and by parthenogenesis. The Louisiana red swamp crayfish (*Procambarus clarkii*), which normally reproduces sexually, has also been suggested to reproduce by parthenogenesis, although no individuals of this species have been reared this way in the lab. *Artemia parthenogenetica* is a species or series of populations of parthenogenetic brine shrimps.

Spiders

At least two species of spiders in the family Oonopidae (goblin spiders), *Heteroonops spinimanus* and *Triaeris stenaspis*, are thought to be parthenogenetic, as no males have ever been collected. Parthenogenetic reproduction has been demonstrated in the laboratory for *T. stenaspis*.

Rotifers

In bdelloid rotifers, females reproduce exclusively by parthenogenesis (obligate parthenogenesis), while in monogonont rotifers, females can alternate between sexual and asexual reproduction (cyclical parthenogenesis). At least in one normally cyclical parthenogenetic species obligate parthenogenesis can be inherited: a recessive allele leads to loss of sexual reproduction in homozygous offspring.

Flatworms

At least two species in the genus *Dugesia*, flatworms in the Turbellaria sub-division of the phylum Platyhelminthes, include polyploid individuals that reproduce by parthenogenesis. This type of parthenogenesis requires mating, but the sperm does not contribute to the genetics of the offspring (the parthenogenesis is pseudogamous, alternatively referred to as gynogenetic). A complex cycle of matings between diploid sexual and polyploid parthenogenetic individuals produces new parthenogenetic lines.

Snails

Several species of parthenogenetic gastropods have been studied, especially with respect to their status as invasive species. Such species include the New Zealand mud snail (*Potamopyrgus antipodarum*), the red-rimmed melania (*Melanoides tuberculata*), and the Quilted melania (*Tarebia granifera*).

Squamata

Most reptiles of the squamata order (lizards and snakes) reproduce sexually, but parthenogenesis has been observed to occur naturally in certain species of whiptails, some geckos, rock lizards,- Komodo dragons and snakes. Some of these like the mourning gecko *Lepidodactylus lugubris*, Indo-Pacific house gecko *Hemidactylus garnotii*, the hybrid whiptails *Cnemidophorus*, Caucasian rock lizards *Darevskia*, and the brahminy blindsnake, *Indotyphlops braminus* are unisexual and obligately parthenogenetic. Others reptiles, such as the Komodo dragon, other monitor lizards, and some species of boas, pythons, filesnakes, gartersnakes and rattlesnakes were previously considered as cases of facultative parthenogenesis, but are in fact cases of accidental parthenogenesis.

Komodo dragon, *Varanus komodoensis*, rarely reproduces offspring via parthenogenesis.

In 2012, facultative parthenogenesis was reported in wild vertebrates for the first time by US researchers amongst captured pregnant copperhead and cottonmouth female pit-vipers. The Komodo dragon, which normally reproduces sexually, has also been found able to reproduce asexually by parthenogenesis. A case has been documented of a Komodo dragon reproducing via sexual reproduction after a known parthenogenetic event, highlighting that these cases of parthenogenesis are reproductive accidents, rather than adaptive, facultative parthenogenesis.

Some reptile species use a ZW chromosome system, which produces either males (ZZ) or females (ZW). Until 2010, it was thought that the ZW chromosome system used by reptiles was incapable of producing viable WW offspring, but a (ZW) female boa constrictor was discovered to have produced viable female offspring with WW chromosomes.

Parthenogenesis has been studied extensively in the New Mexico whiptail in the genus *Cnemidophorus* (also known as *Aspidoscelis*) of which 15 species reproduce exclusively by parthenogenesis. These lizards live in the dry and sometimes harsh climate of the southwestern United States and northern Mexico. All these asexual species appear to have arisen through the hybridization of two or three of the sexual species in the genus leading to polyploid individuals. The mechanism by which the mixing of chromosomes from two or three species can lead to parthenogenetic reproduction is unknown. Recently, a hybrid parthenogenetic whiptail lizard was bred in the laboratory from a cross between an asexual and a sexual whiptail. Because multiple hybridization events can occur, individual parthenogenetic whiptail species can consist of multiple independent asexual lineages. Within lineages, there is very little genetic diversity, but different lineages may have quite different genotypes.

An interesting aspect to reproduction in these asexual lizards is that mating behaviors are still seen, although the populations are all female. One female plays the role played by the male in closely related species, and mounts the female that is about to lay eggs. This behaviour is due to the hormonal cycles of the females, which cause them to behave like males shortly after laying eggs, when levels of progesterone are high, and to take the female role in mating before laying eggs, when estrogen dominates. Lizards who act out the courtship ritual have greater fecundity than those kept in isolation, due to the increase in hormones that accompanies the mounting. So, although the populations lack males, they still require sexual behavioral stimuli for *maximum* reproductive success.

Some lizard parthenogens show a pattern of geographic parthenogenesis, occupying high mountain areas where their ancestral forms have an inferior competition ability. In Caucasian rock lizards of genus *Darevskia*, which have six parthenogenetic forms of hybrid origin hybrid parthenogenetic form *D. "dahli"* has a broader niche than either of its bisexual ancestors and its expansion throughout the Central Lesser Caucasus caused decline of the ranges of both its maternal and paternal species.

Amphibians

Sharks

Parthenogenesis in sharks has been confirmed in at least three species, the bonnethead, the blacktip shark, and the zebra shark, and reported in others.

A bonnethead, a type of small hammerhead shark, was found to have produced a pup, born live on 14 December 2001 at Henry Doorly Zoo in Nebraska, in a tank containing three female hammerheads, but no males. The pup was thought to have been conceived through parthenogenic means. The shark pup was apparently killed by a stingray within days of birth. The investigation of the birth was conducted by the research team from Queen's University Belfast, Southeastern University in Florida, and Henry Doorly Zoo itself, and it was concluded after DNA testing that the reproduction was parthenogenic. The testing showed the female pup's DNA matched only one female who lived in the tank, and that no male DNA was present in the pup. The pup was not a twin or clone of her mother, but rather, contained only half of her mother's DNA ("automictic parthenogenesis"). This type of reproduction had been seen before in bony fish, but never in cartilaginous fish such as sharks, until this documentation.

In the same year, a female Atlantic blacktip shark in Virginia reproduced via parthenogenesis. On 10 October 2008 scientists confirmed the second case of a virgin birth in a shark. The Journal of Fish Biology reported a study in which scientists said DNA testing proved that a pup carried by a female Atlantic blacktip shark in the Virginia Aquarium & Marine Science Center contained no genetic material from a male.

In 2002, two white-spotted bamboo sharks were born at the Belle Isle Aquarium in Detroit. They hatched 15 weeks after being laid. The births baffled experts as the mother shared an aquarium with only one other shark, which was female. The female bamboo sharks had laid eggs in the past. This is not unexpected, as many animals will lay eggs even if there is not a male to fertilize them. Normally, the eggs are assumed to be inviable and are discarded. This batch of eggs was left undisturbed by the curator as he had heard about the previous birth in 2001 in Nebraska and wanted to observe whether they would hatch. Other possibilities had been considered for the birth of the Detroit bamboo sharks including thoughts that the sharks had been fertilized by a male and stored the sperm for a period of time, as well as the possibility that the Belle Isle bamboo shark is a hermaphrodite, harboring both male and female sex organs, and capable of fertilizing its own eggs, but that is not confirmed.

In 2008, a Hungarian aquarium had another case of parthenogenesis after its lone female shark produced a pup without ever having come into contact with a male shark.

The repercussions of parthenogenesis in sharks, which fails to increase the genetic diversity of the offspring, is a matter of concern for shark experts, taking into consideration conservation management strategies for this species, particularly in areas where there may be a shortage of males due to fishing or environmental pressures. Although parthenogenesis may help females who cannot find mates, it does reduce genetic diversity.

In 2011, recurring shark parthenogenesis over several years was demonstrated in a captive zebra shark, a type of carpet shark.

Birds

Parthenogenesis in birds is known mainly from studies of domesticated turkeys and chickens, although it has also been noted in the domestic pigeon. In most cases the egg fails to develop normally or completely to hatching. The first description of parthenogenetic development in a passerine was demonstrated in captive zebra finches, although the dividing cells exhibited irregular nuclei and the eggs did not hatch.

Parthenogenesis in turkeys appears to result from a conversion of haploid cells to diploid; most embryos produced in this way die early in development. Rarely, viable birds result from this process, and the rate at which this occurs in turkeys can be increased by selective breeding, however male turkeys produced from parthenogenesis exhibit smaller testes and reduced fertility.

Mammals

There are no known cases of naturally occurring mammalian parthenogenesis in the wild. Parthenogenetic progeny of mammals would have two X chromosomes, and would therefore be female.

In 1936, Gregory Goodwin Pincus reported successfully inducing parthenogenesis in a rabbit. In April 2004, scientists at Tokyo University of Agriculture used parthenogenesis successfully to create a fatherless mouse. Using gene targeting, they were able to manipulate two imprinted loci H19/IGF2 and DLK1/MEG3 to produce bi-maternal mice at high frequency and subsequently show that fatherless mice have enhanced longevity.

Induced parthenogenesis in mice and monkeys often results in abnormal development. This is because mammals have imprinted genetic regions, where either the maternal or the paternal chromosome is inactivated in the offspring in order for development to proceed normally. A mammal created by parthenogenesis would have double doses of maternally imprinted genes and lack paternally imprinted genes, leading to developmental abnormalities. It has been suggested that defects in placental folding or interdigitation are one cause of swine parthenote abortive development. As a consequence, research on human parthenogenesis is focused on the production of embryonic stem cells for use in medical treatment, not as a reproductive strategy.

Use of an electrical or chemical stimulus can produce the beginning of the process of parthenogenesis in the asexual development of viable offspring.

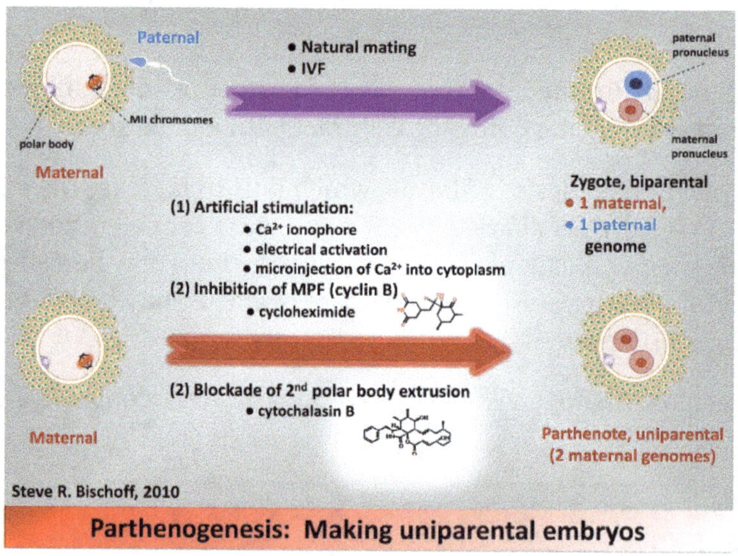

Induction of parthenogenesis in swine. Parthenogenetic development of swine oocytes. High metaphase promoting factor (MPF) activity causes mammalian oocytes to arrest at the metaphase II stage until fertilization by a sperm. The fertilization event causes intracellular calcium oscillations, and targeted degradation of cyclin B, a regulatory subunit of MPF, thus permitting the MII-arrested oocyte to proceed through meiosis. To initiate parthenogenesis of swine oocytes, various methods exist to induce an artificial activation that mimics sperm entry, such as calcium ionophore treatment, microinjection of calcium ions, or electrical stimulation. Treatment with cycloheximide, a non-specific protein synthesis inhibitor, enhances parthenote development in swine presumably by continual inhibition of MPF/cyclin B. As meiosis proceeds, extrusion of the second polar is blocked by exposure to cytochalasin B. This treatment results in a diploid (2 maternal genomes) parthenote. Parthenotes can be surgically transferred to a recipient oviduct for further development, but will succumb by developmental failure after ≈30 days of gestation. The swine parthenote placentae often appears hypo-vascular and is approximately 50% smaller than biparental offspring placentae: see free image (Figure 1) in linked reference.

During oocyte development, high metaphase promoting factor (MPF) activity causes mammalian oocytes to arrest at the metaphase II stage until fertilization by a sperm. The fertilization event causes intracellular calcium oscillations, and targeted degradation of cyclin B, a regulatory subunit of MPF, thus permitting the MII-arrested oocyte to proceed through meiosis.

To initiate parthenogenesis of swine oocytes, various methods exist to induce an artificial activation that mimics sperm entry, such as calcium ionophore treatment, microinjection of calcium ions, or electrical stimulation. Treatment with cycloheximide, a non-specific protein synthesis inhibitor, enhances parthenote development in swine presumably by continual inhibition of MPF/cyclin B. As meiosis proceeds, extrusion of the second polar is blocked by exposure to cytochalasin B. This treatment results in a diploid (2 maternal genomes) parthenote Parthenotes can be surgically transferred to a recipient oviduct for further development, but will succumb to developmental failure after ≈30 days of gestation. The swine parthenote placentae often appears hypo-vascular: see free image (Figure 1) in linked reference.

Humans

On June 26, 2007, International Stem Cell Corporation (ISCC), a California-based stem cell research company, announced that their lead scientist, Dr. Elena Revazova, and her research team were the first to intentionally create human stem cells from unfertilized human eggs using parthenogenesis. The process may offer a way for creating stem cells that are genetically matched to a particular female for the treatment of degenerative diseases that might affect her. In December 2007, Dr. Revazova and ISCC published an article illustrating a breakthrough in the use of parthenogenesis to produce human stem cells that are homozygous in the HLA region of DNA. These stem cells are called HLA homozygous parthenogenetic human stem cells (hpSC-Hhom) and have unique characteristics that would allow derivatives of these cells to be implanted into millions of people without immune rejection. With proper selection of oocyte donors according to HLA haplotype, it is possible to generate a bank of cell lines whose tissue derivatives, collectively, could be MHC-matched with a significant number of individuals within the human population.

On August 2, 2007, after much independent investigation, it was revealed that discredited South Korean scientist Hwang Woo-Suk unknowingly produced the first human embryos resulting from parthenogenesis. Initially, Hwang claimed he and his team had extracted stem cells from cloned human embryos, a result later found to be fabricated. Further examination of the chromosomes of these cells show indicators of parthenogenesis in those extracted stem cells, similar to those found in the mice created by Tokyo scientists in 2004. Although Hwang deceived the world about being the first to create artificially cloned human embryos, he did contribute a major breakthrough to stem cell research by creating human embryos using parthenogenesis. The truth was discovered in 2007, long after the embryos were created by him and his team in February 2004. This made Hwang the first, unknowingly, to successfully perform the process of parthenogenesis to create a human embryon and, ultimately, a human parthenogenetic stem cell line.

Oomycetes

Apomixis can apparently occur in *Phytophthora*, an Oomycete. Oospores derived after an experimental cross were germinated, and some of the progeny were genetically identical to one or other parent, which would imply that meiosis did not occur and the oospores developed by parthenogenesis.

Similar Phenomena

Gynogenesis

A form of asexual reproduction related to parthenogenesis is gynogenesis. Here, offspring are produced by the same mechanism as in parthenogenesis, but with the requirement that the egg merely be stimulated by the *presence* of sperm in order to develop. However, the sperm cell does not contribute any genetic material to the offspring. Since gynogenetic species are all female, activation of their eggs requires mating with males of a closely related species for the needed stimulus. Some salamanders of the genus *Ambystoma* are gynogenetic and appear to have been so for over a million years. It is believed that the success of those salamanders may be due to rare fertilization of eggs by males, introducing new material to the gene pool, which may result from perhaps only one mating out of a million. In addition, the amazon molly is known to reproduce by gynogenesis.

Hybridogenesis

Hybridogenesis is a mode of reproduction of hybrids. Hybridogenetic hybrids (for example AB genome), usually females, during gametogenesis exclude one of parental genomes (A) and produce gametes with unrecombined genome of second parental species (B), instead of containing mixed recombined parental genomes. First genome (A) is restored by fertilization of these gametes with gametes from the first species (AA, sexual host, usually male).

So hybridogenesis is not completely asexual, but instead hemiclonal: half of genome is passed to the next generation clonally, unrecombined, intact (B), other half sexually, recombined (A).

This process continues, so that each generation is half (or hemi-) clonal on the mother's side and has half new genetic material from the father's side.

This form of reproduction is seen in some live-bearing fish of the genus *Poeciliopsis* as well as in some of the *Pelophylax* spp. ("green frogs" or "waterfrogs"):

- *P. kl. esculentus* (edible frog): *P. lessonae* × *P. ridibundus,*
- *P. kl. grafi* (Graf's hybrid frog): *P. perezi* × *P. ridibundus*
- *P. kl. hispanicus* (Italian edible frog) – unknown origin: *P. bergeri* × *P. ridibundus* or *P. kl. esculentus*

and perhaps in *P. demarchii.*

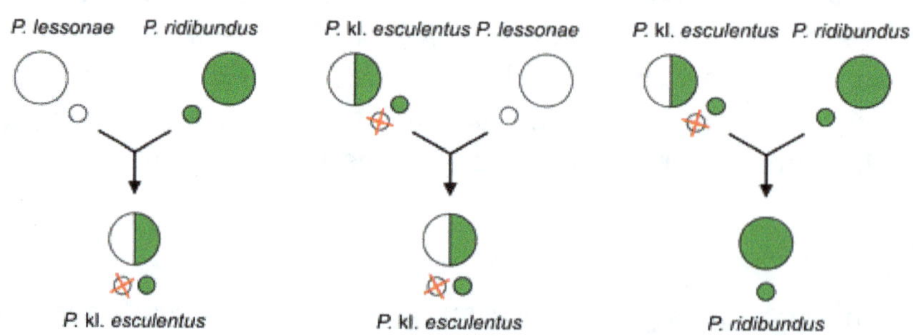

Example crosses between pool frog (*Pelophylax lessonae*), marsh frog (*P. ridibundus*) and their hybrid – edible frog (*P.* kl. *esculentus*). First one is the primary hybridisation generating hybrid, second one is most widespread type of hybridogenesis.

Other examples where hybridogenesis is at least one of modes of reproduction include i.e.

- Iberian minnow *Tropidophoxinellus alburnoides* (*Squalius pyrenaicus* × hypothetical ancestor related with Anaecypris hispanica)
- spined loaches *Cobitis hankugensis* × *C. longicorpus*
- *Bacillus* stick insects B. rossius × Bacillus grandii benazzii

References

- Thorpe, Showick; Thorpe, Edgar (2009). General Studies Manual. Pearson Education India. p. 17. ISBN 9788131721339.
- Berglund A (1997) "Mating systems and sex allocation" Pages 237–265 in JJ Godon, ed. Behavioural ecology of teleost fishes. Oxford University Press. ISBN 0-19-850503-5.
- Birkhead, T.R. & Møller, A.P. (1996) "Monogamy and sperm competition in birds". In J. M. Black (Ed.), Partnerships in Birds: The Study of Monogamy, pp. 323–343, Oxford: Oxford University Press ISBN 0-19-854860-5.
- Barash, D.P. & Lipton, J.E. (2001). The Myth of Monogamy. New York, NY: W.H. Freeman and Company, ISBN 0805071369.
- This section and examples taken from Robert Sapolsky (1998) Why Zebras Don't Get Ulcers, W.H. Freeman and Co., ISBN 0-7167-3210-6, pp. 140–141.
- Trivers, R.L. (1972). Parental investment and sexual selection. In B. Campbell (Ed.), Sexual selection and the descent of man, 1871–1971 (pp. 136–179). Chicago, IL: Aldine. ISBN 0-435-62157-2.
- Virginia Douglass Hayssen; Ari Van Tienhoven (1993). Asdell's Patterns of Mammalian Reproduction: A Compendium of Species-specific Data. Cornell University Press. ISBN 978-0-8014-1753-5. Retrieved 27 September 2013.
- Bear, Mark F. (2007). Neuroscience, exploring the brain. Baltimore, MD: Lippincott Williams and Wilkins. pp. 544–545. ISBN 0781760038.
- Jeffrey Moussaieff Masson (21 October 2009). When Elephants Weep: The Emotional Lives of Animals. Random House Publishing Group. ISBN 978-0-307-57420-6. Retrieved 28 May 2013.
- Jacky Turner; Joyce D'Silva, eds. (2006). Animals, Ethics and Trade: The Challenge of Animal Sentience. Earthscan. ISBN 9781844072545. Retrieved October 26, 2013.
- Balcombe, Jonathan P. (2011). The Exultant Ark: A Pictorial Tour of Animal Pleasure. University of California Press. pp. 89–. ISBN 978-0-520-26024-5.

6

Eating Behavior of Animals

Predation is the process of a predator hunting and eating its prey. They may or may not kill their prey before eating them but the act of predation usually does result in the death of the prey. The alternative eating behavior of animals are hoarding, regurgitation and cannibalism. This chapter helps the readers in developing an in depth understanding the eating behavior of animals.

Predation

In an ecosystem, predation is a biological interaction where a predator (an organism that is hunting) feeds on its prey (the organism that is attacked). Predators may or may not kill their prey prior to feeding on them, but the act of predation often results in the death of the prey and the eventual absorption of the prey's tissue through consumption. Thus predation is often, though not always, carnivory. Other categories of consumption are herbivory (eating parts of plants), fungivory (eating parts of fungi), and detritivory (the consumption of dead organic material (detritus)). All these consumption categories fall under the rubric of consumer-resource systems. It can often be difficult to separate various types of feeding behaviors. For example, some parasitic species prey on a host organism and then lay their eggs on it for their offspring to feed on it while it continues to live in or on its decaying corpse after it has died. The key characteristic of predation however is the predator's direct impact on the prey population. On the other hand, detritivores simply eat dead organic material arising from the decay of dead individuals and have no direct impact on the "donor" organism(s).

A polar bear (*Ursus maritimus*) as the predator feeding on a bearded seal in Svalbard, Norway

Selective pressures imposed on one another often leads to an evolutionary arms race between prey and predator, resulting in various antipredator adaptations. Ways of classifying predation surveyed here include grouping by trophic level or diet, by specialization, and by the nature of the predator's interaction with prey.

Indian python swallowing a small chital deer at Mudumalai National Park

Functional Classification

Classification of predators by the extent to which they feed on and interact with their prey is one way ecologists may wish to categorize the different types of predation. Instead of focusing on what they eat, this system classifies predators by the way in which they eat, and the general nature of the interaction between predator and prey species. Two factors are considered here: How close the predator and prey are (in the latter two cases the term *prey* may be replaced with *host*) and whether or not the prey are directly killed by the predator is considered, with true predation and parasitoidism involving certain death.

True Predation

Predators

Leopard (*Panthera pardus pardus*) killing a young bushbuck (*Tragelaphus sylvaticus*) in Kruger National Park.

A true predator can commonly be known as one that kills and eats another living thing. Whereas other types of predator all harm their prey in some way, this form kills them. Predators may hunt actively for prey in pursuit predation, or sit and wait for prey to approach within striking distance, as in ambush predators. Some predators kill large prey and dismember or chew it prior to eating it, such as a jaguar or a human; others may eat their (usually much smaller) prey whole, as does a bottlenose dolphin swallowing a fish, or a snake, duck or stork swallowing a frog. Some animals that kill both large and small prey for their size (domestic cats and dogs are prime examples) may do either depending upon the circumstances; either would devour a large insect whole but dismember a rabbit. Some predation entails venom that subdues a prey creature before the predator ingests the prey by killing, which the box jellyfish does, or disabling it, found in the behavior of the cone shell. In some cases, the venom, as in rattlesnakes and some spiders, contributes to the digestion of the prey item even before the predator begins eating. In other cases, the prey organism may die in the mouth or digestive system of the predator. Baleen whales, for example, eat millions of microscopic plankton at once, the prey being broken down well after entering the whale. Seed predation and egg predation are other forms of true predation, as seeds and eggs represent potential organisms. Predators of this classification need not eat prey entirely. For example, some predators cannot digest bones, while others can. Some may eat only part of an organism, as in grazing, but still consistently cause its direct death.

Lion and cub eating a Cape buffalo.

Grazing

Grazing organisms may also kill their prey species, but this is seldom the case. While some herbivores like zooplankton live on unicellular phytoplankton and therefore, by the individualized nature of the organism, kill their prey, many only eat a small part of the plant. Grazing livestock may pull some grass out at the roots, but most is simply grazed upon, allowing the plant to regrow once again. Kelp is frequently grazed in subtidal kelp forests, but regrows at the base of the blade continuously to cope with browsing pressure. Animals may also be 'grazed' upon; female mosquitos land on hosts briefly to gain sufficient proteins for the development of their offspring. Starfish may be grazed on, being capable of regenerating lost arms.

Parasitism

Parasites can at times be difficult to distinguish from grazers. Their feeding behavior is similar in many ways, however they are noted for their close association with their host species. While a grazing species such as an elephant may travel many kilometers in a single day, grazing on many plants in the process, parasites form very close associations with their hosts, usually having only one or at most a few in their lifetime. This close living arrangement may be described by the term symbiosis, "living together", but unlike mutualism the association significantly reduces the fitness of the host. Parasitic organisms range from the macroscopic mistletoe, a parasitic plant, to microscopic internal parasites such as cholera. Some species however have more loose associations with their hosts. Lepidoptera (butterfly and moth) larvae may feed parasitically on only a single plant, or they may graze on several nearby plants. It is therefore wise to treat this classification system as a continuum rather than four isolated forms.

Parasitoidism

Parasitoids are organisms living in or on their host and feeding directly upon it, eventually leading to its death. They are much like parasites in their close symbiotic relationship with their host or hosts. Like the previous two classifications parasitoid predators do not kill their hosts instantly. However, unlike parasites, they are very similar to true predators in that the fate of their prey is quite inevitably death. A well-known example of a parasitoids are the ichneumon wasps, solitary insects living a free life as an adult, then laying eggs on or in another species such as a caterpillar. Its larva(e) feed on the growing host causing it little harm at first, but soon devouring the internal organs until finally destroying the nervous system resulting in prey death. By this stage the young wasp(s) are developed sufficiently to move to the next stage in their life cycle. Though limited mainly to the insect order Hymenoptera, Diptera and Coleoptera parasitoids make up as much as 10% of all insect species.

Degree of Specialization

Among predators there is a large degree of specialization. Many predators specialize in hunting only one species of prey. Others are more opportunistic and will kill and eat almost anything (examples: humans, leopards, dogs and alligators). The specialists are usually particularly well suited to capturing their preferred prey. The prey in turn, are often equally suited to escape that predator. This is called an evolutionary arms race and tends to keep the populations of both species in equilibrium. Some predators specialize in certain classes of prey, not just single species. Some will switch to other prey (with varying degrees of success) when the preferred target is extremely scarce, and they may also resort to scavenging or a herbivorous diet if possible.

The land flatworm *Platydemus manokwari* is a predator mainly specialized in land snails

An opportunistic alligator swims with a deer.

Trophic Level

A secondary consumer in action: a mantis (*Tenodera aridifolia*) eating a bee.

Predators are often another organism's prey, and likewise prey are often predators. Though blue jays prey on insects, they may in turn be prey for cats and snakes, and snakes may be the prey of hawks. One way of classifying predators is by trophic level. Organisms that feed on autotrophs, the producers of the trophic pyramid, are known as herbivores or *primary consumers*; those that feed on heterotrophs such as animals are known as *secondary consumers*. Secondary consumers are a type of carnivore, but there are also tertiary consumers eating these carnivores, quartary consumers eating them, and so forth. Because only a fraction of energy is passed on to the next level, this hierarchy of predation must end somewhere, and very seldom goes higher than five or six levels, and may go only as high as three trophic levels (for example, a lion that preys upon large herbivores such as wildebeest, which in turn eat grasses). A predator at the top of any food chain (that is, one that is preyed upon by no organism) is called an apex predator; examples include the orca, sperm whale, anaconda, Komodo dragon, tiger, lion, tiger shark, Nile crocodile, and most eagles and owls—and even omnivorous humans and grizzly bears. An apex predator in one environment may not retain this position as a top predator if introduced to another habitat, such as a dog among alligators, a skunk in the presence of the great horned owl immune to skunk spray, or a snapping turtle among jaguars; a predatory species introduced into an area where it faces no predators, such as a domestic cat or a dog in some insular environments, can become an apex predator by default.

Many organisms (of which humans are prime examples) eat from multiple levels of the food chain and, thus, make this classification problematic. A carnivore may eat both secondary and tertiary consumers, and its prey may itself be difficult to classify for similar reasons. Organisms showing both carnivory and herbivory are known as omnivores. Even herbivores such as the giant panda may supplement their diet with meat. Scavenging of carrion provides a significant part of the diet of some of the most fearsome predators. Carnivorous plants would be very difficult to fit into this classification, producing their own food but also digesting anything that they may trap. Organisms that eat detritivores or parasites would also be difficult to classify by such a scheme.

Predation as Competition

An alternative view offered by Richard Dawkins is of predation as a form of competition: the genes of both the predator and prey are competing for the body (or 'survival machine') of the prey organism. This is best understood in the context of the gene centered view of evolution. Another manner in which predation and competition are connected is throughout intraguild predation. Intraguild predators are those that kill and eat other predators of different species at the same trophic level, and thus that are potential competitors.

Ecological Role

Predators may increase the biodiversity of communities by preventing a single species from becoming dominant. Such predators are known as keystone species and may have a profound influence on the balance of organisms in a particular ecosystem. Introduction or removal of this predator, or changes in its population density, can have drastic cascading effects on the equilibrium of many other populations in the ecosystem. For example, grazers of a grassland may prevent a single dominant species from taking over.

The elimination of wolves from Yellowstone National Park had profound impacts on the trophic pyramid. Without predation, herbivores began to over-graze many woody browse species, affecting the area's plant populations. In addition, wolves often kept animals from grazing in riparian areas, which protected beavers from having their food sources encroached upon. The removal of wolves had a direct effect on beaver populations, as their habitat became territory for grazing. Furthermore, predation keeps hydrological features such as creeks and streams in normal working order. Increased browsing on willows and conifers along Blacktail Creek due to a lack of predation caused channel incision because they helped slow the water down and hold the soil in place.

Adaptations and Behavior

The act of predation can be broken down into a maximum of four stages: Detection of prey, attack, capture and finally consumption. The relationship between predator and prey is one that is typically beneficial to the predator, and detrimental to the prey species. Sometimes, however, predation has indirect benefits to the prey species, though the individuals preyed upon themselves do not benefit. This means that, at each applicable stage, predator and prey species are in an evolutionary arms race to maximize their respective abilities to obtain food or avoid being eaten. This interaction has resulted in a vast array of adaptations in both groups.

Camouflage of the dead leaf mantis makes it less visible to both its predators and prey.

One adaptation helping both predators and prey avoid detection is camouflage, a form of crypsis where species have an appearance that helps them blend into the background. Camouflage consists of not only color but also shape and pattern. The background upon which the organism is seen can be both its environment (e.g., the praying mantis to the right resembling dead leaves) or other organisms (e.g., zebras' stripes blend in with each other in a herd, making it difficult for lions to focus on a single target). The more convincing camouflage is, the more likely it is that the organism will go unseen.

Mimicry in *Automeris io*.

Mimicry is a related phenomenon where an organism has a similar appearance to another species. One such example is the drone fly, which looks a lot like a bee, yet is completely harmless as it cannot sting at all. Another example of batesian mimicry is the io moth, (*Automeris io*), which has markings on its wings that resemble an owl's eyes. When an insectivorous predator disturbs the moth, it reveals its hind wings, temporarily startling the predator and giving it time to escape. Predators may also use mimicry to lure their prey, however. Female fireflies of the genus *Photuris*, for example, copy the light signals of other species, thereby attracting male fireflies, which are then captured and eaten.

Predator

While successful predation results in a gain of energy, hunting invariably involves energetic costs as well. When hunger is not an issue, in general most predators will not seek to attack prey since the costs outweigh the benefits. For instance, a large predatory fish like a shark that is well fed in an aquarium will typically ignore the smaller fish swimming around it (while the prey fish take advantage of the fact that the apex predator is apparently uninterested). Surplus killing represents a deviation from this type of behaviour. The treatment of consumption in terms of cost-benefit analysis is known as optimal foraging theory, and has been quite successful in the study of animal behavior. Some adaptations of successful predators include speed- built for speed, weapons- sharp teeth and claws, camouflage- to avoid being seen by prey, and depth perception- eyes to the front of the head to judge size and distance.

A juvenile red-tailed hawk eating a California vole

Social predation allows predators to kill creatures larger than those that members of the species could overpower singly. Lions, hyenas, wolves, dholes, African wild dogs, and piranhas can kill large herbivores that single animals of the same species usually don't dispatch. Social predation allows some animals to organize hunts of creatures that would easily escape a single predator; thus chimpanzees can prey upon colobus monkeys, and Harris's hawks can cut off all possible escapes for a doomed rabbit. Social predation is often complex behavior, and not all social creatures perform it. Even without complex intelligence, some ant species can destroy much larger creatures.

Great blue heron with prey

Size-selective predation involves predators preferring prey of a certain size. Large prey may prove troublesome for a predator, while small prey might prove hard to find and in any case provide less of a reward. This has led to a correlation between the size of predators and their prey. Size may also act as a refuge for large prey, for example adult elephants are, in general, safe from predation by lions, but juveniles are vulnerable.

Antipredator Adaptations

Many antipredator adaptations have evolved in prey populations due to the selective pressures of predation over long periods of time. Some species mob predators cooperatively. Others such as Thomson's gazelle stot to signal to predators such as cheetahs that they will have an unprofitable chase. Many prey animals are aposematically colored or patterned as a warning to predators that they are distasteful or able to defend themselves. Such distastefulness or toxicity is brought about by chemical defenses, found in a wide range of prey, especially insects, but the skunk is a dramatic mammalian example.

Thomson's gazelle stotting to advertise its ability to escape

Population Dynamics

It is fairly clear that predators tend to lower the survival and fecundity of their prey, but on a higher level of organization, populations of predator and prey species also interact. It is obvious that predators depend on prey for survival, and this is reflected in predator populations being affected by changes in prey populations. It is not so obvious, however, that predators affect prey populations. Eating a prey organism may simply make room for another if the prey population is approaching its carrying capacity.

The population dynamics of predator–prey interactions can be modelled using the Lotka–Volterra equations. These provide a mathematical model for the cycling of predator and prey populations. Predators tend to select young, weak, and ill individuals.

Evolution of Predation

Predation appears to have become a major selection pressure shortly before the Cambrian period—around 550 million years ago—as evidenced by the almost simultaneous development of calcification in animals and algae, and predation-avoiding burrowing. However, predators had been grazing on micro-organisms since at least 1,000 million years ago.

Humans and Predation

Humans are to some extent predatory, fishing, hunting and trapping animals using weapons and tools. They also use other predatory species, such as dogs, cormorants, and falcons to catch prey for food or for sport.

Humans and dogs as a predatory team

In biological pest control, predators from a pest's natural range are introduced to control populations, at the risk of causing unforeseen problems. Besides their use in conservation biology, predators are also important for controlling pests in agriculture. Natural predators are an environmentally friendly and sustainable way of reducing damage to crops, and are one alternative to the use of chemical agents such as pesticides.

Carnivore

A carnivore meaning 'meat eater' is an organism that derives its energy and nutrient requirements from a diet consisting mainly or exclusively of animal tissue, whether through predation or scavenging. Animals that depend solely on animal flesh for their nutrient requirements are called obligate carnivores while those that also consume non-animal food are called facultative carnivores. Omnivores also consume both animal and non-animal food, and apart from the more general definition, there is no clearly defined ratio of plant to animal material that would distinguish a facultative carnivore from an omnivore. A carnivore that sits at the top of the food chain is termed an apex predator.

Plants that capture and digest insects (and, at times, other small animals) are called carnivorous plants. Similarly, fungi that capture microscopic animals are often called carnivorous fungi.

Classification

The word "carnivore" sometimes refers to the mammalian order Carnivora, but this is somewhat misleading. While many Carnivora meet the definition of being meat eaters, not all do, and even fewer are true obligate carnivores. For example, most species of bears are actually omnivorous, except for the giant panda, which is almost exclusively herbivorous, and the exclusively meat-eating polar bear, which lives in the Arctic, where few plants grow. In addition, there are plenty of carnivorous species that are not members of Carnivora.

Outside the animal kingdom, there are several genera containing carnivorous plants and several phyla containing carnivorous fungi. The former are predominantly insectivores, while the latter prey mostly on microscopic invertebrates, such as nematodes, amoebae and springtails.

The Venus flytrap, a well known carnivorous plant

Carnivores are sometimes characterized by the type of prey that they consume. For example, animals that eat insects and similar invertebrates primarily or exclusively are called insectivores, while those that eat fish primarily or exclusively are called piscivores. The first tetrapods, or land-dwelling vertebrates, were piscivorous amphibians known as labyrinthodonts. They gave rise to insectivorous vertebrates and, later, to predators of other tetrapods.

Carnivores may alternatively be classified according to the percentage of meat in their diet. The diet of a hypercarnivore consists of more than 70% meat, that of a mesocarnivore 50–70%, and that of a hypocarnivore less than 30%, with the balance consisting of non-animal foods such as fruits, other plant material, or fungi.

Obligate Carnivores

Obligate carnivores or "true" carnivores depend on the nutrients found only in animal flesh for their survival. While they may consume small amounts of plant material, they lack the physiology required for the efficient digestion of vegetable matter and, in fact, some carnivorous mammals eat vegetation specifically as an emetic. For instance, felids including the domestic cat are obligate carnivores requiring a diet of primarily animal flesh and organs. Specifically, cats have high protein requirements and their metabolisms appear unable to synthesize certain essential nutrients (including retinol, arginine, taurine, and arachidonic acid), and thus they rely on animal flesh in their diet to supply these nutrients. Synthetic forms of essential nutrients such as taurine in the lab have allowed feed manufacturers to formulate foods for carnivores including domestic pets and zoo animals with varying amounts of plant material.

This Bengal tiger's sharp teeth and strong jaws are the classical physical traits expected from carnivorous mammalian predators

Characteristics of Carnivores

Characteristics commonly associated with carnivores include organs for capturing and disarticulating prey (teeth and claws serve these functions in many vertebrates) and status as a predator. In truth, these assumptions may be misleading, as some carnivores do not hunt and are scavengers (though most hunting carnivores will scavenge when the opportunity exists). Thus they do not have the characteristics associated with hunting carnivores. Carnivores have comparatively short digestive systems, as they are not required to break down tough cellulose found in plants. Many animals that hunt other animals have evolved eyes that face forward, thus making depth perception possible. This is almost universal among mammalian predators. Other predators, like crocodiles, as well as most reptiles and amphibians, have sideways facing eyes and hunt by ambush rather than pursuit.

Prehistoric Carnivores

The first vertebrate carnivores were fish, and then amphibians that moved on to land. Early tetrapods were large amphibious piscivores. Some scientists assert that *Dimetrodon* "was the first terrestrial vertebrate to develop the curved, serrated teeth that enable a predator to eat prey much larger than itself." While amphibians continued to feed on fish and later insects, reptiles began exploring two new food types, tetrapods (carnivory), and later, plants (herbivory). Carnivory was a natural transition from insectivory for medium and large tetrapods, requiring minimal adaptation (in contrast, a complex set of adaptations was necessary for feeding on highly fibrous plant materials).

Carnivoramorphs are currently the dominant carnivorous mammals, and have been so since the Miocene. In the early to mid-Cenozoic, however, hyaenodonts, oxyaenid, entelodonts, ptolemaiidans, "arctocyonids" and "mesonychians" were dominant instead, representing a very high diversity of eutherian carnivores in the northern continents and Africa. In South America, sparassodonts were dominant instead, while Australia saw the presence of several marsupial predators, such as the dasyuromorphs and thylacoleonids.

In the Mesozoic, while theropod dinosaurs were the larger carnivores, several carnivorous mammal groups were already present. Most notable are the gobiconodontids, the triconodontid *Jugulator*, the deltatheroideans and *Cimolestes*. Many of these, such as *Repenomamus*, *Jugulator* and *Cimolestes*, were among the largest mammals in their faunal assemblages, capable of attacking dinosaurs.

Most carnivorous mammals, from dogs to *Deltatheridium*, share several adaptations in common, such as carnassialiforme teeth, long canines and even similar tooth replacement patterns. Most aberrant are thylacoleonids, which bear a diprodontan dentition completely unlike that of any mammal, and "eutriconodonts" like gobioconodontids and *Jugulator*, by virtue of their cusp anatomy, though they still worked in the same way as carnassials.

Some theropod dinosaurs such as *Tyrannosaurus rex* that existed during the Mesozoic Era were probably obligate carnivores.

Herbivore

A herbivore is an animal anatomically and physiologically adapted to eating plant material, for example foliage, for the main component of its diet. As a result of their plant diet, herbivorous animals typically have mouthparts adapted to rasping or grinding. Horses and other herbivores have wide flat teeth that are adapted to grinding grass, tree bark, and other tough plant material.

A large percentage of herbivores have mutualistic gut flora that help them digest plant matter, which is more difficult to digest than animal prey. This gut flora is made up of cellulose-digesting protozoans or bacteria living in the herbivores' intestines.

A deer and two fawns feeding on foliage

A caterpillar feeding on a leaf

Etymology

Herbivore is the anglicized form of a modern Latin coinage, *herbivora,* cited in Charles Lyell's 1830 *Principles of Geology*. Richard Owen employed the anglicized term in an 1854 work on fossil teeth and skeletons. *Herbivora* is derived from the Latin *herba* meaning a small plant or herb, and *vora,* from *vorare,* to eat or devour.

Definition and Related Terms

Herbivory is a form of consumption in which an organism principally eats autotrophs such as plants, algae and photosynthesizing bacteria. More generally, organisms that feed on autotrophs in general are known as primary consumers. *Herbivory* usually refers to animals eating plants; fungi, bacteria and protists that feed on living plants are usually termed plant pathogens (plant diseases), and microbes that feed on dead plants are saprotrophs. Flowering plants that obtain nutrition from other living plants are usually termed parasitic plants. There is however no single exclusive and definitive ecological classification of consumption patterns; each textbook has its own variations on the theme.

Evolution of Herbivory

A fossil *Viburnum lesquereuxii* leaf with evidence of insect herbivory; Dakota Sandstone (Cretaceous) of Ellsworth County, Kansas. Scale bar is 10 mm.

Our understanding of herbivory in geological time comes from three sources: fossilized plants, which may preserve evidence of defence (such as spines), or herbivory-related damage; the observation of plant debris in fossilised animal faeces; and the construction of herbivore mouthparts.

Although herbivory was long thought to be a Mesozoic phenomenon, fossils have shown that within less than 20 million years after the first land plants evolved, plants were being consumed by arthropods. Insects fed on the spores of early Devonian plants, and the Rhynie chert also provides evidence that organisms fed on plants using a "pierce and suck" technique.

During the next 75 million years, plants evolved a range of more complex organs, such as roots and seeds. There is no evidence of any organism being fed upon until the middle-late Mississippian, 330.9 million years ago. There was a gap of 50 to 100 million years between the time each organ evolved and the time organisms evolved to feed upon them; this may be due to the low levels of oxygen during this period, which may have suppressed evolution. Further than their arthropod status, the identity of these early herbivores is uncertain. Hole feeding and skeletonisation are recorded in the early Permian, with surface fluid feeding evolving by the end of that period.

Herbivory among four-limbed terrestrial vertebrates, the tetrapods developed in the Late Carboniferous (307 - 299 million years ago). Early tetrapods were large amphibious piscivores. While amphibians continued to feed on fish and insects, some reptiles began exploring two new food types, tetrapods (carnivory) and plants (herbivory). The entire dinosaur order ornithischia was composed with herbivores dinosaurs. Carnivory was a natural transition from insectivory for medium and large tetrapods, requiring minimal adaptation. In contrast, a complex set of adaptations was necessary for feeding on highly fibrous plant materials.

Arthropods evolved herbivory in four phases, changing their approach to it in response to changing plant communities. Tetrapod herbivores made their first appearance in the fossil record of their jaws near the Permio-Carboniferous boundary, approximately 300 million years ago. The earliest evidence of their herbivory has been attributed to dental occlusion, the process in which teeth from the upper jaw come in contact with teeth in the lower jaw is present. The evolution of dental occlusion led to a drastic increase in plant food processing and provides evidence about feeding strategies based on tooth wear patterns. Examination of phylogenetic frameworks of tooth and jaw morphologes has revealed that dental occlusion developed independently in several lineages tetrapod herbivores. This suggests that evolution and spread occurred simultaneously within various lineages.

Food Chain

Herbivores form an important link in the food chain; because they consume plants in order to digest the carbohydrates photosynthetically produced by a plant. Carnivores in turn consume herbivores for the same reason, while omnivores can obtain their nutrients from either plants or animals. Due to a herbivore's ability to survive solely on tough and fibrous plant matter, they are termed the primary consumers in the food cycle (chain). Herbivory, carnivory, and omnivory can be regarded as special cases of Consumer-Resource Systems.

Leaf miners feed on leaf tissue between the epidermal layers, leaving visible trails

Feeding Strategies

Two herbivore feeding strategies are grazing (e.g. cows) and browsing (e.g. moose). Although the exact definition of the feeding strategy may depend on the writer, most authors agree that to define a grazer at least 90% of the forage has to be grass, and for a browser at least 90% tree leaves and/or twigs. An intermediate feeding strategy is called "mixed-feeding". In their daily need to take up energy from forage, herbivores of different body mass may be selective in choosing their food. "Selective" means that herbivores may choose their forage source depending on, e.g., season or food availability, but also that they may choose high quality (and consequently highly nutritious) forage before lower quality. The latter especially is determined by the body mass of the herbivore, with small herbivores selecting for high quality forage, and with increasing body mass animals are less selective. Several theories attempt to explain and quantify the relationship between animals and their food, such as Kleiber's law, Holling's disk equation and the marginal value theorem.

Kleiber's law describes the relationship between an animal's size and its feeding strategy, saying that larger animals need to eat less food per unit weight than smaller animals. Kleiber's law states that the metabolic rate (q_0) of an animal is the mass of the animal (M) raised to the 3/4 power: $q_0=M^{3/4}$ Therefore, the mass of the animal increases at a faster rate than the metabolic rate.

Herbivores employ numerous types of feeding strategies. Many herbivores do not fall into one specific feeding strategy, but employ several strategies and eat a variety of plant parts.

Types of feeding strategies

Feeding Strategy	Diet	Example
Algivores	Algae	krill, crabs, sea snail, sea urchin, parrotfish, surgeonfish, flamingo
Frugivores	Fruit	Ruffed lemurs
Folivores	Leaves	Koalas
Nectarivores	Nectar	Honey possum
Granivores	Seeds	Hawaiian honeycreepers

Palynivores	Pollen	Bees
Mucivores	Plant fluids, i.e. sap	Aphids
Xylophages	Wood	Termites

Optimal Foraging Theory is a model for predicting animal behavior while looking for food or other resource, such as shelter or water. This model assesses both individual movement, such as animal behavior while looking for food, and distribution within a habitat, such as dynamics at the population and community level. For example, the model would be used to look at the browsing behavior of a deer while looking for food, as well as that deer's specific location and movement within the forested habitat and its interaction with other deer while in that habitat.

This model has been criticized as circular and untestable. Critics have pointed out that its proponents use examples that fit the theory, but do not use the model when it does not fit the reality. Other critics point out that animals do not have the ability to assess and maximize their potential gains, therefore the optimal foraging theory is irrelevant and derived to explain trends that do not exist in nature.

Holling's disk equation models the efficiency at which predators consume prey. The model predicts that as the number of prey increases, the amount of time predators spend handling prey also increases and therefore the efficiency of the predator decreases. In 1959, S. Holling proposed an equation to model the rate of return for an optimal diet: Rate (R) = Energy gained in foraging (Ef)/ (time searching (Ts) + time handling (Th))

Where s = cost of search per unit time f = rate of encounter with items, h = handling time, e = energy gained per encounter In effect, this would indicate that a herbivore in a dense forest would spend more time handling (eating) the vegetation because there was so much vegetation around than a herbivore in a sparse forest, who could easily browse through the forest vegetation. According to the Holling's disk equation, a herbivore in the sparse forest would be more efficient at eating than the herbivore in the dense forest

The marginal value theorem describes the balance between eating all the food in a patch for immediate energy, or moving to a new patch and leaving the plants in the first patch to regenerate for future use. The theory predicts that absent complicating factors, an animal should leave a resource patch when the rate of payoff (amount of food) falls below the average rate of payoff for the entire area. According to this theory, locus should move to a new patch of food when the patch they are currently feeding on requires more energy to obtain food than an average patch. Within this theory, two subsequent parameters emerge, the Giving Up Density (GUD) and the Giving Up Time (GUT). The Giving Up Density (GUD) quantifies the amount of food that remains in a patch when a forager moves to a new patch. The Giving Up Time (GUT) is used when an animal continuously assesses the patch quality.

Attacks and Counter-attacks

Herbivore Offense

The myriad defenses displayed by plants means that their herbivores need a variety of skills to overcome these defenses and obtain food. These allow herbivores to increase their feeding and use of a host plant. Herbivores have three primary strategies for dealing with plant defenses: choice, herbivore modification, and plant modification.

Aphids are fluid feeders on plant sap.

Feeding choice involves which plants a herbivore chooses to consume. It has been suggested that many herbivores feed on a variety of plants to balance their nutrient uptake and to avoid consuming too much of any one type of defensive chemical. This involves a tradeoff however, between foraging on many plant species to avoid toxins or specializing on one type of plant that can be detoxified.

Herbivore modification is when various adaptations to body or digestive systems of the herbivore allow them to overcome plant defenses. This might include detoxifying secondary metabolites, sequestering toxins unaltered, or avoiding toxins, such as through the production of large amounts of saliva to reduce effectiveness of defenses. Herbivores may also utilize symbionts to evade plant defences. For example, some aphids use bacteria in their gut to provide essential amino acids lacking in their sap diet.

Plant modification occurs when herbivores manipulate their plant prey to increase feeding. For example, some caterpillars roll leaves to reduce the effectiveness of plant defenses activated by sunlight.

Plant Defense

A plant defense is a trait that increases plant fitness when faced with herbivory. This is measured relative to another plant that lacks the defensive trait. Plant defenses increase survival and/or reproduction (fitness) of plants under pressure of predation from herbivores.

Defense can be divided into two main categories, tolerance and resistance. Tolerance is the ability of a plant to withstand damage without a reduction in fitness. This can occur by diverting herbivory to non-essential plant parts or by rapid regrowth and recovery from herbivory. Resistance refers to the ability of a plant to reduce the amount of damage it receives from a herbivore. This can occur via avoidance in space or time, physical defenses, or chemical defenses. Defenses can either be constitutive, always present in the plant, or induced, produced or translocated by the plant following damage or stress.

Physical, or mechanical, defenses are barriers or structures designed to deter herbivores or reduce intake rates, lowering overall herbivory. Thorns such as those found on roses or acacia trees are one example, as are the spines on a cactus. Smaller hairs known as trichomes may cover leaves or stems and are especially effective against invertebrate herbivores. In addition, some plants have waxes or resins that alter their texture, making them difficult to eat. Also the incorporation of silica into cell walls is analogous to that of the role of lignin in that it is a compression-resistant structural component of cell walls; so that plants with their cell walls impregnated with silica are thereby afforded a measure of protection against herbivory.

Chemical defenses are secondary metabolites produced by the plant that deter herbivory. There are a wide variety of these in nature and a single plant can have hundreds of different chemical defenses. Chemical defenses can be divided into two main groups, carbon-based defenses and nitrogen-based defenses.

1. Carbon-based defenses include terpenes and phenolics. Terpenes are derived from 5-carbon isoprene units and comprise essential oils, carotenoids, resins, and latex. They can have a number of functions that disrupt herbivores such as inhibiting adenosine triphosphate (ATP) formation, molting hormones, or the nervous system. Phenolics combine an aromatic carbon ring with a hydroxyl group. There are a number of different phenolics such as lignins, which are found in cell walls and are very indigestible except for specialized microorganisms; tannins, which have a bitter taste and bind to proteins making them indigestible; and furanocumerins, which produce free radicals disrupting DNA, protein, and lipids, and can cause skin irritation.

2. Nitrogen-based defenses are synthesized from amino acids and primarily come in the form of alkaloids and cyanogens. Alkaloids include commonly recognized substances such as caffeine, nicotine, and morphine. These compounds are often bitter and can inhibit DNA or RNA synthesis or block nervous system signal transmission. Cyanogens get their name from the cyanide stored within their tissues. This is released when the plant is damaged and inhibits cellular respiration and electron transport.

Plants have also changed features that enhance the probability of attracting natural enemies to herbivores. Some emit semiochemicals, odors that attract natural enemies, while others provide food and housing to maintain the natural enemies' presence, e.g. ants that reduce herbivory. A given plant species often has many types of defensive mechanisms, mechanical or chemical, constitutive or induced, which allow it to escape from herbivores.

Herbivore–plant Interactions per Predator–prey Theory

According to the theory of predator–prey interactions, the relationship between herbivores and plants is cyclic. When prey (plants) are numerous their predators (herbivores) increase in numbers, reducing the prey population, which in turn causes predator number to decline. The prey population eventually recovers, starting a new cycle. This suggests that the population of the herbivore fluctuates around the carrying capacity of the food source, in this case the plant.

Several factors play into these fluctuating populations and help stabilize predator–prey dynamics. For example, spatial heterogeneity is maintained, which means there will always be pockets of plants not found by herbivores. This stabilizing dynamic plays an especially important role for specialist herbivores that feed on one species of plant and prevents these specialists from wiping out their food source. Prey defenses also help stabilize predator–prey dynamics, and for more information on these relationships see the section on Plant Defenses. Eating a second prey type helps herbivores' populations stabilize. Alternating between two or more plant types provides population stability for the herbivore, while the populations of the plants oscillate. This plays an important role for generalist herbivores that eat a variety of plants. Keystone herbivores keep vegetation populations in check and allow for a greater diversity of both herbivores and plants. When an invasive herbivore or plant enters the system, the balance is thrown off and the diversity can collapse to a monotaxon system.

The back and forth relationship of plant defense and herbivore offense can be seen as a sort of "adaptation dance" in which one partner makes a move and the other counters it. This reciprocal change drives coevolution between many plants and herbivores, resulting in what has been referred to as a "coevolutionary arms race". The escape and radiation mechanisms for coevolution, presents the idea that adaptations in herbivores and their host plants, has been the driving force behind speciation.

While much of the interaction of herbivory and plant defense is negative, with one individual reducing the fitness of the other, some is actually beneficial. This beneficial herbivory takes the form of mutualisms in which both partners benefit in some way from the interaction. Seed dispersal by herbivores and pollination are two forms of mutualistic herbivory in which the herbivore receives a food resource and the plant is aided in reproduction.

Impacts

Herbivorous fish and marine animals are an indispensable part of the coral reef ecosystem. Since algae and seaweeds grow much faster than corals they can occupy spaces where corals could have settled. They can outgrow and thus outcompete corals on bare surfaces. In the absence of plant-eating fish, seaweeds deprive corals of sunlight. They can also physically damage corals with scrapes.

The impact of herbivory can be seen in areas ranging from economics to ecological, and both. For example, environmental degradation from white-tailed deer (Odocoileus virginianus) in the US alone has the potential to both change vegetative communities through over-browsing and cost forest restoration projects upwards of $750 million annually. Agricultural crop damage by the same species totals approximately $100 million every year. Insect crop damages also contribute largely to annual crop losses in the U.S. Herbivores affect economics through the revenue generated by hunting and ecotourism. For example, the hunting of herbivorous game species such as white-tailed deer, cottontail rabbits, antelope, and elk in the U.S. contributes greatly to the billion-dollar annually hunting industry. Ecotourism is a major source of revenue, particularly in Africa, where many large mammalian herbivores such as elephants, zebras, and giraffes help to bring in the equivalent of millions of US dollars to various nations annually.

Fungivore

Fungivory or mycophagy is the process of organisms consuming fungi. Many different organisms have been recorded to gain their energy from consuming fungi, including birds, mammals, insects, plants, amoeba, gastropods, nematodes, bacteria and other fungi. Some of these, which only eat fungi are called fungivores whereas others eat fungi as only part of their diet, being omnivores.

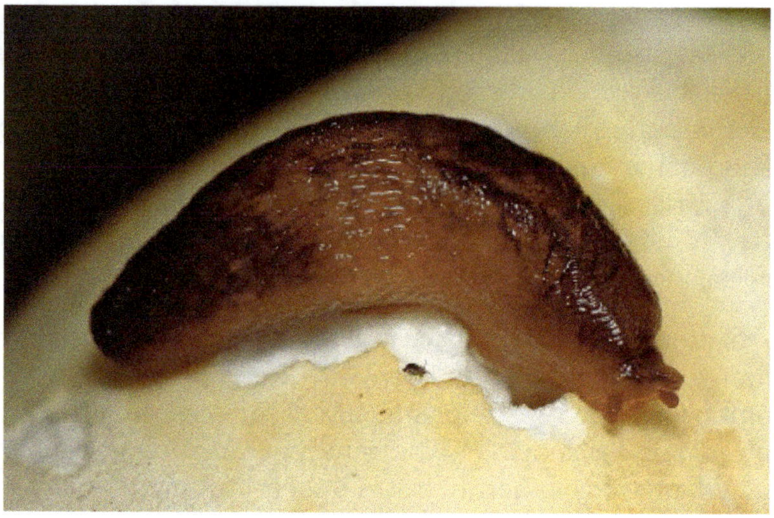

A slug (*Lehmannia nyctelia*) feeding on a mushroom

Animals

Mammals

Many mammals eat fungi, but only a few feed exclusively on fungi, most are opportunistic feeders and fungi only make up part of their diet. At least 22 species of primate, including humans, bonobos, colobines, gorillas, lemurs, macaques, mangabeys, marmosets and vervet monkeys are known to feed on fungi. Most of these species spend less than 5% of the time they spend feeding, eating fungi and they therefore form only a small part of their diet. Some species spend longer foraging for fungi and they account for a greater part of their diet; buffy-tufted marmosets spend up to 12% of their time consuming sporocarps, Goeldi's monkeys spend up to 63% of their time doing so and the Yunnan snub-nosed monkey spends up to 95% of its feeding time eating lichens. Fungi are comparatively very rare in tropical rainforests compared to other food sources such as fruit and leaves and they are also distributed more sparsely and appear unpredictably, making them a challenging source of food for Goeldi's monkeys.

Fungi are renowned for their poisons to deter animals from feeding on them: even today humans die from eating poisonous fungi. A natural consequence of this is the virtual absence of obligate vertebrate fungivores. One of the few extant vertebrate fungivores is the northern flying squirrel, but it is believed that in the past there were numerous vertebrate fungivores and that toxin development greatly lessened their number and forced these species to abandon fungi or diversify. Although some monkeys still eat fungi today, there are no completely fungivorous primates, though their dentition is very suitable for eating fungi.

Mollusks

Many terrestrial gastropod mollusks are known to feed on fungi. It is the case in several species of slugs from distinct families. Among them are the Philomycidae (e. g. *Philomycus carolinianus* and *Phylomicus flexuolaris*) and *Ariolimacidae* (*Ariolimax californianus*), which respectively feed on slime molds (myxomycetes) and mushrooms (basidiomycetes). Species of mushroom producing fungi used as food source by slugs include milk-caps, *Lactarius* spp., the oyster mushroom,

Pleurotus ostreatus and the penny bun, *Boletus edulis*. Other species pertaining to different genera, such as *Agaricus*, *Pleurocybella* and *Russula*, are also eaten by slugs. Slime molds used as food source by slugs include *Stemonitis axifera* and *Symphytocarpus flaccidus*. Some slugs are selective towards certain parts or developmental stages of the fungi they eat, though this behavior varies greatly. Depending on the species and other factors, slugs eat only fungi at specific stages of development. Moreover, in other cases, whole mushrooms can be eaten, without any trace of selectivity.

A banana slug feeding on *Amanita*

Insects

In 2008, *Euprenolepis procera* a species of ant from the rainforests of South East Asia was found to harvest mushrooms from the rainforest. Witte & Maschwitz found that their diet consisted almost entirely of mushrooms, representing a previously undiscovered feeding strategy in ants. Several beetle families, including the Erotylidae, Endomychidae, and certain Tenebrionidae also are specialists on fungi, though they may eat other foods occasionally. Other insects, like fungus gnats and scuttle flies, utilize fungi at their larval stage. Feeding on fungi is crucial for dead wood eaters as this is the only way to acquire nutrients not avilable in nutritionally scarce dead wood.

Euprenolepis procera, the only species of ant known to harvest mushrooms, feeding on a *Pleurotus* mushroom

Birds

Jays (*Perisoreus*) are believed to be the first birds in which mycophagy was recorded. Canada jays (*P. canadensis*), Siberian jays (*P. infaustus*) and Oregon jays (*P. obscurus*) have all been recorded to eat mushrooms, with the stomachs of Siberian jays containing mostly fungi in the early winter. The ascomycete, *Phaeangium lefebvrei* found in north Africa and the Middle East is eaten by migrating birds in winter and early spring, mainly be species of lark (Alaudidae). Bedouin hunters have been reported to use *P. lefebvrei* as bait in traps to attract birds.

Fungi are known to form an important part of the diet of the southern cassowary (*Casuarius casuarius*) of Australia. Bracket fungi have been found in their droppings throughout the year, and Simpson in the *Australasian Mycological Newsletter* suggested it is likely they also eat species of Agaricales and Pezizales but these have not been found in their droppings since they disintegrate when they are eaten. Emus (*Dromaius novaehollandiae*) will eat immature *Lycoperdon* and *Bovista* fungi if presented to them as will brush turkeys (*Alectura lathami*) if offered *Mycena*, suggesting that species of Megapodiidae may feed opportunistically on mushrooms.

Microbial

Fungi

Mycoparasitism occurs when any fungus feeds on other fungi, a form of parasitism, our knowledge of it in natural environments is very limited. *Collybia* grow on dead mushrooms.

The fungal genus, *Trichoderma* produces enzymes such as chitinases which degrade the cell walls of other fungi. They are able to detect other fungi and grow towards them, they then bind to the hyphae of other fungi using lectins on the host fungi as a receptor, forming an appressorium. Once this is formed, *Trichoderma* inject toxic enzymes into the host and probably peptaibol antibiotics, which create holes in the cell wall, allowing *Trichoderma* to grow inside of the host and feed. *Trichoderma* are able to digest sclerotia, durable structures which contain food reserves, which is important if they are to control pathogenic fungi in the long term. *Trichoderma* species have been recorded as protecting crops from *Botrytis cinerea, Rhizoctonia solani, Alternaria solani, Glomerella graminicola, Phytophthora capsici, Magnaporthe grisea* and *Colletotrichum lindemuthianum*; although this protection may not be entirely due to *Trichoderma* digesting these fungi, but by them improving plant disease resistance indirectly.

Bacteria

Bacterial mycophagy was a term coined in 2005, to describe the ability of some bacteria to "grow at the expense of living fungal hyphae". In a 2007 review in the *New Phytologist* this definition was adapted to only include bacteria which play an active role in gaining nutrition from fungi, excluding those that feed off passive secretions by fungi, or off dead or damaged hyphae. The majority of our knowledge in this area relates to interactions between bacteria and fungi in the soil and in or around plants, little is known about interactions in marine and freshwater habitats, or those occurring on or inside animals. It is not known what affects bacterial mycophagy has on the fungal communities in nature.

There are three mechanisms by which bacteria feed on fungi; they either kill fungal cells, cause them to secrete more material out of their cells or enter into the cells to feed internally and they are categorised according to these habits. Those that kill fungal cells are called nectrotrophs, the molecular mechanisms of this feeding are thought to overlap considerably with bacteria that feed on fungi after they have died naturally. Necrotrophs may kill the fungi through digesting their cell wall or by producing toxins which kill fungi, such as tolaasin produced by *Pseudomonas tolaasii*. Both of these mechanisms may be required since fungal cell walls are highly complex, so require many different enzymes to degarde them, and because experiments demonstrate that bacteria that produce toxins cannot always infect fungi. It is likely that these two systems act synergistically, with the toxins killing or inhibiting the fungi and exoenzymes degrading the cell wall and digesting the fungus. Examples of necrotrophs include *Staphylococcus aureus* which feed on *Cryptococcus neoformans*, *Aeromonas caviae* which feed on *Rhizoctonia solani*, *Sclerotium rolfsii* and *Fusarium oxysporum*, and some myxobacteria which feed on *Cochliobolus miyabeanus* and *Rhizoctonia solani*.

Bacteria which manipulate fungi to produce more secretions which they in turn feed off are called extracellular biotrophs; many bacteria feed on fungal secretions, but do not interact directly with the fungi and these are called saprotrophs, rather than biotrophs. Extracellular biotrophs could alter fungal physiology in three ways; they alter their development, the permeability of their membranes (including the efflux of nutrients) and their metabolism. The precise signalling molecules that are used to achieve these changes are unknown, but it has been suggested that auxins (better known for their role as a plant hormone) and quorum sensing molecules may be involved. Bacteria have been identified that manipulate fungi in these ways, for example mycorhizzal helper bacteria (MHBs) and *Pseudomonas putida*, but it remains to be demonstrated whether the changes they make are cause are directly beneficial to the bacteria. In the case of MHBs, which increase infection of plant roots by mycorrhizal fungi, they may benefit, because the fungi gain nutrition from the plant and in turn the fungi will secrete more sugars.

The third group, that enter into living fungal cells are called endocellular biotrophs. Some of these are transmitted vertically whereas others are able to actively invade and subvert fungal cells. The molecular interactions involved in these interactions are mostly unknown. Many endocellular biotrophs, for example some *Burkholderia* species, belong to the β-proteobacteria which also contains species which live inside the cells of mammals and amoeba. Some of them, for example *Candidatus* Glomeribacter gigasporarum, which colonises the spores of *Gigaspora margarita*, have reduced genome sizes indicating that they have become entirely dependent on the metabolic functions of the fungal cells in which they live. When all the endocellular bacteria inside *G. margarita* were removed, the fungus grew differently and was less fit, suggesting that some bacteria may also provide services to the fungi they live in.

Ciliates

The Grossglockneridae family of ciliates, including the species *Grossglockneria acuta*, feed exclusively on fungi. *G. acuta* first attaches themselves to a hyphae or sporangium via a feeding tube and then a ring-shaped structure, around 2 μm in diameter is observed to appear on the fungus, possibly consisting of degraded cell wall material. *G. acuta* then feeds through the hole in the cell wall for, on average, 10 minutes, before detaching itself and moving away. The precise mechanism of feeding is not known, but it conceivably involves enzymes including acid phosphatases, cellulases and chitinases. Microtubules are visible in the feeding tube, as are possible reserves of cell membrane, which may be used to form food vacuoles filled with the cytoplasm of the fungus, via endocytosis, which are then transported back into *G. acuta*. The holes made by *G. acuta* bear some similarities to those made by amoeba, but unlike amoeba *G. acuta* never engulfs the fungus.

Plants

Around 90% of land plants live in symbiosis with mycorrhizal fungi, where fungi gain sugars from plants and plants gain nutrients from the soil via the fungi. Some species of plant have evolved to manipulate this symbiosis, so that they no longer give fungi sugars that they produce and instead gain sugars from the fungi, a process called myco-heterotrophy. Some plants are only dependent on fungi as a source of sugars during the early stages of their development, these include most of the orchids as well as many ferns and lycopods. Others are dependent on this food source for their entire lifetime, including some orchids and Gentianaceae, and all species of Monotropaceae and Triuridaceae. Those that are dependent on fungi, but still photosynthesise are called mixotrophs since they gain nutrition in more than one way, by gaining a significant amount of sugars from fungi, they are able to grow in the deep shade of forests. Examples include the orchids *Epipactis*, *Cephalanthera* and *Plantanthera* and the Pyroleae tribe of the Ericaceae family. Others, such as *Monotropastrum humile*, no longer photosynthesise and are totally dependent on fungi for nutrients. Around 230 such species exist, and this trait is thought to have evolved independently on five occasions outside of the orchid family. Some individuals of the orchid species *Cephalanthera damasonium* are mixotrophs, but others do not photosynthesise. Because the fungi that myco-heterotrophic plants gain sugars from in turn gain them from plants that do photosynthesise, they are considered indirect parasites of other plants. The relationship between orchids and orchid mycorrhizae has been suggested to be somewhere between predation and parasitism.

Monotropastrum humile, a myco-heterotroph dependent on fungi throughout its lifetime

The precise mechanisms by which these plants gain sugars from fungi are not known and has not been demonstrated scientifically. Two pathways have been proposed; they may either degrade fungal biomass, particularly the fungal hyphae which penetrate plant cells in a similar manner to in arbuscular mycorrhizae, or absorb sugars from the fungi by disrupting their cell membranes, through mass flow. To prevent the sugars returning to the fungi, they must compartmentalise the sugars or convert them into forms which the fungi cannot use.

Fungal Farming

Insects

Three insect lineages, the beetles, ants and termites, independently evolved the ability to farm fungi between 40 and 60 million years ago. In a similar way to the way that human societies became more complex after the development of plant-based agriculture, the same occurred in these insect lineages when they evolved this ability and these insects are now of major importance in ecosystems. The methods that insects use to farm fungi share fundamental similarities with human agriculture. Firstly, insects inoculate a particular habitat or substrate with fungi, much in the same as humans plant seeds in fields. Secondly, they cultivate the fungi by regulating the growing environment to try to improve the growth of the fungus, as well as protecting it from pests and diseases. Thirdly they harvest the fungus when it is mature and feed on it. Lastly they are dependent on the fungi they grow, in the same way that humans are dependent on crops.

Beetles

Ambrosia beetles, for example *Austroplatypus incompertus*, farm ambrosia fungi inside of trees and feed on them. The mycangia (organs which carry fungal spores) of ambrosia beetles contain various species of fungus, including species of *Ambrosiomyces*, *Ambrosiella*, *Ascoidea*, *Ceratocystis*, *Dipodascus*, *Diplodia*, *Endomycopsis*, *Monacrosporium* and *Tuberculariella*. The ambrosia fungi are only found in the beetles and their galleries, suggesting that they and the beetles have an obligate symbiosis.

Gallery of *Xylosandrus crassiusculus* split open, with larvae and black fungus

Termites

Around 330 species of termites in twelve genera of the subfamily Macrotermitinae cultivate a specialised fungus in the *Termitomyces* genus. The fungus is kept in a specialised part of the nest in fungus cones. Worker termites eat plant matter, producing faecal pellets which they continuously place on top of the cone. The fungus grows into this material and soon produces immature mushrooms, a rich source of protein, sugars and enzymes, which the worker termites eat. The nodules also contain indigestible asexual spores, meaning that the faecal pellets produced by the workers always contain spores of the fungus that colonise the plant material that they defaecate. The *Termitomyces* also fruits, forming mushrooms above ground, which mature at the same time that the first workers emerge from newly formed nests. The mushrooms produce spores that are wind dispersed, and through this method, new colonies acquire a fungal strain. In some species, the genetic variation of the fungus is very low, suggesting that spores of the fungus are transmitted vertically from nest to nest, rather than from wind dispersed spores.

Termitomyces mushrooms growing out of a termite nest

Ants

Around 220 described species, and more undescribed species of ants in the tribe Attini cultivate fungi. They are only found in the New World and are thought to have evolved in the Amazon Rainforest, where they are most diverse today. For these ants, farmed fungi are the only source of food on which their larvae are raised on and are also an important food for adults. Queen ants carry a small part of fungus in small pouches in their mouthparts when they leave the nest to mate, allowing them to establish a new fungus garden when they form a new nest. Different lineages cultivate fungi on different substrates, those that evolved earlier do so on a wide range of plant matter, whereas leaf cutter ants are more selective, mainly using only fresh leaves and flowers. The fungi are members of the Lepiotaceae and Pterulaceae families. Other fungi in the *Escovopsis* genus parasitise the gardens and antibiotic-producing bacteria also inhabit the gardens.

Humans

Gastropods

The marine snail *Littoraria irrorata*, which lives in the salt marshes of the southeast of the United States feeds on fungi that it encourages to grow. It creates and maintains wounds on the grass, *Spartina alterniflora* which are then infected by fungi, probably of the *Phaeosphaeria* and *Mycosphaerella* genera, which are the preferred diet of the snail. They also deposit faeces on the wounds that they create, which encourage the growth of the fungi because they are rich in nitrogen and fungal hyphae. Juvenile snails raised on uninfected leaves do not grow and are more likely to die, indicating the importance of the fungi in the diet of *L. irrorata*.

Hoarding (Animal Behavior)

Hoarding or caching in animal behavior is the storage of food in locations hidden from the sight of both conspecifics (animals of the same or closely related species) and members of other species. Most commonly, the function of hoarding or caching is to store food in times of surplus for times when food is less plentiful. However, there is evidence that some amount of caching or hoarding is done in order to ripen the food, called ripening caching. The term hoarding is most typically used for rodents, whereas caching is more commonly used in reference to birds, but the behaviors in both animal groups are quite similar.

Western scrub jays cache food such as acorns and insects.

Hoarding is done either on a long-term basis – cached on a seasonal cycle, with food to be consumed months down the line – or on a short term basis, in which case the food will be consumed over a period of one or several days.

Some common animals that cache their food are rodents such as hamsters and squirrels, and many different bird species, such as rooks and woodpeckers. The western scrub jay is noted for its particular skill at caching. There are two types of caching behavior: larder-hoarding, where a species creates a few large caches which it often defends, and scatter-hoarding, where a species will create multiple caches, often with each individual food item stored in a unique place. Both types of caching have their advantage.

Function

Caching behavior is typically a way to save excess edible food for later consumption - either soon to be eaten food, such as when a jaguar hangs a partially eaten prey in a tree to be eaten within a few days, or long term where the food is hidden and retrieved many months later. Caching is a common adaptation to seasonal changes in food availability. In regions where winters are harsh, food availability typically becomes low, and caching food during the times of high food availability in the warmer months provides a significant survival advantage. This phenomenon is referenced in the fable The Ant and the Grasshopper.

However, in ripening caching behavior, animals collect and cache food which is immediately inedible but will become "ripe" and edible after a short while. For instance, tayras (a Central American weasel) have been observed to harvest whole green plantains, hide them, and then come back to eat them after they have ripened. Crocodilians such as the alligator are predators with ineffective teeth action - they can bite prey to kill it but can't tear flesh or chew. Small enough prey are swallowed whole while for larger prey, such as a deer, the carcass is cached underwater and left to rot (or "ripen") until it is easy to eat. Leafcutter ants harvest pieces of inedible leaves and then cache them in underground chambers to ripen with a fungus which is the main food for the colony.

Hoard Distribution and Size

Scatter hoarding is the formation of a large number of small hoards. This behavior is present in both birds (especially the Gray jay) and small mammals, mainly squirrels and other rodents, such as the eastern gray squirrel, fox squirrel, and wood mouse. Specifically, those who do not migrate to warmer climates or hibernate for winter are most likely to scatter hoard. This behavior plays an important part in seed dispersal, as those seeds that are left uneaten will have a chance to germinate, thus enabling plants to spread their populations effectively.

Cache spacing is the primary technique that scatter hoarders use to protect food from pilferers. By spreading the food supply around geographically, hoarders discourage competitors who happen upon a cache from conducting area-restricted searching for more of the supply. Despite cache spacing, however, hoarders are still unable to completely eliminate the threat of pilferage. However, having multiple cache sites is costly because it requires a good memory. Scatter-hoarders generally have a large hippocampus.

In larder hoarding, the hoard is large and is found in a single place termed a larder, which usually also serves as the nest where the animal lives. Hamsters are famous larder hoarders. Indeed, the German verb "hamstern" (to hoard) is derived from the noun "Hamster" which refers to the rodent; similar verbs are found in various related languages (Dutch *hamsteren*, and Swedish *hamstra*). Other languages also draw a clear connection between hamsters and hoarding: Polish *chomikować*, from *chomik* – hamster; Hebrew *hamster*; *oger* (אוגר) comes from *to hoarde*; *le'egor* (לאגור). A disadvantage of larder hoarding is that if a cache is raided, this is far more problematic for the animal than if it were a scatter hoarder. While the hoard is much easier to remember the location of, these larger hoards must also be more staunchly defended.

Related Behaviors

Guarding

Most species are particularly wary of onlooking individuals during caching and ensure that the cache locations are secret. Not all caches are concealed however, for example shrikes store prey items on thorns on branches in the open.

Shared or individual hoarding

Although a small handful of species share food stores, food hoarding is a solo endeavor for most species, including almost all rodents and birds. For example, a number of jays live in large family groups, but they don't demonstrate sharing of cached food. Rather, they hoard their food supply selfishly, caching and retrieving the supply in secret.

There are only two species in which kin selection has resulted in a shared food store, i.e. beavers (*Castor canadensis*) and acorn woodpeckers (*Melanerpes formicivorous*); the former live in family groups and construct winter larders of submerged branches, while the latter are unusual in that they construct a conspicuous communal larder.

Pilfering

Pilferage occurs when one animal takes food from another animal's larder. Some species experience high levels of cache pilferage, up to 30% of the supply per day. Models of scatter hoarding suggested the value of cached food is equal to the hoarders ability to retrieve it.

Reciprocal pilfering

It has been observed that members of certain species, such as rodents and chickadees, act as both hoarder and pilferer. In other words, pilfering can be reciprocal and, thus, tolerable. Although this kind of food caching system seems cooperative, it has been suggested that it is actually driven by the selfish interests of the individual.

Recaching

Animals recache the food that they've pilfered from other animal's caches. For example, 75% percent of mildly radioactive (thus traceable) Jeffrey pine seeds cached by yellow pine chipmunks were found in two cache sites, 29% of the seeds were found in three sites, 9.4% were found in four sites and 1.3% were found in five sites over a 3-month period. These results, and those from other studies, demonstrate the dynamic nature of the food supplies of scatter hoarding animals.

Deception

Group-foraging common ravens, (*Corvus corax*), scatter hoard their food and also raid the caches made by others. Cachers withdraw from conspecifics when hiding their food and most often place their caches behind structures, obstructing the view of potential observers. Raiders watch inconspicuously and keep at a distance to cachers close to their cache sites. In response to the presence of potential raiders or because of their initial movements towards caches, the cachers frequently interrupt caching, change cache sites, or recover their food items. These behaviors suggest that ravens are capable of withholding information about their intentions, which may qualify as tactical deception.

Similarly, Eurasian jays (*Garrulus glandarius*) when being watched by another jay, prefer to cache food behind an opaque barrier rather than a transparent barrier, suggesting they may opt to cache in out-of-view locations to reduce the likelihood of other jays pilfering their caches.

Eating Behavior in Insects

Insects are among the most diverse groups of animals on the planet, including more than a million described species and representing more than half of all known living organisms. The number of extant species is estimated at between six and ten million, found in nearly all environments, although only a small number of species occur in the oceans. This large extant means that the dietary habits of taxa include a large variety of behaviors.

Insect Mouthparts

The insect mouthparts consist of the maxilla, labium, and in some species the mandibles. The labrum is a simple fused sclerite, often called the upper lip, and moves longitudinally, which is hinged to the clypeus. The mandibles (jaws) are a highly sclerotized pair of structures that move at right angles to the body; used for biting, chewing and severing food. The maxillae are paired structures that can also move at right angles to the body and possess segmented palps. The labium (lower lip) is the fused structure that moves longitudinally and possesses a pair of segmented palps.

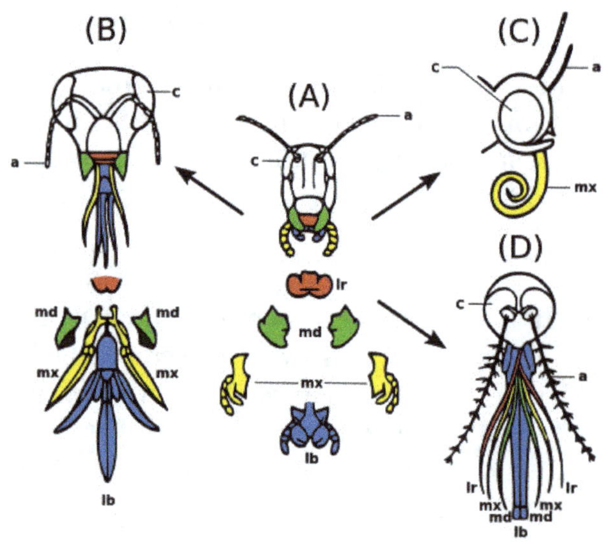

The development of insect mouthparts from the primitive chewing mouthparts of a grasshopper in the centre (A), to the lapping type (B) and the siphoning type (C).

Legend: a - antennae

c - compound eye

lb -labium

lr - labrum

md - mandibles

mx - maxillae.

The mouthparts, along with the rest of the head, can be articulated in at least three different positions: prognathous, opisthognathous and hypognathous. In species with prognathous articulation, the head is positioned vertically aligned with the body, such as species of Formicidae; while in a hypognathous type, the head is aligned horizontally adjacent to the body. In an opisthognathous head, it was positioned diagonally, such as species of Blattodea and some Coleopterans. The mouthparts very greatly among insects of different orders but there are two main functional groups: mandibulate and haustellate. Haustellate mouthparts are those used for sucking liquids and can be further classified, by the presence of stylets, which include: piercing-sucking, sponging, and siphoning. The stylets are needle-like projections used to penetrate plant and animal tissue. The stylets and the feeding tube form the modified mandibles, maxilla, and hypopharynx.

- Mandibulate: These forms of mouthparts are among the most common in insects, which are used for biting and grinding solid foods.
- Piercing-sucking: These forms of mouthparts have stylets, and are used to penetrate solid tissue and then suck up liquid food.
- Sponging: These forms of mouthparts are used to sponge and suck liquids and lack stylets (e.g. most Diptera).
- Siphoning: These forms of mouthparts lack stylets and are used to suck liquids, which are commonly found among species of Lepidoptera.

Mouthparts that are mandibular are found in species of Odonata, Blattodea, adult Neuroptera, Coleoptera, Hymenoptera, Orthoptera, and Lepidoptera. However most adult Lepidoptera have siphoning mouthparts, while the larvae (commonly called caterpillars) are the ones with the mandibles.

Measurement

Drosophila melanogaster has a long history of use as a model organism for genetic studies. However, as of 2014 the quantitative analysis of feeding behavior remains "challenging" and it is "often ignored or poorly characterized". Among many methods, the most commonly applied are capillary feeder (CAFE), radioactive tracer labeling in food, dye tracer labeling in food and counting of proboscis extension (PE) events.

Regurgitation (Digestion)

Regurgitation is the expulsion of material from the pharynx, or esophagus, usually characterized by the presence of undigested food or blood.

Flesh fly, from the Sarcophagidae family "blowing a bubble". One explanation for this behaviour is that it concentrates the fly's meal by evaporation. The diet of the flesh fly is very high in water content. The fly regurgitates the liquid portion of the food, holds it while evaporation reduces the water content and the fly then swallows a much more concentrated meal without the water content. This continues until sufficient amount of liquid is left for the fly. - *Australian Museum*

Regurgitation is used by a number of species to feed their young. This is typically in circumstances where the young are at a fixed location and a parent must forage or hunt for food, especially under circumstances where the carriage of small prey would be subject to robbing by other predators or the whole prey is larger than can be carried to a den or nest. Some bird species also occasionally regurgitate pellets of indigestible matter such as bones and feathers.

It is in most animals a normal and voluntary process unlike the complex vomiting reflex in response to toxins. Honey is produced by a process of regurgitation by honey bees, which is stored in the beehive as a primary food source.

Humans

In humans it can be voluntary or involuntary, the latter being due to a small number of disorders. Regurgitation of a person's meals following ingestion is known as rumination syndrome, a rarely diagnosed eating disorder. It may be a symptom of gastroesophageal reflux disease (GERD).

Some people are able to regurgitate without using any external stimulation or drug, by means of muscle control. Practitioners of yoga have also been known to do this. Professional regurgitators perfect the ability to such a degree as being able to exploit it as entertainment.

Other Animals

For birds that transport food to their mates and/or their young over long distances — especially seabirds — it is impractical to carry food in their bills because of the risk that it would be stolen by other birds, such as frigatebirds, skuas and gulls. Such birds often employ a regurgitative feeding strategy. Many species of gulls have an orange to red spot near the end of the bill (called a "subterminal spot") that the chicks peck in order to stimulate regurgitation.

Moorhen chick being fed regurgitated food by an adult

All of the Suliformes employ a regurgitative strategy to feed their young. In some species — such as the blue-footed booby, masked booby, and the Nazca booby — a brood hierarchy exists, in which the older chick is fed before the younger, subordinate chick. In times when food is scarce, siblicide may occur, where the dominant chick kills its younger sibling in order to sequester all of the resources of the parents.

Some birds, such as fulmars, employ regurgitation as a defense when threatened.

Bubbling fly

List of Feeding Behaviours

Feeding is the process by which organisms, typically animals, obtain food.

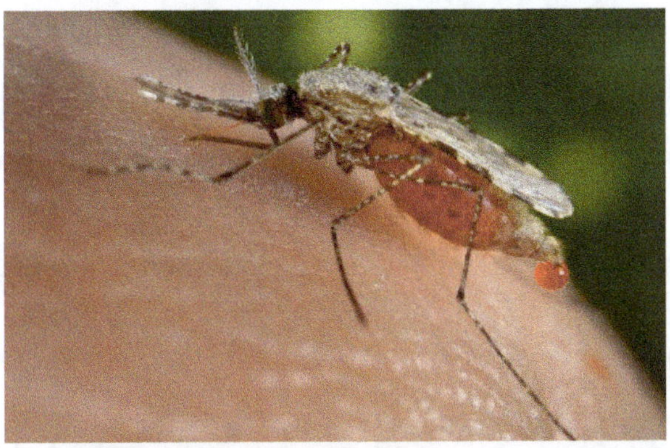

A mosquito drinking blood (hematophagy) from a human (note the droplet of blood being expelled as a surplus)

A rosy boa eating a mouse whole

Evolutionary History

The evolution of different feeding strategies is varied with some feeding strategies evolving several times in independent lineages. In terrestrial vertebrates, the earliest forms were large amphibious piscivores 400 million years ago. While amphibians continued to feed on fish and later insects, reptiles began exploring two new food types, other tetrapods (carnivory), and later, plants (herbivory). Carnivory was a natural transition from insectivory for medium and large tetrapods, requiring minimal adaptation (in contrast, a complex set of adaptations was necessary for feeding on highly fibrous plant materials).

Evolutionary Adaptations

The specialization of organisms towards specific food sources is one of the major causes of evolution of form and function, such as:

- mouth parts and teeth, such as in whales, vampire bats, leeches, mosquitos, predatory animals such as felines and fishes, etc.
- distinct forms of beaks in birds, such as in hawks, woodpeckers, pelicans, hummingbirds, parrots, kingfishers, etc.
- specialized claws and other appendages, for apprehending or killing (including fingers in primates)
- changes in body colour for facilitating camouflage, disguise, setting up traps for preys, etc.
- changes in the digestive system, such as the system of stomachs of herbivores, commensalism and symbiosis

Classification

By Mode of Ingestion

There are many modes of feeding that animals exhibit, including:

- Filter feeding: obtaining nutrients from particles suspended in water
- Deposit feeding: obtaining nutrients from particles suspended in soil
- Fluid feeding: obtaining nutrients by consuming other organisms' fluids
- Bulk feeding: obtaining nutrients by eating all of an organism.
- Ram feeding and suction feeding: ingesting prey via the fluids around it.

By Mode of Digestion

- Extra-cellular digestion: excreting digesting enzymes and then reabsorbing the products
- Myzocytosis: one cell pierces another using a feeding tube, and sucks out cytoplasm
- Phagocytosis: engulfing food matter into living cells, where it is digested

By Food Type

Polyphagy is the ability of an animal to eat a variety of food, whereas monophagy is the intolerance of every food except of one specific type (see generalist and specialist species).

Another classification refers to the specific food animals specialize in eating, such as:

- Carnivore: the eating of animals
    - Araneophagy: eating spiders
    - Avivore: eating birds
    - Durophagy: eating hard-shelled or exoskeleton bearing organisms
    - Haematophagy: eating blood
    - Insectivore: eating insects
        - Myrmecophagy: eating ants and/or termites
    - Invertivore: eating invertebrates
    - Keratophagy or Ceratophagy: eating horny material, such as wool by cloths moths, or snakes eating their own skin after ecdysis.
    - Lepidophagy: eating fish scales
    - Molluscivore: eating molluscs
    - Mucophagy: eating mucus
    - Ophiophagy: eating snakes
    - Oophagy: eating eggs (but also see "Intrauterine cannibalism" below), also Ovivore
    - Piscivore: eating fish
    - Anurophagy: eating frogs
    - Spongivore: eating sponges
    - Teuthophagore: eating mainly squid and other cephalopods
    - Vermivore: eating worms
- Herbivore: the eating of plants
    - Exudativore: eating plant and/or insect exudates (gum, sap, lerp, etc.)
        - Gumivore: eating tree gum
    - Folivore: eating leaves
    - Florivore: eating flower tissue prior to seed coat formation
    - Frugivore: eating fruits
    - Graminivore: eating grasses

    - Granivore: eating seeds
    - Nectarivore: eating nectar
    - Palynivore: eating pollen
    - Xylophagy: eating wood
- Omnivore: the eating of both plants, animals, fungi, bacteria etc. The term means "all-eater".
- Fungivore: the eating of fungus
- Bacterivore: the eating of bacteria
- Planktivore: the eating of plankton

The eating of non-living or decaying matter:

- Coprophagy: eating faeces
- Detritivore: eating decomposing material
- Geophagy: eating inorganic earth
- Osteophagy: eating bones
- Saprophagy: eating decaying organic matter
- Scavenger: eating carrion

There are also several unusual feeding behaviours, either normal, opportunistic, or pathological, such as:

- Cannibalism: feeding on members of the same species
    - Anthropophagy: the practice of eating human flesh
    - Intrauterine cannibalism
        - Oophagy or Ovophagy: the embryo/foetus eats sibling eggs
        - Embryophagy: the foetus eats sibling embryos
    - Filial cannibalism
    - Self-cannibalism: feeding on parts of one's own body
    - Sexual cannibalism: cannibalism after mating
- Kleptoparasitism: stealing food from another animal
- Lignophagia: eating wood, typically a pathological condition in some domestic animals
- Paedophagy: eating young animals
- Pica: appetite for largely non-nutritive substances, e.g. clay or hair, sometimes in pregnancy or in pathological states, typically a medical or veterinary concern.

- Placentophagy: eating placenta
- Trophallaxis: eating food regurgitated by another animal
- Zoopharmacognosy: self-medication by eating plants, soils, and insects to treat and prevent disease.

An opportunistic feeder sustains itself from a number of different food sources, because the species is behaviourally sufficiently flexible.

Storage Behaviours

Some animals exhibit hoarding and caching behaviours in which they store or hide food for later use.

Others

Alcohol – it is widely believed that some animals eat rotting fruit for this to ferment and make them drunk, however, this has been refuted in the case of at least elephants.

Filter Feeder

Filter feeders are a sub-group of suspension feeding animals that feed by straining suspended matter and food particles from water, typically by passing the water over a specialized filtering structure. Some animals that use this method of feeding are clams, krill, sponges, baleen whales, and many fish (including some sharks). Some birds, such as flamingos and certain species of duck, are also filter feeders. Filter feeders can play an important role in clarifying water, and are therefore considered ecosystem engineers.

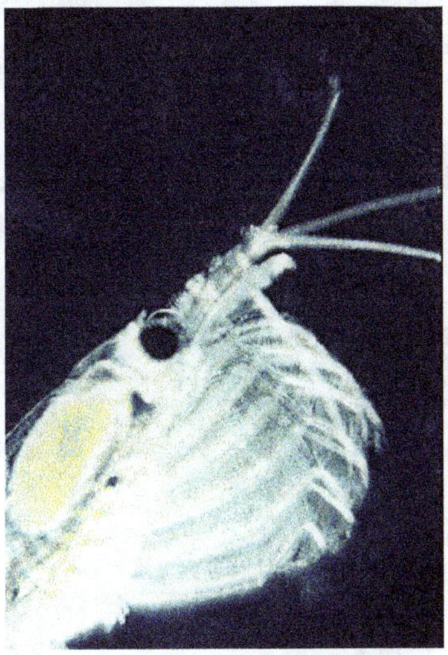

Krill feeding under high phytoplankton concentration (slowed down by a factor of 12)

Fish

Most forage fish are filter feeders. For example, the Atlantic menhaden, a type of herring, lives on plankton caught in midwater. Adult menhaden can filter up to four gallons of water a minute and play an important role in clarifying ocean water. They are also a natural check to the deadly red tide.

In addition to these bony fish, four types of cartilaginous fishes are also filter feeders. The whale shark sucks in a mouthful of water, closes its mouth and expels the water through its gills. During the slight delay between closing the mouth and opening the gill flaps, plankton is trapped against the dermal denticles which line its gill plates and pharynx. This fine sieve-like apparatus, which is a unique modification of the gill rakers, prevents the passage of anything but fluid out through the gills (anything above 2 to 3 mm in diameter is trapped). Any material caught in the filter between the gill bars is swallowed. Whale sharks have been observed "coughing" and it is presumed that this is a method of clearing a build up of food particles in the gill rakers. The megamouth shark has luminous organs called photophores around its mouth. It is believed they may exist to lure plankton or small fish into its mouth. The basking shark is a passive filter feeder, filtering zooplankton, small fish, and invertebrates from up to 2,000 tons of water per hour. Unlike the megamouth and whale sharks, the basking shark does not appear to actively seek its quarry; but it does possess large olfactory bulbs that may guide it in the right direction. Unlike the other large filter feeders, it relies only on the water that is pushed through the gills by swimming; the megamouth shark and whale shark can suck or pump water through their gills. Manta rays can time their arrival at the spawning of large shoals of fish and feed on the free-floating eggs and sperm. This stratagem is also employed by whale sharks.

Crustaceans

Filter basket of a mysid

Mysidacea are small crustaceans that live close to shore and hover above the sea floor, constantly collecting particles with their filter basket. They are an important food source for herring, cod, flounder, and striped bass. Mysids have a high resistance to toxins in polluted areas, and may contribute to high toxin levels in their predators. Antarctic krill manages to directly utilize the minute phytoplankton cells, which no other higher animal of krill size can do. This is accomplished through filter feeding, using the krill's developed front legs, providing for a very efficient filtering apparatus: the six thoracopods form a very effective "feeding basket" used to collect phytoplankton from the open water. In the animation at the top of this page, the krill is hovering at a 55° angle on the spot. In lower food concentrations, the feeding basket is pushed through the water for over half a meter in an opened position, and then the algae are combed to the mouth opening with special setae on the inner side of the thoracopods. Porcelain crab species have feeding appendages covered with setae to filter food particles from the flowing water. Most species of barnacles are filter feeders, using their highly modified legs to sift plankton from the water.

Baleen Whales

The baleen whales (Mysticeti), one of two suborders of the Cetacea (whales, dolphins, and porpoises), are characterized by having baleen plates for filtering food from water, rather than teeth. This distinguishes them from the other suborder of cetaceans, the toothed whales (Odontoceti). The suborder contains four families and fourteen species. Baleen whales typically seek out a concentration of zooplakton, swim through it, either open-mouthed or gulping, and filter the prey from the water using their baleens. A baleen is a row of a large number of keratin plates attached to the upper jaw with a composition similar to those in human hair or fingernails. These plates are triangular in section with the largest, inward-facing side bearing fine hairs forming a filtering mat. Right whales are slow swimmers with large heads and mouths. Their baleen plates are narrow and very long — up to 4 m (13 ft) in bowheads — and accommodated inside the enlarged lower lip which fits onto the bowed upper jaw. As the right whale swims, a front gap between the two rows of baleen plates lets the water in together with the prey, while the baleens filter out the water. Rorquals such as the blue whale, in contrast, have smaller heads, are fast swimmers with short and broad baleen plates. To catch prey, they widely open their lower jaw — almost 90° — swim through a swarm gulping, while lowering their tongue so that the head's ventral grooves expand and vastly increase the amount of water taken in. Baleen whales typically eat krill in polar or subpolar waters during summers, but can also take schooling fish, especially in the Northern Hemisphere. All baleen whales except the gray whale feed near the water surface, rarely diving deeper than 100 m (330 ft) or for extended periods. Gray whales live in shallow waters feeding primarily on bottom-living organisms such as amphipods.

Baleen of a right whale

Bivalves

Bivalves are aquatic molluscs which have two-part shells. Typically both shells (or valves) are symmetrical along the hinge line. The class has 30,000 species, including scallops, clams, oysters and mussels. Most bivalves are filter feeders (although some have taken up scavenging and predation), extracting organic matter from the sea in which they live. Nephridia, the shell fish version of kidneys, remove the waste material. Buried bivalves feed by extending a siphon to the surface. For example, oysters draw water in over their gills through the beating of cilia. Suspended food (phytoplankton, zooplankton, algae and other water-borne nutrients and particles) are trapped in the mucus of a gill, and from there are transported to the mouth, where they are eaten, digested and expelled as feces or pseudofeces. Each oyster filters up to five litres of water per hour. Scientists believe that the Chesapeake Bay's once-flourishing oyster population historically filtered the estuary's entire water volume of excess nutrients every three or four days. Today that process would take almost a year, and sediment, nutrients, and algae can cause problems in local waters. Oysters filter these pollutants, and either eat them or shape them into small packets that are deposited on the bottom where they are harmless.

Bivalve shellfish recycle nutrients that enter waterways from human and agricultural sources. Nutrient bioextraction is "an environmental management strategy by which nutrients are removed from an aquatic ecosystem through the harvest of enhanced biological production, including the aquaculture of suspension-feeding shellfish or algae." Nutrient removal by shellfish, which are then harvested from the system, has the potential to help address environmental issues including excess inputs of nutrients (eutrophication), low dissolved oxygen, reduced light availability and impacts on eelgrass, harmful algal blooms, and increases in incidence of paralytic shellfish poisoning (PSP). For example, the average harvested mussel contains: 0.8–1.2 % nitrogen and 0.06–0.08 % phosphorus Removal of enhanced biomass can not only combat eutrophication and also support the local economy by providing product for animal feed or compost. In Sweden, environmental agencies utilize mussel farming as a management tool in improving water quality conditions, where mussel bioextraction efforts have been evaluated and shown to be a highly effective source of fertilizer and animal feed In the U.S., researchers are investigating potential to model the use of shellfish and seaweed for nutrient mitigation in certain areas of Long Island Sound.

Bivalve are also largely used as bioindicators to monitor the health of an aquatic environment, either fresh- or seawater. Their population status or structure, physiology, behaviour or their content of certain elements or compounds can reveal the contamination status of any aquatic ecosystem. They are extremelly useful as they are sessile - which means they are closely representative of the environment where they are sampled or placed (caging) -, and they are breathing water all along the day, exposing their gills and internal tissues: bioaccumulation. One of the most famous project in that field is the Mussel Watch Programme in U.S. but today they are used worldwide for that purpose (ecotoxicology).

Sponges

Sponges have no true circulatory system; instead, they create a water current which is used for circulation. Dissolved gases are brought to cells and enter the cells via simple diffusion. Metabolic wastes are also transferred to the water through diffusion. Sponges pump remarkable amounts of water. *Leuconia*, for example, is a small leuconoid sponge about 10 cm tall and 1 cm in diameter. It is estimated that water enters through more than 80,000 incurrent canals at a speed of 6 cm per minute. However, because *Leuconia* has more than 2 million flagellated chambers whose combined diameter is much greater than that of the canals, water flow through chambers slows to 3.6 cm per hour. Such a flow rate allows easy food capture by the collar cells. Water is expelled through a single osculum at a velocity of about 8.5 cm/second: a jet force capable of carrying waste products some distance away from the sponge.

Tube sponges attracting small reef fish

The arcuate bill of this lesser flamingo is well adapted to bottom scooping

Cnidarians

The moon jellyfish has a grid of fibres which are slowly pulled through the water. The motion is so slow that copepods cannot sense it and do not react with an escape response.

Other filter-feeding cnidarians include sea pens, sea fans, plumose anemones, and *Xenia*.

An undulating live *Aurelia* in the Baltic Sea showing the grid in action.

Flamingos

Flamingos filter-feed on brine shrimp. Their oddly shaped beaks are specially adapted to separate mud and silt from the food they eat, and are uniquely used upside-down. The filtering of food items is assisted by hairy structures called lamellae which line the mandibles, and the large rough-surfaced tongue.

Pterosaurs

Traditionally, Ctenochasmatoidea as a group has been listed as filter-feeders, due to their long, multiple slender teeth, clearly well adapted to trap prey. However, only *Pterodaustro* showcases a proper pumping mechanism, having up-turned jaws and powerful jaw and tongue musculature. Other ctenochasmatoids lack these, and are now instead thought to have been spoonbill-like catchers, using their specialised teeth simply to offer a larger surface area. Tellingly, these teeth, while small and numerous, are comparatively unspecialised to the baleen-like teeth of *Pterodaustro*.

Boreopterids are thought to have relied on a kind of rudimentary filter feeding, using their long, slender teeth to trap small fish, though probably lacking the pumping mechanism of *Pterodaustro*. In essence, their foraging mechanism was similar to that of modern young *Platanista* "dolphins".

Marine Reptiles

Filter feeding habits are conspicuously rare among Mesozoic marine reptiles, the main filter feeding niche being seemingly instead occupied by pachycormid fish. However, some sauropsids have been suggested to have engaged in filter feeding. *Henodus* was a placodont with unique baleen-like denticles and features of the hyoid and jaw musculature comparable to those of flamingos. Combined with its lacustrine environment, it might have occupied a similar ecological niche. In particular, it was probably a herbivore, filtering out algae and other small-sized flora from the substrates. Stomatosuchidae is a family of freshwater crocodylomorphs with rorqual-like jaws and minuscule teeth, and the unrelated Cenozoic *Mourasuchus* shares similar adaptations. Hupehsuchia is a lineage of bizarre Triassic reptiles adapted for suspension feeding. Some plesiosaurs might have had filter-feeding habits.

Other Filter Feeders

Other filter-feeding organisms include Brachiopoda, Branchiopoda, Bryozoa, Canalipalpata, Ctenophora, Echiura Entoprocta, *Lobodon carcinophagus*, Holothuroidea, Phoronida, Sipuncula, and Urochordata.

Earthworms are a good example of soil-dwelling detritivores.

Detritivores, also known as detrivores, detritophages, detritus feeders, or detritus eaters, are heterotrophs that obtain nutrients by consuming detritus (decomposing plant and animal parts as well as feces). There are many kinds of invertebrates, vertebrates and plants that carry out coprophagy. By doing so, all these detritivores contribute to decomposition and the nutrient cycles. They should be distinguished from other decomposers, such as many species of bacteria, fungi and protists, which are unable to ingest discrete lumps of matter, but instead live by absorbing and metabolizing on a molecular scale (saprotrophic nutrition). However, the terms detritivore and decomposer are often used interchangeably.

Two Adonis blue butterflies lap at a small lump of feces lying on a rock.

Detritivores are an important aspect of many ecosystems. They can live on any soil with an organic component, including marine ecosystems, where they are termed interchangeably with bottom feeders.

Typical detritivorous animals include millipedes, springtails, woodlice, dung flies, slugs, many terrestrial worms, sea stars, sea cucumbers, fiddler crabs, and some sedentary polychaetes such as amphitrites (*Amphitritinae*, worms of the family Terebellidae) and other terebellids.

Scavengers are typically not thought to be detritivores, as they generally eat large quantities of organic matter, but both detritivores and scavengers are specific cases of consumer-resource systems. The eating of wood, whether live or dead, is known as xylophagy. The activity of animals feeding only on dead wood is called sapro-xylophagy and those animals, sapro-xylophagous.

Ecology

In food webs, detritivores generally play the roles of decomposers. Detritivores are often eaten by consumers and therefore commonly play important roles as recyclers in ecosystem energy flow and biogeochemical cycles.

Fungi are the primary decomposers in most environments, illustrated here *Mycena interrupta*. Only fungi produce the enzymes necessary to decompose lignin, a chemically complex substance found in wood.

Many detritivores live in mature woodland, though the term can be applied to certain bottom-feeders in wet environments. These organisms play a crucial role in benthic ecosystems, forming essential food chains and participating in the nitrogen cycle.

Fungi, acting as decomposers, are important in today's terrestrial environment. During the Carboniferous period, fungi and bacteria had yet to evolve the capacity to digest lignin, and so large deposits of dead plant tissue accumulated during this period, later becoming the fossil fuels.

A decaying tree trunk in Canada's boreal forest. Decaying wood fills an important ecological niche, providing habitat and shelter, and returning important nutrients to the soil after undergoing decomposition.

By feeding on sediments directly to extract the organic component, some detritivores accidentally concentrate toxic pollutants.

Aquatic Predation

Aquatic predation presents a special difficulty as compared to predation on land, because the density of water is about the same as that of the prey, so that the prey tends to be pushed away. This problem was first identified by Robert McNeill Alexander. As a result, underwater predators, especially bony fish, have evolved a number of specialized feeding mechanisms, such as filter feeding, ram feeding, suction feeding, protrusion, and pivot feeding.

Grouper capture their prey by sucking them into their mouth

Most underwater predators combine more than one of these basic principles. For example, a typical generalized predator, such as the cod, combines suction with some amount of protrusion and pivot feeding.

Suction Feeding

Suction feeding is a method of ingesting a prey item in fluids by sucking the prey into the predator's mouth. This is typically accomplished by the predator expanding the volume of its oral cavity and/or throat, resulting in a pressure difference between the inside of the mouth and the outside environment. When the mouth is opened, the pressure difference causes water to flow into the predator's mouth, carrying the prey item in with the fluid flow. Bony fish have evolved ingenious mechanisms, so-called mechanical linkages for rapid and forceful expansion of the buccal cavity.

Ram Feeding

Ram feeding, also known as lunge feeding, is a method of feeding underwater in which the predator moves forward with its mouth open, engulfing the prey along with the water surrounding it. During ram feeding, the prey remains fixed in space, and the predator moves its jaws past the prey to capture it. The motion of the head may induce a bow wave in the fluid which pushes the prey away from the jaws, but this can be avoided by allowing water to flow through the jaw. This can be accomplished by means of an expandable throat, as in snapping turtles and baleen whales, or by allowing water to flow out through the gills, as in sharks and herring. A number of species have evolved narrow snouts, as in gar fish and water snakes .

Foraging *Manta alfredi* ram feeding, swimming against the tidal current with its mouth open and sieving zooplankton from the water

Herrings often hunt copepods. If they encounter copepods schooling in high concentrations, the herrings switch to ram feeding. They swim with their mouth wide open and their opercula fully expanded. Every several feet, they close and clean their gill rakers for a few milliseconds (filter feeding). The fish all open their mouths and opercula wide at the same time (the red gills are visible in the photo below—click to enlarge). The fish swim in a grid where the distance between them is the same as the jump length of the copepods.

Herring ram feeding on a school of copepods

School of adult Indian mackerel ram feeding on macroplankton

Pivot Feeding

Pivot feeding is a method to transport the mouth towards the prey by an upward turning of the head, which is pivoting on the neck joint. Pipefish such as sea horses and sea dragons are specialized on this feeding mechanism. With prey capture times of down to 5 ms (shrimpfish *Centriscus scutatus*) this method is used by the fastest feeders in the animal kingdom.

Seahorses rely on stealth to ambush small prey such as copepods. They use pivot feeding to catch the copepod, which involves rotating their snout at high speed and then sucking in the copepod.

The secret of the speed of pivot feeding is in a locking mechanism, in which the hyoid arch is folded under the head and is aligned with the urohyal which connects to the shoulder girdle. A four-bar linkage at first locks the head in a ventrally bent position by the alignment of two bars. The release of the locking mechanism jets the head up and moves the mouth toward the prey within 5-10 ms. The trigger mechanism of unlocking is debated, but is probably in lateral adduction.

Protrusion

Protrusion is the extension of the mouth or premaxilla towards the prey, via mechanical linkages. Protrusion is only known in modern bony fishes, which possess many forms of coupled linkages in their head. Remarkable examples are the slingjaw wrasse and the sand eel which can protrude their mouth by several centimeters. Another example of protrusion is seen in dragonfly larvae, or nymphs, which have hydraulic lower mandibles, protruding forward to catch prey and bring it to the top jaw.

Filter and Suspension Feeding

This is the selection of food particles from a water flow, for example by the gill rakers of fish, the baleens of whales, or the ostia of sponges.

Filter Feeding

In filter feeding, the water flow is primarily generated by the organism itself, for example by creating a pressure, by active swimming, or by ciliary movements.

Suspension Feeding

In suspension feeding, the water flow is primarily external or if the particles themselves move with respect to the ambient water, such as in sea lilies.

Lunge Feeding

Baleen whales feed on plankton by a technique called lunge feeding. Lunge feeding could be regarded as a kind of inverted suction feeding, during which a whale takes a huge gulp of water, which is then filtered through the baleens. Biomechanically this is a unique and extreme feeding method, for which the animal at first must accelerate to gain enough momentum to fold its elastic throat (buccal cavity) around the volume of water to be swallowed. Subsequently, the water flows back through the baleens keeping back the food particles. The highly elastic and muscular buccal rills are a specialized adaptation to this feeding mode.

References

- Begon, M., Townsend, C., Harper, J. (1996). Ecology: Individuals, populations and communities (Third edition). Blackwell Science, London. ISBN 0-86542-845-X, ISBN 0-632-03801-2, ISBN 0-632-04393-8.
- Godfray, H.C.J. (1994). Parasitoids: Behavioral and Evolutionary Ecology. Princeton University Press, Princeton. ISBN 0-691-03325-0, ISBN 0-691-00047-6.
- Alcock, John (1998). Animal Behavior: An Evolutionary Approach (6th ed.). Sunderland, Mass.: Sinauer Associates. ISBN 0-87893-009-4.
- Molles, Manuel C., Jr. (2002). Ecology: Concepts and Applications (International ed.). New York: The McGraw-Hill Companies, Inc. ISBN 0-07-112252-4.
- Richard J. King (1 October 2013). The Devil's Cormorant: A Natural History. University of New Hampshire Press. pp. 9–. ISBN 978-1-61168-225-0.
- Flint, Maria Louise & Dreistadt, Steve H. (1998). Clark, Jack K., ed. Natural Enemies Handbook: The Illustrated Guide to Biological Pest Control. University of California Press. ISBN 978-0-520-21801-7.
- Ullrey, D. E. (2004). "Mammals: Carnivores". In Pond, Wilson. Encyclopedia of Animal Science. CRC Press. p. 591. ISBN 978-0-8247-5496-9.
- Ullrey, D. E. (2004). "Nutrient Requirements: Carnivores". In Pond, Wilson. Encyclopedia of Animal Science. CRC Press. p. 670. ISBN 978-0-8247-5496-9.
- Abraham, Martin A. A. Sustainability Science and Engineering, Volume 1. page 123. Publisher: Elsevier 2006. ISBN 978-0444517128
- Thomas, Peter & Packham, John. Ecology of Woodlands and Forests: Description, Dynamics and Diversity. Publisher: Cambridge University Press 2007. ISBN 978-0521834520
- Steven L. Stephenson (21 April 2010). The Kingdom Fungi: The Biology of Mushrooms, Molds, and Lichens. Timber Press. pp. 200–. ISBN 9780881928914. Retrieved 10 February 2011.
- Chapman, A. D. (2006). Numbers of living species in Australia and the World. Canberra: Australian Biological Resources Study. pp. 60pp. ISBN 978-0-642-56850-2.

7

Abnormal Behavior in Animals

Agonistic behavior is the behavior that animals represent during a fight. It also includes threats, displays, placation and conciliation. Animal psychopathology, broodiness, licking, feather pecking and vent pecking are some of the topics explained in the chapter. This section on abnormal behavior in animals offers an insightful focus, keeping in mind the subject matter.

Agonistic Behaviour

Agonistic behaviour is any social behaviour related to fighting. The term has broader meaning than aggressive behaviour because it includes threats, displays, retreats, placation, and conciliation. The term was coined by Scott and Fredericson in 1951. Agonistic behaviour is seen in many animal species because resources including food, shelter, and mates are often limited.

Ritualized agonistic behaviour between *Zygoballus sexpunctatus* males

Some forms of agonistic behaviour are between contestants who are competing for access to the same resources, such as food or mates. Other times, it involves tests of strength or threat display that make animals look large and more physically fit, a display that may allow it to gain the resource before an actual battle takes place. Although agonistic behaviour varies among species, agonistic interaction consists of three kinds of behaviours: threat, aggression, and submission. These three behaviours are functionally and physiologically interrelated with aggressive behaviour yet fall outside the narrow definition of aggressive behaviour. While any one of these divisions of behaviours may be seen alone in an interaction between two animals, they normally occur in sequence from start to end. Depending on the availability and importance of a resource, behaviours can range from a fight to the death or a much safer ritualistic behaviour, though ritualistic or display behaviours are the most common form of agonistic behaviours.

Prediction of Winning

The type of agonistic behaviour observed, whether it be aggressive or submissive, all depend on the likelihood of winning. For instance, size is usually a good predictor of fighting success, and many animals will display to flaunt their size. Animals are better able to assess their next form of agonistic action by judging the opponent's size and if they are likely to win a fight if a physical fight were to occur.

Example: Stalk-eyed Flies (Diopsidae)

In aggressive behaviour by male stalk-eyed flies the males "square off" by displaying their eyes. Females show a strong preference for mating with males with longer eyestalks. Due to the female preference, males have evolved to compete with each other for mating rights. In the threat display the two flies face each other head-to-head, with their forelegs spread outward and parallel to the eyestalks. This behaviour allows each individual to judge the distance between its competitor's eyes. Eyestalk length increases with body size, and males with shorter eyestalks will usually retreat. A further distance between the eyes conveys a bigger body size, and a better chance of winning.

Avoidance

Physical fighting is actually rare between animals. It would seem that normally the more aggressive an animal is, the more it has to gain. However, in a normal scenario if an animal is too aggressive it might face an unacceptably high cost such as severe injury or death. Unless an animal has a sure indication that they will win without injury, or the resources are valuable enough for the risk of death, animals usually avoid fighting. An animal must weigh the relative costs and benefits of fighting. If the costs are too high, avoiding a fight is preferable.

Ritual Display

For animals, display is any behaviour modified by evolution that is used to convey information. Animals display particular signs, which recipients can use to infer something about the mental and physical state of the first animal. To avoid the heavy cost of fighting, animals have evolved sophisticated rituals, which they use to bluff their opponents into backing down or fleeing. The cost-benefit model of display makes three assumptions: (1) type of display varies depending on the cost; (2) the risk of the display increases as the effectiveness of display increases; and (3) the value of resource being disputed over determines the choice of display used. Animals have evolved to use their physical attributes as a display of ability. If contests can be resolved with ritual display, fighting is not needed. Display can be used to dispute for mates, territory, and food through symbolic gestures instead of battles to the death. If an animal can display without fighting that he is more physically fit than his opponent, he will have gained more than he would have if he had fought and in the process possibly been injured.

Example: Male Grey Catbird (Dumetella Carolinensis)

Male grey catbirds fluff their feathers and spread their lower tails to defend their territory when threatened by another male. The bird that is capable of puffing up and appearing to be the biggest will win the territory.

Example: Western Gorilla (Gorilla Gorilla)

Male western gorillas display a wide range of both vocal and gestural communications when threatened by an opponent. A silverback (alpha male) will start hooting, throwing, chest pounding, leg kicks, and sideways running when approached by another male. This is done to intimidate the opponent and show physical abilities without actually making any physical contact.

Threats

Threat behaviour is any behaviour that signifies hostility or intent to attack another animal. Threat behaviour is meant to cause the opponent to back down and leave. While ritual display can be used for an array of reasons or communicative purposes, threat distinctly is meant for hostility and is the last step before fighting or submission. Threat does not involve physical contact with another animal. Any threat behaviour most often elicits other agonistic behaviour in the recipient. This initiation of threat will result in a display of physical attributes, a fight, or submission; the behaviour or sequence of behaviours depends on what resources are being fought over and each individual's chance of winning against his opponent. In any animal species, threat always contains components of attack and fleeing, which expresses an animal's readiness and likelihood of winning. An intimidation display with a means to threat are exhibited through: hair bristling, feather ruffling, raising skin folds and crest, teeth displaying, horn displaying, making sound, etc.

Two domestic cats threatening each other. Note the more flattened ears of the cat on the right

Example: Frill-necked lizard (Chlamydosaurus Kingii)

Chlamydosaurus kingii, an Australian agamid lizard, uses its frill as a way to display size and aggression to opponents. It is one of the largest and most notable displays seen in the animal kingdom. In comparison to its body size, the frill can flare out to make the lizards head look several times bigger, and it displays bright orange and red scales. Males of *C. kinggi* fight and display frills often during the mating seasons. The male ritualistic display includes repeated partial erections of the frill, head bobbing, tail lashing, and waving of forelimbs.

Agonistic Fighting

Actual fighting in contests is rare because of the risk of injury to both participants. It is most likely to occur when individuals are similarly sized, or when the contested resource is essential for reproduction or survival. Even when agonistic behaviour escalates to fighting, restraint may be used. Fish such as Oreochromis mossambicus often exhibit aggressive displays, but rarely fight to the point of injury or bodily harm. This is also the case in fights among some male venomous snakes; they wrestle, but refrain from biting.

Agonistic behaviour in a zoo between two chickens

Example: Black Mamba (Dendroaspis Polylepis)

Agonistic fighting for black mambas involves a wrestling match in which opponents attempt to pin each other's head repeatedly to the ground. Fights normally last a few minutes but can extend to over an hour. The purpose of fighting is to secure mating rights to receptive females nearby during the breeding season.

Submissive Behaviour

Submissive behaviour involves an individual indicating by an act or posture that it will not challenge a dominant individual in a social group. Submissive behaviours are part of the maintenance of a dominance hierarchy of cooperating individuals in a social group that have overlapping but not entirely coincident interests.

Example: Bearded Dragon (Pogona Vitticeps)

Communication between animals is often achieved by adding a succession of behaviours to a display. Social interactions among bearded dragons (*Pogona vitticeps*) consist of a unique set of movements or visual signals. Waving is one of the most visible signs of submission one lizard can display to another. The lizard rests on three of its legs, raises one of the front arms and then slowly waves the arm in a circular motion. This circular motion, along with the dragon puffing up slightly, shows submission. This display is seen between opponents, as well as adolescents towards adults.

Animal Psychopathology

Animal psychopathology is the study of mental or behavioral disorders in non-human animals.

Historically, there has been an anthropocentric tendency to emphasize the study of animal psychopathologies as models for human mental illnesses. But animal psychopathologies can, from an evolutionary point of view, be more properly regarded as non-adaptive behaviors due to some sort of a cognitive disability, emotional impairment or distress. This article provides a non-exhaustive list of animal psychopathologies.

Eating Disorders

Animals in the wild appear to be relatively free from eating disorders although their body composition fluctuates depending on seasonal and reproductive cycles. However, domesticated animals including farm, laboratory and pet animals are prone to disorders. Evolutionary fitness drives feeding behavior in wild animals. The expectation is that farm animals also display this behavior, but questions arise if the same principles apply to laboratory and pet animals.

Activity Anorexia

Activity anorexia (AA) is a condition where rats begin to exercise excessively while simultaneously cutting down on their food intake, similar to human anorexia nervosa or hypergymnasia. When given free access to food and an exercise wheel, rats normally develop a balanced routine between exercise and food intake, which turns them into fit rats. However, if food intake is restricted and wheel access is unrestricted, rats begin to exercise more and eat less, resulting in excessive weight loss and, ultimately, death. The running cycles shift so that most of the running is done in hours before feeding is scheduled. In other conditions, AA does not develop. Unrestricted food access and restricted wheel access will not cause any significant change in either feeding or exercise routine. Also, if rats are restricted both in food intake and wheel access, they will adjust accordingly. In fact, if rats are first trained to the feeding schedule and then given unrestricted access to a running wheel, they will not develop AA behavior. Results support the notion that the running interferes with adaptation to the new feeding schedule and is associated with the reward system in the brain. One theory is that running simulates foraging, a natural behavior in wild rats. Laboratory rats therefore run (forage) more in response to food shortages. The effect of semi-starvation on activity has also been studied in primates. Rhesus macaque males become hyperactive in response to long-term chronic food restriction.

Thin Sow Syndrome

Thin Sow Syndrome (TSS) is a behavior observed in stalled sows that is similar to AA where some sows after early pregnancy are extremely active, eat little and waste away resulting very often in death. They suffer from emaciation, hypothermia, a depraved appetite, restlessness and hyperactivity. The syndrome may mainly be related to social and environmental stressors. Studies on the effects of overcrowding were pioneered by John B. Calhoun in the 1940s by placing pregnant Norway rats in a room with plenty of water and food and observing the population growth. The population reached a number of individuals and did not grow after; overcrowding produced stress and psychopathologies. Even though there was plenty of water and food the rats stop eating and stop reproducing. These effects have also been observed in jam-packed beetles. When overcrowding occurs female beetles destroy their eggs and turn cannibalistic, eating each other. Male beetles lose interest in the females and although there is plenty of water and food there is no population growth. Similar effects have been observed in overcrowded situations in jack rabbits in Minnesota, and deer in the Chesapeake Bay. Stress in stalled sows is often perceived as the consequence of the restraint of animals that happens in intensive production units. The sows that suffer the most restraining conditions are those lactating or pregnant as they have very little room to move around because they are kept in barred gestation crates or tethered for the 16 weeks of pregnancy which prevents natural and social behaviors. However, increased movement and freedom is also stressful for adult sows, which is usually the case after weaning. When placed into groups they fight vigorously with one dominant sow emerging that eats greedily and becomes large and fat. It is also very likely that two subservient sows make up part of the group who actively avoid competitive feeding situations and are bullied around by the dominant sow. Affected sows have poor appetite but often show PICA and excessive water intake and are anemic.

Pica

Pica is the ingestion of non-nutritive substances and has so far been poorly documented. In non-human animals in the laboratory it has been examined through the ingestion of kaolin (a clay mineral) by rats. Rats were induced to intake kaolin by administering various emetic stimuli such as copper sulfate, apomorphine, cisplatin, and motion. Rats are unable to vomit when they ingest a substance that is harmful thus pica in rats is analogous to vomiting in other species; it is a way for rats to relieve digestive distress. In some animals pica seems to be an adaptive trait but in others it seems to be a true psychopathology like in the case of some chickens. Chickens can display a type of pica when they are feed-deprived (feeding restriction has been adopted by the egg industry to induce molting). They increase their non-nutritive pecking, such as pecking structural features of their environment like wood or wire on fences or the feathers of other birds. It is a typical response that occurs when feeding is restricted or is completely withdrawn. Some of the non-nutritive pecking may be due to a redirection of foraging related behavior. Another animal that has displayed a more complex pica example are cattle. Cattle eat bones when they have a phosphorus deficiency. However, in some cases they persist on eating bones even after their phosphorus levels have stabilized and they are getting adequate doses of phosphorus in their diet. In this case evidence supports both a physical and psychological adaptive response. Cattle that continue to eat bones after their phosphorus levels are adequate do it because of a psychological reinforcer. "The persistence of pica in the seeming absence of a physiological cause might be due to the fortuitous acquisition of a conditioned illness during the period of physiological insult."

Cats also display pica behavior in their natural environments and there is evidence to support that this behavior has a psychological aspect to it. Some breeds (such as the Siamese cat) are more predisposed to showing this type of behavior than other breeds, but several types of breeds have been documented to show pica. Cats have been observed to start by chewing and sucking on non-nutritive substances like wool, cotton, rubber, plastic and even cardboard and then progress into ingestion of these substances. This type of behavior occurs through the first four years of a cat's life but it is primarily observed during the first two months of life when cats are introduced into new homes is most common. Theories explaining why this behavior becomes active during this time suggest that early weaning and stress as a consequence of separation from the mother and litter-mates and exposure to a new environment are to blame. Eating wool or other substances may be a soothing mechanism that cats develops to cope with the changes. Pica is also observed predominately during 6–8 months of a cat's life when territorial and sexual behaviors emerge. Pica may be induced by these social stressors. Other theories contemplated include pica as a redirection of prey-catching/ ingestion behavior as a result of indoor confinement, especially common among oriental breeds due to risk of theft. In natural environments pica has been observed in parrots (such as macaws) and other birds and mammals. Charles Munn has been studying Amazon macaws lick clay from riverbeds in the Amazon to detoxify the seeds they eat. Amazon macaws spend two to three hours a day licking clay. Munn has found that clay helps counter the tannin and alkaloid in the seeds the macaws ingest, a strategy that is also used by native cultures in the Andes Mountains in Peru.

Pica also affects domesticated animals. While drugs like Prozac are often able to diminish troublesome behaviors in pet dogs, they don't seem to help with this eating disorder. The following story about Bumbley, a wire fox terrier who appeared on *20/20* as a result of his eating disorder, is taken from a book by Dr. Nicholas Dodman.

> *"This dog's presenting problem was light chasing (otherwise known as shadow chasing). It chased shadows for hours on end, even excavating through plasterboard walls to pursue its will-o'-the-wisp illusions...The one thing that didn't come across clearly in the show was that Bumbley ate everything in sight and the house had to be "Bumbley-proofed" against his relentless ingestion of anything his owners left around...He had already had surgery to relieve intestinal obstructions resulting from his habit and, each day, his owners reentered their house with trepidation after work, fearing that Bumbley might have eaten something else."*

Dodman talks about new research relating bulimia and compulsive overeating to seizural behavior in human patients. He suggests that anti-epileptic medication might be a possible treatment for some cases of pica in animals.

Behavioral Disorders

Behavioral disorders are difficult to study in animal models because it is difficult to know what animals are thinking and because animal models used to assess psychopathologies are experimental preparations developed to study a condition. Can a monkey effectively communicate that he is sad or that he feels overwhelmed? Lacking the ability to use language to study behavioral disorders like depression and stress questions the validity of those studies conducted. It can be difficult to attribute human afflictions to non-human animals.

Obsessive Compulsive Disorder (OCD)

Obsessive compulsive behavior in animals can be defined as a specific, unnecessary action (or series of actions) repeated more often than would normally be expected. It is unknown whether animals are able to 'obsess' in the same way as humans, and because the motivation for compulsive acts in non-human animals is unknown, the term "abnormal repetitive behaviour" is less misleading.

A wide variety of animals exhibit behaviors that can be considered abnormally repetitive.

Ritualized and Stereotyped Behaviors

Though obsessive-compulsive behaviors are often considered to be pathological or maladaptive, some ritualized and stereotyped behaviors are beneficial. These are usually known as fixed action patterns. These behaviors sometimes share characteristics with obsessive-compulsive behavior, including a high degree of similarity in form and use among many individuals and a repetitive dimension.

There are many observable animal behaviors with characteristic, highly conserved patterns. One example is grooming behavior in rats. This behavior is defined by a specific sequence of actions that does not normally differ between individual rats. The rat first begins by stroking its whiskers, then expands the stroking motion to include the eyes and the ears, finally moving on to lick both sides of its body. Other behaviors may be added to the end of this chain, but these four actions themselves are fixed. Its ubiquity and high degree of stereotypy suggest that this is a beneficial behavior pattern which has been maintained throughout evolutionary history.

Pathological stereotyped behaviors also exist in animals, but they do not necessarily provide a similar model to OCD in humans. Feather picking in orange-winged amazon parrots has both a genetic component, with the behavior being more likely in one sibling if the other does it, and more common in parrots close to a door when they were housed in groups. The same study found that feather picking was more common in females and that there was no social transmission of the behavior; neighbors of feather picking birds were only more likely to show the behavior as well if they were related.

An Evolutionary Basis

Some researchers believe that disadvantageous obsessive compulsive behaviors can be thought of as a normally beneficial process gone too far. Brüne (2006) suggests that change of various origin in striatal and frontal brain circuits, which play a role in predicting needs and threats that may arise in the future, may result in a hyperactive cognitive harm avoidance system, in which a person becomes consciously and unreasonably fearful of an unlikely or impossible event. This may also be true in other animals.

Genetic Factors

Looking for a genetic link, Hannes Lohi, a professor of canine genomics at the University of Helsinki has observed that canine compulsions are typically more common in certain breeds and behavioral dispositions are often shared within the same litter. This suggests that there is a genetic factor to the disorder. Lohi hoped to prove this through a questionnaire to dog owners and a blood sample of 181 dogs from four breeds: miniature and standard bull terriers, German shepherds, and Staffordshire bull terriers, all breeds shown to be more susceptive to compulsive and repetitive behaviors used to release anxiety. J.A. Miller suggests that the more we learn through studying Obsessive Compulsive Disorder in dogs, the more we can come to understand about human biology and the genetics involved in the heredity of susceptibility to disorders such as Obsessive Compulsive Disorder. Research done by the Behavior Service at the Cummings School of Veterinary Medicine, the Program in Medical Genetics at the University of Massachusetts Medical School, and the Broad Institute at the Massachusetts Institute of Technology have supposedly located a chromosome in dogs that confers a high risk of susceptibility to Obsessive Compulsive Disorder. Canine chromosome 7 has been found to be most significantly associated with obsessive compulsive disorder in dogs, or more specifically, canine compulsive disorder (CCD). This breakthrough helped further relate OCD in humans to CCD in canines. Canine chromosome 7 is expressed in the hippocampus of the brain, the same area that Obsessive Compulsive Disorder is expressed in human patients. Similar pathways are involved in drug treatment responses for both humans and dogs, offering more research that the two creatures exhibit symptoms and respond to treatment in similar ways. This data can help scientists to discover more effective and efficient ways to treat OCD in humans through the information they find by studying CCD in dogs.

Animal Models

Animals exhibiting obsessive and compulsive behaviors that resemble OCD in humans have been used as a tool for elucidating possible genetic influences on the disease, potential treatments, and to better understand the pathology of this behavior in general. While such models are useful, they are also limited; it is unclear whether the behavior is ego dystonic in animals. That is, it is difficult to evaluate whether an animal is aware that its behavior is excessive and unreasonable and whether this awareness is a source of anxiety.

One study done by Simon Vermeier used neuroimaging to investigate serotonergic and dopaminergic neurotransmission in 9 dogs with Canine Compulsive Disorder (CCD) to measure the serotonin 2A receptor availability. When compared to the 15 non-compulsive dogs used as a control group, the dogs with CCD were found to have lower receptor availability as well as lower subcortical perfusion and hypothalamic availability . The results of this study provide evidence that there are imbalanced serotonergic and dopaminergic pathways in dogs. Similarities between other studies about human OCD provide construct validity for this study, which suggests that the research will be valid and useful in continuing to investigate brain activity and drug treatment in Obsessive Compulsive Disorder.

Some treatment has been given to dogs with CCD to observe their reactions and how they are similar or different from how humans would react to the same pharmaceutical or behavioral treatment. A combination of the two approaches has been found to be most effective in lowering the intensity and regularity of OCD in both canines and humans. Pharmaceutically, clomipramine was found to be more effective than an alternative chemical, amitriptyline, in treatments for dogs. One study by Karen Overall discovered that by combining behavioral therapy with the more effective clomipramine, the symptoms of Canine Compulsive Disorder decreased by over 50% for all of the dogs involved in the study. Overall acknowledges that OCD is not something that can be completely cured, but studies like this are still important because Obsessive Compulsive Disorder can be controlled effectively enough so it does not interfere with one's life, a valuable and commonly sought after thing for those who suffer from the disorder.

Alicia Graef's article makes several bold claims that dogs are the future in understanding how to better diagnose, recognize, and treat Obsessive Compulsive Disorder in humans. There is evidence supporting her statements, but the connection between CCD and OCD is not clearly understood. So far, studies have proved that effective treatments in dogs are similarly effective for humans, but there are still so many things unknown. Obsessive Compulsive Disorder is a unique mental disorder that cannot be fully cured. It can be controlled and understood, and one possible way of better doing that might be through studying CCD in canines. Studying dogs that exhibit compulsive behaviors has led scientists to genetic breakthroughs in understanding more how biology and genetics factor into Obsessive Compulsive Disorder. By observing and studying how CCD manifests in the brain activity, behaviors, and genes of diagnosed canines, scientists have been able to use their newfound information to develop better diagnostic tests and more readily recognize symptoms and susceptible humans. The similar brain functions and behaviors of dogs with CCD and humans with OCD suggests they have a connection, not only in behavior and symptoms, but in reacting to treatments. Understanding Canine Compulsive Disorder in dogs has helped scientists to better understand and apply their learning to developing new and more effective ways to treat Obsessive Compulsive Disorder in humans.

Some examples of ways in which rats and mice, two of the most common animal models, have been used to represent human OCD are provided below.

Lever Pressing in Rats

Certain laboratory rat strains that have been created by controlled breeding for many generations show a higher tendency towards compulsive behaviors than other strains. Lewis rats show more compulsive lever pressing behavior than Sprague Dawley or Wistar rats and are less responsive to the anti-compulsive drug paroxetine. In this study, rats were taught to press a lever to receive food in an operant conditioning task. Once food was no longer provided when they pressed the lever, rats were expected to stop pressing it. Lewis rats pressed the lever more often than the other two types, even though they had presumably learned that they would not receive food, and continued to press it more often even after treatment with the drug. An analysis of the genetic differences between the three rat strains might help to identify genes that might be responsible for the compulsive behavior.

Rats have also been used to test the possibility of a problem with dopamine levels in the brains of animals that exhibit compulsive checking behavior. After treating rats with quinpirole, a chemical that specifically blocks dopamine D2/D3 receptors, compulsive checking of certain locations in an open field increased. Some components of the checking behavior, such as the level of stereotypy in the path animals took to checked locations, the number of checks, and the length of the checks indicated and increase in compulsivity as doses of quinpirole increased; other components, such as the time taken to return from the checked location to the starting point and the time taken to make that trip remained constant after the initial injection throughout the experiment. This means that there might be both all-or-none and a sensitization aspects in the biology of the dopamine deficiency model of OCD. In addition, quinpirole might reduce a sense of satisfaction in the rats after they check a location, causing them to return to that location again and again.

Estrogen Deficiency in Male Mice

Based on findings of changes in OCD symptoms in menstruating women and differences in the development of the disease between men and women, Hill and colleagues set out to research the effect of estrogen deprivation on the development of compulsive behavior in mice. Male mice with an aromatase Gene knockout who were unable to produce estrogen showed excessive grooming and wheel running behaviors, but female mice did not. When treated with 17β-estradiol, which replaced estrogen in these mice, the behaviors disappeared. This study also found that COMT protein levels decreased in mice that did not produce estrogen and increased in the hypothalamus after estrogen-replacement treatment. Briefly, the COMT protein is involved in degrading some neurotransmitters, including dopamine, norepinephrine and epinephrine. This data suggests that there may be a hormonal component and a hormone-gene interaction effect that may contribute to obsessive behaviors.

Pets

Dr. Nicholas Dodman describes a wide variety of OCD-like behaviors in his book Dogs Behaving Badly. Such behaviors typically appear when the dog is placed in a stressful situation, including an environment that is not very stimulating or a history of abuse. Different breeds of dog seem to display different compulsions. Lick granuloma, or licking repeatedly until ulcers form on the skin, affects more large dogs, like Labradors, golden retrievers, Great Danes, and Dobermans, while bull terriers, German shepherds, Old English sheepdogs, Rottweilers, wire-haired fox terriers, and springer spaniels are more likely to snap at imaginary flies or chase light and shadows. These associations probably have an evolutionary basis, although Dodman does not clearly explain that aspect of the behaviors.

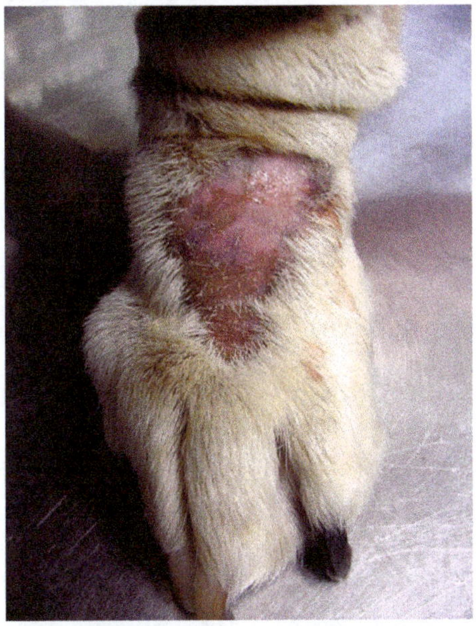

Lick granuloma from excessive licking

Louis Shuster and Nicholas Dodman noticed that dogs often demonstrate obsessive and compulsive behaviors similar to humans. Canine Compulsive Disorder (CCD) is not only specific only to certain breeds of dogs, but the breed may affect the specific types of compulsions. For example, bull terriers frequently exhibit obsessively predatory or aggressive behaviors. Breed may factor into the types of compulsions, but some behaviors are more common across the canine spectrum. Most commonly, CCD is seen in canines as they repeat behaviors such as chasing their tails, compulsively chewing on objects, or licking their paws excessively, similar to the common hand-washing compulsion many people with Obsessive Compulsive Disorder have. Hallucinating and attacking the air around their head, as if there were a bug there, is another compulsion that has been seen in some dogs. Circling, hair biting, staring, and sometimes even barking are other examples of behaviors that are considered compulsions in dogs when taken to extreme, repetitive actions.

Treatment (Pharmaceutical)

Dodman advocates the use of exercised, an enriched environment (like providing noises for dogs to listen to while owners are at work), and often Prozac (an SSRI used to treat OCD in humans) as treatments.

Shuster and Dodman tested pharmaceutical treatment on canines with CCD to see if it would work as effectively as it does in humans. They used glutamate receptor blockers (memantine) and fluoxetine, commonly known as the antidepressant Prozac, to treat and observe the reactions of 11 dogs with compulsions. Seven of the 11 dogs significantly reduced their compulsions in intensity and frequency after receiving medication.

Dodman includes a story about Hogan, a castrated deaf male Dalmatian, and his compulsive behavior. Hogan had a history of neglect and abuse before he was adopted by Connie and Jim, who attempted to improve his behavior by teaching him to respond to American Sign Language. The following are some excerpts from Hogan's file.

> *"All was well for a year and a half when suddenly, one March morning, he woke up and started pawing everything in sight, and just wouldn't stop. He pawed rugs and blankets, hardwood floors and linoleum, grass and dirt surfaces...The similarity between what he was doing and prey-seeking behavior was remarkable."*

> *"I do believe...that Hogan was under some kind of psychological pressure at the time the compulsive pawing behavior developed... Connie and Jim were compelled to leave him for some eight hours a day while they went to work...The pendulum was set and ready to swing. The actual compulsion that develops under such circumstances is less relevant than the fact that one 'does' develop."*

> *"The "three R's" of rehabilitation are exercise, nutrition, and communication. First, I advised Connie to step up Hogan's exercise to a minimum of thirty minutes of aerobic activity a day. In addition, I advised that Hogan should be fed a low-protein, preservative-free diet. Completing the rehabilitation checklist, I exhorted Connie to work even harder with the sign-language and instructed her on a new sign to use when Hogan started digging. The sign was a piece of card with the letter 'H' written on it in thick black pen. Connie was to show Hogan this sign as soon as possible after he engaged in a bout of unwanted pawing and then leave the room. The idea was to let him know that the behavior was not wanted by signaling to him that Connie was about to leave the room...Call me a coward, but I didn't think that alone would cut it because of previous experiences with canine compulsive disorders so, employing a belt-and-suspenders strategy, I also advised medicating Hogan with the tricyclic antidepressant Elavil. Theoretically, Elavil wouldn't be that good in obsessive-compulsive behavior but, limited for reasons of expense, and bearing in mind the possible contribution of separation anxiety, Elavil was my best shot."*

> *"It took six months before Hogan was over the hump of treatment success...At this time Hogan only engaged in occasional pawing of significantly reduced intensity, and the pawing only occurred in moments of stress. Connie reported that stresses particularly likely to induce pawing included being unable to find her and sensing that he was about to be left alone...Hogan continued to improve and reached a point at which he was almost pawing-free - but not quite. That seems to be the way with compulsive disorders in man and beast. They can be reduced to the level of permitting affectees to lead relatively normal lives, but there are occasional relapses."*

Addiction

Sugar addiction has been examined in laboratory rats and it develops in the same way that drug addiction develops. Eating sugary foods causes the brain to release natural chemicals called opioids and dopamine in the limbic system. Tasty food can activate opioid receptors in the ventral tegmental area and thereby stimulate cells that release dopamine in the nucleus accumbens (NAc). The brain recognizes the intense pleasure derived from the dopamine and opioids release and learns to crave more sugar. Dependence is created through these natural rewards, the sugary treats, and the opioid and dopamine released into the synapses of the mesolimbic system. The hippocampus, the insula and the caudate activate when rats crave sugar, which are the same areas that become active when drug addicts crave the drug. Sugar is good because its provides energy but when the body becomes dependent on the sugar intake and the nervous system goes through a change somatic signs of withdrawal begin to appear like chattering teeth, forepaw tremors and head shakes when sugar is not ingested. Morphine tolerance, a measure of addiction, was observed in rats and their tolerance on Morphine was attributed to environmental cues and the systemic effects of the drug. Morphine tolerance does not depend merely on the frequency of pharmacological stimulation, but rather on both the number of pairings of a drug-predictive cue with the systemic effects of the drug. Rats became significantly more tolerant to morphine when they had been exposed to a paired administration than those rats that were not administered morphine along with a drug-predictive cue.

Depression

Using dogs, Martin Seligman and his colleagues pioneered the study of depression in the animal model of learned helplessness at the University of Pennsylvania. Dogs were separated into three groups, the control group, group A had control over when they were being shocked and group B had no control over when they were being shocked. After the shocking condition the dogs were tested in a shuttle box where they could escape shock by jumping over a partition. To eliminate an interference effect - that the dogs did not learn responses while being shocked that would interfere with their normal escape behavior - the dogs were immobilized using curare, a paralyzing drug while they were being shocked. Both the control group and group A tended to jump over the partition to escape shock while group B dogs did not jump and would passively take the shock. The dogs in group B perceived that the outcome was not related to their efforts. Consequently, a theory emerged that attributed the behavior of the animals to the effects of the shock as a stressor so extreme that it depleted a neurochemical needed by the animals for movement. After the dogs study the effects of helplessness have been tested in species from fish to cats. Most recently learned helplessness has been studied in rhesus macaques using inescapable shock, evoked through stress situations like forced swimming, behavioral despair tasks, tails suspension and pinch induced catalepsy; situations that render the monkey incapable of controlling the environment.

Depression and low mood were found to be of a communicative nature. They signal yielding in a hierarchy conflict or a need for help. Low mood or extreme low mood (depression) can regulate a pattern of engagement and foster disengagement from unattainable goals. "Low mood increases an organism's ability to cope with the adaptive challenges characteristic of unpropitious situations in which effort to pursue a major goal will likely result in danger, loss, bodily damage, or wasted effort." Being apathetic can have a fitness advantage for the organism. Depression has also been studied as a behavioral strategy used by vertebrates to increase their personal or inclusive fitness in the threat of parasites and pathogens.

The lack of neurogenesis has been linked to depression. Animals with stress (isolated, cortisol levels) show a decrease in neurogenesis and antidepressants have been discovered to promote neurogenesis. Rene Hen and his colleagues at Columbia University ran a study on rats in which they blocked neurogenesis by applying radiation to the hippocampal area to test the efficacy of antidepressants. Results suggested that antidepressants failed to work when neurogenesis was inhibited.

Stress

Robert Sapolsky has extensively studied baboons in their natural environment in the Serengeti in Africa. He noticed that baboons have very similar hierarchies in their society as do humans. They spend very few hours searching for food and fulfilling their primary needs leaving them with time to develop their social network. In primates mental stresses show up in the body. Primates experience psychological stresses that can elicit physiological responses that overtime can make them sick. Sapolsky observed the baboon's ranks, personalities and social affiliations then collected blood samples of the baboons to control the cortisol (stress hormone) levels of the baboons then matched social position to cortisol levels. Most of the data have been collected from male baboons because at any given time 80 percent of the females were pregnant. Three factors influenced a baboon's cortisol levels; friendships, perspective and rank. Baboons that played with infants and cultivated friendships, could tell if a situation was a real threat and could tell if they were going to win or lose, and were top ranking had lower levels of cortisol. Cortisol levels rise with age and hippocampal cells express fewer hormone receptors on their surface to protect themselves from excess, making it harder to control stress levels. Cortisol levels are elevated in half of people suffering from major depression, it is the hippocampal region that is affected by both. Stress can have negative effects on gastrointestinal function causing ulcers, and it can also decrease sex drive, effect sleeping patterns and elevate blood pressure but it can also stimulate and motivate. When animals experience stress they are generally more alert than when they are not stressed. It may help them be better aware of unfamiliar environments and possible threats to their life in these environments. Yerkes and Dodson developed a law that explains the empirical relationship between arousal and performance illustrated by and inverted U-shape graph. According to the Yerkes-Dodson Law performance increases as does cognitive arousal but only to a certain point. The downward part of the U-shape is caused by stress and as stress increases so does efficiency and performance but to a certain point. When stress becomes too great performance and efficiency decline. Sapolsky has also studied stress in rats and his results indicate that early experiences in young rats have strong, lasting effects. Rats that were exposed to human handling (stressful situation) had finely tuned stress responses that may have lowered their lifetime exposure to stress hormones compared to those that were not handled. In short stress can be adaptive. The more exposure to stressful situations the better the rat can handle that situation.

Stereotypies

Stereotypies are repetitive, sometimes abnormal behaviors like pacing on the perch for birds. There are adaptive stereotypic behaviors such as grooming in cats and preening in birds. Captive parrots commonly perform a range of stereotypies. These behaviors are repeated identically and lack any function or goal. Captive parrots perform striking oral and locomotor stereotypies like pacing on the perch or repetitive play with a certain toy. Feather picking and loud vocalizations can be stereotypies but are not as rigid and may be reactions to confinement, stress, boredom and loneliness as studies have shown that parrots that are in cages closest to the door are the most prone to feather pick or scream. Feather picking is not a true stereotypy and is more like hair pulling in human and loud vocalizations or screaming can be a stereotypy but vocalization is part of a parrot's natural behavior. Captive parrots lack sufficient stimulation. Presumably they suffer from lack of companionship and opportunities to forage. Stereotypies can evolve from the social environment for example the presence or absence of certain social stimuli, social isolation, low feeder space and high stocking density (especially for tail biting in pigs). These behaviors can also be transmitted through social learning. Bank voles, pigeons and pigs when housed next to animals that show stereotypies, pick them up as well as through stimulus enhancement which is what happens in tail biting in pigs and feather pecking by hens.

Stereotypies may be coping mechanisms as results suggest from study on tethered and stalled sows. Sows that are tethered and stalled exhibited more stereotypies like licking and rubbing than sows that are in groups outdoors. This abnormal behavior seems to be related to opioid (related to the reward system) receptor density. In sows, prolonged confinement, being tethered or being in gestation crates, results in abnormal behaviors and stereotypies. Mu and Kappa receptors are associated with aversion behaviors and Mu receptor density is greater in tethered sows than sows that are in groups outdoors. However, sows with stereotypy behaviors experienced a decrease both in Mu and Kappa receptor density in the brain suggesting that inactivity increases Mu receptor density and stereotypy development decrease both kappa and Mu receptor density.

Self-aggression

Rhesus macaques have been observed to display self-aggression (SA) including self-biting, self-clasping, self-slapping, self-rubbing and threatening of body parts. The rhesus macaques observed were individually caged and free of disease. Their self-aggression level rose in stressful and stimulating conditions such as moving from one cage to another. Stump-tailed macaques were studied to examine the source of their SA. SA increased in an impoverished environment and results support that SA may increase sensory input in poor environments. Captive macaques do not socialize the way wild macaques do which may affect SA. When allowed to socialize by putting another macaque in the cage or not putting them in a cage, SA levels in macaques decrease. Results indicate that SA is a form of redirected social aggression. SA is related to frustration and social status, especially in macaques that have an intermediate dominance rank.

Broodiness

Broodiness is the action or behavioral tendency to sit on a clutch of eggs to incubate them, often requiring the non-expression of many other behaviors including feeding and drinking. Being broody has been defined as "Being in a state of readiness to brood eggs that is characterized by cessation of laying and by marked changes in behavior and physiology". Broody birds often pluck feathers from their chest and abdomen, using them to cover the eggs. As a consequence of this, they develop one or several patches of bare skin on the ventral surface. These reddish, well-vascularized areas of skin are usually called brood patches, and improve heat transfer to the eggs. Broodiness is usually associated with female birds, although males of some bird species become broody and some non-avian animals also show broodiness.

A brooding White tern (*Gygis alba*).

In Wild Birds

In wild birds, egg incubation is a normal and essential phase in the process of reproduction, and in many families of birds, e.g. pigeons, the eggs are incubated by both male and female parents.

Broodiness in Males

In all the species of phalaropes the males become broody rather than the female. The females leave the nest after finishing laying to let the males incubate the eggs and take care of the young. Male emus (*Dromaius novaehollandiae*) become broody after their mates start laying, and begin to incubate the eggs before the laying period is complete.

Non-broodiness

A small number of atypical birds such as *Passeriformes* of the genus *Molothrus* do not become broody but lay their eggs in the nests of other species for incubation, known as brood parasitism. The Australian Brushturkey *(Alectura lathami)* also does not become broody, rather, it covers the eggs with a large mound of vegetable matter which decomposes keeping the eggs warm until hatching. The Crab-plover, (*Dromas ardeola*), which live on the coasts and islands of the Indian Ocean, let their eggs incubate primarily by the heat of the sun and will leave their nests unattended, occasionally for days at a time.

In Domestic Poultry

Broody hens can be recognized by their behaviour. They sit firmly over the eggs, and when people approach or try to remove the eggs, threaten the person by erecting their feathers, emitting a characteristic sound like *clo-clo-clo* and will peck aggressively. When broody, hens often temporarily cease eating or reduce their feed consumption.

A brooding domestic hen.

Letting eggs accumulate in a relatively dark place near the floor often stimulates hens to become broody. Placing artificial eggs into nests also stimulates broodiness. Keeping hens in dark places with warm temperatures and in view of vocalising orphan chicks can induce broodiness, even in breeds that normally do not go broody.

Some environmental conditions stimulate broodiness. In heavy breeds of chickens, warm weather tends to bring about broodiness. Removing eggs each day, out of the sight of the hens, helps avoid broodiness not only in domestic poultry but also in some wild species in captivity. This continued egg laying means more eggs are laid than would occur under natural conditions. Poultry farming in battery cages also helps to avoid broodiness.

In Commercial Egg-laying

An egg incubator.

Because hens stop laying when they become broody, commercial poultry breeders perceive broodiness not as a normal physiological process, but as a drawback to egg and poultry meat production. With domestication, it has become more profitable to incubate eggs artificially, while keeping hens in full egg production. To help achieve this, there has been intense artificial selection for non-broodiness in commercial egg laying chickens Gallus gallus domesticus and parent stock of poultry. As a result of this artificial selection, broodiness has been greatly reduced to very low levels in present breeds of commercial fowl, both among egg-laying and meat-producing breeds.

Physiological Basis

Broodiness is due to the secretion of prolactine hormone by the front lobe of the hypophysis. Prolactine injection in hens provokes egg laying to stop within a few days, vitellum reabsorbtion, ovary regression (hens only have a left ovary) and finally broodiness. However, attempts to stop broodiness by the administration of several hormones have failed because this state, once evoked, requires time to revert.

Prolactine injections inhibit the production of gonadotropin hormone, a hormone that stimulates ovarian follicles which is produced in the frontal lobe of hypophysis.

Castrated males can go broody with baby chicks, showing that broodiness is not limited to females, however, castrated males do not incubate eggs.

Contrary to common opinion, the temperature of broody hens barely differs from that of laying hens. Broody hens pluck feathers from their chest, using them to cover the eggs. As a consequence of this, they develop one or several patches of bare skin on the ventral surface. These reddish, well-vascularized areas of skin are usually called brood patches. which improve heat transfer to the eggs.

Genetic Basis

Broodiness is more common in some chicken breeds than others, indicating that it is an inherited characteristic. Breeds such as Cochin, Cornish and Silkie exhibit a tendency to broodiness, including brooding eggs from other species such as quails, pheasants, turkeys and geese. In some breeds such as the White Leghorn, broodiness is extremely rare.

Some studies on crosses of chicken breeds point to the hypothesis of complementary genes acting on broodiness. Other results point to the hypothesis of sex-linked genes, or, inheritance through the maternal chromosome. Although these studies have been made on different breeds of chickens, their results are not contradictory. There is common agreement that artificial selection for egg production succeeded in reducing the incidence of broody hens in chicken populations.

Chicken Breeds that Commonly Exhibit Broodiness

- Cochin
- Manx Rumpy
- Plymouth Rock

- Silkie
- Saipan Jungle Fowl
- Orpington
- Kraienköppe
- Belgian Bearded d'Anvers
- Icelandic chicken
- Java (chicken)
- Philippine Native Chicken
- Belgian Bearded d'Uccle
- Iowa Blue
- Nankin
- Delaware
- Booted Bantam
- New Hampshire
- Pekin
- Dutch Bantam
- Indian Game (Cornish)

Chicken Breeds that Rarely Exhibit Broodiness

- Leghorn
- Polish
- Norwegian Jærhøne
- Brabanter
- Orloff
- Sultan
- Scots Grey
- Sussex
- Sebright
- Menorca
- Wyandotte
- Egyptian Fayoumi

Broodiness in Non-avian Animals

There is some evidence that non-avian dinosaurs also practiced brooding. A specimen of the Mongolian oviraptorid *Citipati osmolskae* was discovered in a chicken-like brooding position in 1993, which may indicate that they had begun using an insulating layer of feathers to keep the eggs warm.

Several deinonychosaur and oviraptorosaur specimens have also been found preserved on top of their nests, likely brooding in a bird-like manner.

Lungless salamanders in the family *Plethodontidae* lay a small number of eggs in a cluster among damp leaf litter. The female salamander often broods the eggs and in the genus *Ensatinas*, she has been observed to coil around them and press her throat area against them, effectively massaging them with a mucous secretion. The black mountain salamander mother broods her eggs, guarding them from predation as the larvae feed on the yolks of their eggs. They eventually break their way out of the egg capsules and disperse. Some species of Gymnophiona (caecilians, with long, cylindrical, limbless bodies) brood their eggs.

A brooding female python.

Most pythons coil around their egg-clutches and remain with them until they hatch. A female python will not leave the eggs, except to occasionally bask in the sun or drink water. She will even "shiver" to generate heat to incubate the eggs.

Some cichlid fish lay their eggs in the open, on rocks, leaves, or logs. Male and female parents usually engage in differing brooding roles. Most commonly, the male patrols the pair's territory and repels intruders, while females fan water over the eggs, removing the infertile and leading the fry while foraging. However, both sexes are able to perform the full range of parenting behaviours.

Mouthbrooding

Mouthbrooding, also known as oral incubation, refers to the care given by some groups of animals to fertilized eggs or their offspring by holding them in the mouth of the parent for extended periods of time. Although it has been observed in a variety of animals, most mouthbrooders are fish. The parent performing this behavior invariably feeds less often and afterwards will be underweight, requiring a period of feeding and restoring the depleted energy reserves.

Others

Marsupial frogs are so-called because they possess a dorsal brood pouch. In some species the eggs are fertilized on the female's lower back, and are inserted in her pouch with the aid of the male's toes. The eggs remain in contact with the female's vascular tissue, which provides them oxygen.

Some animals have a common name that includes the word 'brood' or its derivatives, although it is arguable whether the animals show 'broodiness' per se. For example, the female gastric-brooding frog (*Rheobatrachus sp.*) from Australia, now probably extinct, swallows her fertilized eggs, which then develop inside her stomach. She ceases to feed and stops secreting stomach acid and the tadpoles rely on the yolks of the eggs for nourishment. After six or seven weeks the mother opens her mouth wide and regurgitates the tadpoles which hop away from her mouth. The Brooding sea anemone (*Epiactis prolifera*) is a colonial hermaphrodite that fertilizes and incubates its eggs internally. The motile larvae, after swimming out of the mouth, migrate down to the disk and become fixed there until they become little anemones, ready to move and feed independently.

In Darwin's Frog (*Rhinoderma darwinii*), the female lays about 30 eggs and then the male guards them for about two weeks, until they hatch. The male then takes all the survivors and carries around the developing young in his vocal pouch. When the tiny tadpoles have developed they hop out and swim away. In this animal, the parents hold the hatched young rather than eggs in their mouths, so is arguably not showing 'broodiness'.

Cribbing (Horse)

Cribbing or crib biting involves a horse grasping a solid object such as the stall door or fence rail with its incisor teeth, then arching its neck, and contracting the lower neck muscles to retract the larynx. This coincides with an in-rush of air into the oesophagus producing the characteristic cribbing grunt. Usually, air is not swallowed but returns to the pharynx. Wind-sucking is a related behavior whereby the horse arches its neck and sucks air into the windpipe but does so without grasping an object. Wind-sucking is thought to form part of the mechanism of cribbing, rather than being defined as an entirely separate behavior.

A horse cribbing on a wooden fence, note anti-cribbing collar intended to reduce this behavior and tension in neck muscles

Cribbing is considered to be an abnormal, compulsive behavior or stereotypy seen in some horses, and is often labelled a stable vice.

Cribbing was mentioned in the literature as early as 1578 and occurs in 2.4-8.3% of horses depending on breed and management.

There is evidence that stomach ulcers may lead to a horse becoming a cribber, and that cribbing may be a coping mechanism in response to stress.

A 1998 study found that cribbing increased endorphins and found no evidence that cribbing generally impairs the health of affected horses, but later studies reported that cribbing and wind-sucking were related to a history of colic or the subsequent development of colic.

A similar but unrelated behavior, wood-chewing or lignophagia, is another undesirable habit observed in horses, but it does not involve sucking in air; the horse simply gnaws on wood rails or boards as if they were food.

Description

Cribbing, or crib biting, involves a horse grasping a solid object such as the stall door or fence rail with its incisor teeth, arching its neck, and contracting the lower neck muscles to retract the larynx caudally. This movement is coincided with an in-rush of air through the crico-pharynx into the oesophagus producing the characteristic cribbing sound or grunt. Usually, air is not swallowed but returns to the pharynx. It is considered to be an abnormal, compulsive behavior or stereotypy, and often labelled as a stable vice.

Wind-sucking is a related behavior whereby the horse arches its neck and sucks air into the windpipe but does so without grasping an object. Wind-sucking is thought to form part of the mechanism of cribbing, rather than being defined as an entirely separate behavior.

Wood-chewing

A similar, but unrelated behavior, wood-chewing (lignophagia), is another undesirable behavior sometimes observed in horses. The horse gnaws on wood rails or boards as if they were food, but it does not involve sucking in air.

Prevalence and Incidence

It is reported that 2.4–8.3% of horses in Europe and Canada are cribbers and occupies 15-65% of an individual horse's daily time budget. A postal survey in 2009 found that an average of 4.4% horses in the US are cribbers, but 13.3% of Thoroughbreds perform the behavior. Young Thoroughbred and part-Thoroughbred horses fed concentrated food after weaning are four times more likely to become cribbers than foals not fed concentrate. In several studies, Thoroughbreds consistently have the greatest prevalence of cribbing compared to other breeds.

Wind-sucking occurs in 3.8% of non-racing horses in the US. One study shows that stereotypies in general, including cribbing, are more prevalent in dressage horses compared to several other uses.

Geldings and stallions are more likely to exhibit cribbing than mares and the behavior has been reported as occurring in horses on pasture.

Negative Consequences

A study in 1998 found that cribbing was correlated to an increase in endorphins and a reduced stress level in horses, finding no studies with evidence that cribbing generally impairs the health of affected horses. A 2008 study found cribbing and wind-sucking as risk factors for certain types of colic. A 2010 study found a correlation between cribbing and colic, but found no clear association between degree of risk or any particular type of colic.

It has been anecdotally claimed that horses can learn to copy these behaviors from other horses, although this has not been substantiated by scientific study. A study in 2009 found that 48.8% of US horse owners believed that cribbing could be learned by observation, but research demonstrated that only 1.0% of horses developed cribbing after being housed in sight of an affected horse.

Causes

Boredom, stress, habit and addiction are all possible causes of cribbing and wind-sucking. It was proposed in a 2002 study that the link between intestinal conditions such as gastric inflammation or colic and abnormal oral behavior was attributable to environmental factors. There is evidence that stomach ulcers may be correlated to a horse becoming a cribber.

Researchers now generally agree that cribbing and wind-sucking occur most often in stabled horses, although once established in an individual horse, the horse may exhibit these behaviors in other places. Recent studies indicate cribbing occurs more frequently in horses that were stable-weaned as foals than in those that were pasture-weaned. In the same study, feeding concentrates after weaning was associated with a fourfold increase in the rate of development of cribbing.

Because Thoroughbreds are so consistently the most prevalent cribbers, this suggests there may be a genetic component, however, this may be confounded by different uses and management of different horse breeds.

Functions

Stereotypies are sometimes considered to be a coping mechanism for animals experiencing stress. A physiological stress response can be induced by injecting an animal with ACTH and the animal's ability to cope with this stress can be monitored by measuring salivary cortisol. In a 2015 study, after ACTH injection, cribbers had higher cortisol levels than non-cribbers. Furthermore, cribbers which did not perform the stereotypy during the 3-hrs of testing had higher cortisol levels than non-cribters, whereas those performing the stereotypy did not. The researchers concluded that cribbing is a coping mechanism to stressful situations and that because of this, it should not be prevented.

Cribbing and wind-sucking may cause a sensation of pleasure by releasing endorphins in the horse's brain. It has also been suggested that the increase in saliva produced during wind-sucking could be a mechanism for neutralizing stomach conditions in stable-kept, grain-fed horses. Stereotypies have been defined as "repetitive, invariant behaviour patterns with no obvious goal or function", therefore, if cribbing and wind-sucking have one of the above possible functions, it may be inappropriate to label them as a stereotypy. However, as the causes and resulting reinforcement for these behaviors are probably multifactorial and they remain abnormal behaviors, this indicates that husbandry changes are needed for animals that exhibit cribbing or wind-sucking.

Treatment

Several methods have been devised to prevent cribbing once the behavior has started. However, some ethologists have argued that prevention of the behavior without addressing the causes is not a cure and may result in cribbing being expressed in a modified form, or may interfere with an animal's attempt to adapt to its environment.

Dietary and Management

It has been shown that feeding cribbing horses an antacid diet can significantly reduce the frequency of the behavior. Current research indicates that the prevention of cribbing and related behavior is based upon management conditions which allow daily free movement and feeding practices that provide higher amounts of roughage and limited amounts of concentrates. A growing body of work suggests that fat and fiber-based diets may also result in calmer patterns of behavior.

One study investigated the effects of providing a feeder that delivered small amounts of concentrate feed when activated by the animal. The feeder increased the feeding time of both cribbers and non-cribbers, however, although the feeder decreased cribbing, it increased again once the feeder was removed.

Physical Devices

There are a number of traditional devices used to minimize or prevent cribbing and wind-sucking. However, the effectiveness of these methods is arguable since they do not address the underlying causal factors. One method involves the horse wearing a collar-like device that stops a horse from arching and swelling its neck to suck in air. However, one of the only studies of the equipment showed that although wearing such a collar for 24 hours reduced cribbing in six of eight horses, once the collar was removed, cribbing returned to greater levels than before. The authors concluded cribbing has a function and that preventing this by using anti-cribbing collars may compromise the horse's welfare.

Covering exposed edges with metal or wire or painting surfaces with bitter substances such as carbolineum or a commercial "chew stop" product may reduce chewing-related damage to surfaces, though this does not prevent edges from being gripped by the teeth.

Surgical and Others

Other methods to prevent cribbing have included surgery, acupuncture, use of pharmaceuticals, operant feeding, and environmental enrichment.

One surgical technique is the modified Forssell's procedure in which muscles and nerves in the ventral neck region are cut as well as some muscle tissue being removed. This makes it more difficult for a horse to contract the larynx and exhibit cribbing. An adaptation of this technique using a laser has proved successful in preventing some cribbers from exhibiting the behavior, although this was less successful in horses which had been cribbers for more than three years prior to the surgery.

Licking

Licking is the action of passing the tongue over a surface, typically either to deposit saliva onto the surface, or to collect liquid, food or minerals onto the tongue for ingestion, or to communicate with other animals. Many animals both groom themselves and eat or drink by licking.

A tiger licking its paw

In Animals

Grooming: Animals commonly clean themselves through licking. In mammals, licking helps keep the fur clean and untangled. The tongues of many mammals have a rough upper surface that acts like a brush when the animal licks its fur. Certain reptiles, such as geckos, clean their eyes by licking them.

Mammals typically lick their offspring clean immediately after birth; in many species this is necessary to free the newborn from the amniotic sac. The licking not only cleans and dries the offspring's fur, but also stimulates its breathing and digestive processes.

Food and water acquisition: Hummingbirds are often said to "sip" nectar, but in fact they lap up nectar on their long tongues. Their tongues have fringed edges, which help both in nectar-eating and in catching tiny insects. Mother hummingbirds also lick their chicks after a rainstorm to dry them by licking water droplets from the coats of the chicks to avoid them chilling. Many animals drink by licking. While young mammals drink milk from their mothers' teats by sucking, the typical method of drinking for adult mammals involves dipping the tongue repeatedly into water and using it to scoop water into the mouth. This method of drinking relies in part on the water adhering to the surface of the tongue and in part on muscular control of the tongue to form it into a spoon-like shape. Cattle, horses and other animals lick rocks, salt licks or other objects to obtain mineral nutrients.

A humming bird licking for nectar

Gustation: Animals also use their tongues to enhance their sense of smell. By licking a surface or extending the tongue beyond the mouth, molecules are transferred via the tongue to the olfactory receptors in the nose and in some animals, to the vomeronasal organ. In some mammals, the tongue is used to "lick" the air during the flehmen response to assist transfer of pheremones. Similarly, snakes use smell to track their prey. They smell by using their forked tongues to collect airborne particles, then passing them to the vomeronasal organ. They keep their tongues constantly in motion, sampling particles from the air, ground, and water, analyzing the chemicals found, and determining the presence of prey or predators in the local environment.

A kangaroo licking its wrists to thermoregulate

Communication: Dogs and cats use licking both to clean and to show affection among themselves or to humans, typically licking their faces. Many animals use licking as a submissive or appeasement signal in dominance hierarchies.

Thermoregulation: Some animals use licking to cool themselves. Cats do not sweat the way humans do and the saliva deposited by licking provides a similar means of evaporative cooling. Some animals spread saliva over areas of the body with little or no fur to maximise heat loss. For example, kangaroos lick their wrists and rats lick their testicles.

In Primates

Ring-tailed lemurs lick each other's babies as a means of collective grooming and of reinforcing social cohesion within the community. Macaques and other primates lick leaves for water in addition to dipping their arms into tree crevices and licking the water off. Chimpanzees use licking in a variety of ways: licking objects, such as dead trees, that others in their community have licked, licking each other's body parts for grooming and sex and licking rocks for salt. Gorillas use licking in addition to other senses to determine the nature of an object.

In Humans

Compared to most other mammals, licking has a minor role for humans. The human tongue is relatively short and inflexible, and is not well adapted for either grooming or drinking. Instead, humans prefer to wash themselves using their hands and drink by sucking fluid into their mouth. Humans have much less hair over their skin than most other mammals, and much of that hair is in places which they cannot reach with their own mouth. The presence of sweat glands all over the human body makes licking as a cooling method unnecessary.

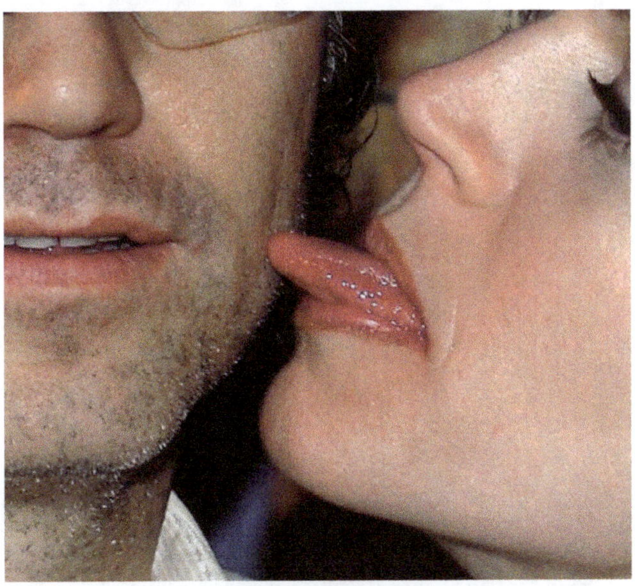

A woman licks a man's face.

Nonetheless, licking does play a role for humans. Even though humans cannot effectively drink water by licking, the human tongue is quite sufficient for licking more viscous fluids. Some foods are sold in a form intended to be consumed mainly by licking, e.g. ice cream cones and lollipops.

Some people in the Afar tribe of Ethiopia have been reported to have used their tongues to lick other humans, as a way of cleaning them from the dust that accumulates on them in a very water-scarce region.

Humans use licking for a number of other purposes. For example, licking can moisten the adhesive surfaces of stamps or envelopes. Many people lick a finger to help turning a page, taking a sheet of paper from the top of a pile or opening a plastic bag. This is often considered unhygienic and it is questioned whether there really is any necessity to do so, although the people who do it claim that, for example, in certain situations turning a page is difficult and that the process goes much easier after licking the top of the finger used to turn that page for some extra grip. In sewing, thread ends are commonly wet by licking to make the fibres stick together and thus make threading them through the eye of a needle easier. Another practice considered uncivilized is licking one's hand and using it to groom one's hair.

Humans also use their tongues for sexual purposes, such as during cunnilingus, anilingus, foot licking, and whilst French kissing, where two people lick each other's tongues.

Abnormal Licking

Self-licking can sometimes become abnormally frequent occasionally resulting in a lick granuloma. The most common cause of lick granuloma appears to be psychological, related to stress, anxiety, separation anxiety, boredom, or compulsiveness. Lick granulomae are especially seen in active dogs left alone for long periods of time. One theory is that excessive licking causes endorphin release, which reduces pain and makes the dog feel temporarily euphoric. This provides the animal with positive feedback from the licking, and subsequent addiction to the behaviour.

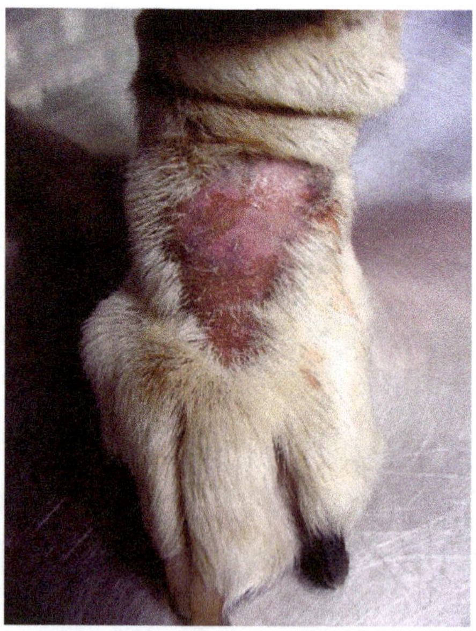

Lick granuloma on a dog's paw

Animals in captivity sometimes develop a licking stereotypy during which surfaces (walls, bars, gates, etc.) are repeatedly licked for no apparent reason. This has been observed in captive giraffes and camels.

Feather Pecking

Feather pecking is a behavioural problem that occurs most frequently amongst domestic hens reared for egg production, although it does occur in other poultry such as pheasants, turkeys, ducks and is sometimes seen in farmed ostriches. Feather pecking occurs when one bird repeatedly pecks at the feathers of another. The levels of severity are recognised as gentle and severe. Gentle feather pecking is considered to be normal investigatory behaviour where the feathers of the recipient are hardly disturbed and therefore does not represent a problem. In severe feather pecking, however, the feathers of the recipient are grasped, pulled at and sometimes removed. This is painful for the receiving bird and can lead to trauma of the skin or bleeding, which in turn can lead to cannibalism and death.

Feather pecking amongst laying hens. In the lower right of the picture, the white hen has lost her tail feathers and the brown hen has been feather pecked on the thigh and wing.

Feather pecking is one of the major problems facing the egg industry in non-cage systems and is set to become an even greater issue with the EU legislation (Council Directive 1999/74/EC) ban on the keeping of laying hens in barren battery cages which came into force in 2012, and the prospect of a ban on beak-trimming. Reducing feather pecking without resorting to beak-trimming is an important goal for the poultry industry.

Motivational Basis

Feather pecking is considered to be re-directed behaviour, developing either from ground pecking or pecking during dustbathing, although the former hypothesis is now the more favoured. Captive birds are very often kept in barren environments with limited foraging opportunities and in addition, are usually fed a nutrient-dense diet which can be eaten in a few minutes rather than the hours it would require to acquire during normal foraging. In combination, these cause the birds' foraging activity to be re-directed to the feathers of their conspecifics.

- Feather pecking is *not* aggression. During aggressive encounters, hens peck exclusively at the top of the head or the comb, whereas during feather pecking, the areas of the body that are usually targeted are the base of the tail over the uropygial or preen gland, the back, the tail feathers and the wing feathers.

- Although feather pecking activity may be related to dominance relationships or the pecking order, formation of the dominance hierarchy is not involved in the causation of feather pecking.
- Feather pecking is also distinct from another psychopathological behaviour called feather-plucking or feather-picking. In feather-plucking, birds, often housed in isolation, remove feathers from their own body; in feather pecking, however, birds peck at each other's feathers.

Sometimes, feathers that are removed are then eaten, in which case the behaviour is termed "feather eating". Whilst there may be a positive association between feather pecking and eating, at least in the individual bird, this is likely due to an overall higher pecking motivation. Eating feathers increases gut transit indicating that feather pecking and feather eating have a different motivational basis.

Development

Early experience can influence severe feather pecking in later life. Commercial egg-laying hens have often already begun feather pecking when they are transferred to the egg laying farm from the rearing farm at approximately 16–20 weeks of age, and plumage quality can then rapidly deteriorate until peak lay at approximately 25 weeks of age. Severe feather pecking can either begin or persist beyond this age although it rarely begins after 40 weeks of age.

Although there are links between gentle feather pecking and severe feather pecking, it is still not clear whether the gentle form leads to the severe form.

Some areas of the body are targeted for feather pecking and there is a pattern in the development of which areas are pecked. The rump area over the uropygial gland and the tail are often the first body regions to show signs of plumage damage due to feather pecking, followed by the neck, wings and back, although it should be noted that in the ostrich which has a similar pattern of feather pecking development, the uropygial gland is absent.

Prevalence

Although feather pecking occurs in all commercial housing systems used for egg laying hens, it is often more prevalent or severe in loose flock systems because it is less easy to control and can spread more rapidly. Prevalence figures range between 57 to 86% of free-range flocks and 99% of hens within a flock can be affected. The UK national flock of egg laying hens is currently (2011) approximately 33 million birds of which approximately 10 million are free-range. This indicates that 5.5 million free-range hens/year are likely to be affected by feather pecking. It has been estimated that 4% of hens on free-range farms die because of feather pecking, representing 220,000 deaths each year in the UK alone due to this behavioural problem. EU legislation (Council Directive 1999/74/EC) will ban battery or conventional cages in 2012 meaning that many producers will change to using free-range systems, possibly exacerbating this welfare problem until we learn effective methods of its control.

Risk Factors

Feather pecking is a multifactorial problem and a large number of risk factors have been identified for commercial flocks.

Factors likely to *reduce* feather pecking are:

Diet

- Minimal number of diet changes
- Ad libitum feeding
- Mashed feed rather than pelleted
- Diet balanced for protein and methionine
- Dietary tryptophan

Genetics

- White breeds such as the Amberlink compared to pigmented breeds
- Less flighty breeds

Housing and Husbandry

- Dark brooders
- Purchasing the hens at an earlier age and allowing them on the range earlier
- Delaying the onset of lay
- Maintaining a uniform flock (purchase single flocks and do not mix)
- Pan feeders rather than chain feeders
- Nipple drinkers rather than bell drinkers
- Good litter quality
- Good air quality (low levels of ammonia and carbon dioxide)
- Low light and noise levels
- House temperature above 20 °C
- Multiple persons inspecting the hens
- Minimal light changes for inspection
- Avoiding using lights in nest boxes

Hen Behaviour

- Increased use of the range (e.g. smaller flocks, increasing shelter, cockerels)
- Reduced fearfulness

Health

- Good health, especially avoiding egg peritonitis and infectious bronchitis

Methods of Control

Beak-trimming

Beak-trimming, sometimes misleadingly termed debeaking, is perhaps most accurately described as 'partial beak-amputation'. It is performed on poultry to reduce the incidence or damage caused by feather pecking or cannibalism and involves amputating the distal one to two thirds of the bird's beak by either a blade or infra-red beam. Beak-trimming causes welfare concerns because the internal tissue of the beak contains many nerves which are transected during the process - it is only the surface and extreme tip of the beak that is keratinised, dead tissue. This can lead to neuromas (abnormal nerve regeneration) developing in the amputated beak stump from which there might be abnormal spontaneous neural discharges similar to the discharges originating from stump neuromas in human amputees and implicated in phantom limb pain.

It has been shown that domestic hens have iron mineral deposits in the dendrites in the upper beak and are capable of magnetoreception. Because hens use directional information from the magnetic field of the earth to orient in relatively small areas, this raises the possibility that beak-trimming impairs the ability of hens to orient in extensive systems, or move in and out of buildings in free-range systems.

A further negative aspect of beak-trimming is that it leaves birds less able to groom themselves effectively, thus beak-trimmed hens have greater ectoparasite burdens than hens with intact beaks.

Light Manipulations

A widely used method of reducing feather pecking is to reduce light intensity, but because a minimum of 5 lux is necessary to maintain egg laying, intensities of 10 lux or more are recommended. At these low intensities it becomes difficult for humans to inspect the hens properly, especially in the more crowded densely populated housing systems, and human colour vision is hindered making the detection of blood almost impossible. Low light intensites may be associated with other welfare costs to the hens as they prefer to eat in brightly lit environments and prefer brightly lit areas for active behaviour but dim (<10 lux) for inactive behaviour. Dimming the lights can also cause problems when the intensity is then abruptly increased temporarily to inspect the hens; this has been associated as a risk factor of increased feather pecking and the birds can become frightened resulting in panic-type ("hysteria") reactions which can increase the risk of injury. In turkeys, low light intensities (perhaps in combination with long light phases) can cause retinal detachment and buphthalmia, a distortion of the eye morphology that can lead to blindness. This does not appear to have been investigated for layer hens under modern lighting patterns. Gradual changes in light intensity simulating a dawn and dusk at the beginning and end of the light phase rather than switching off lights abruptly enables birds to feed in anticipation of the dark period and to move safely to roosts, rather than moving in the dark and risking injury which is possibly more important in furnished systems. Many producers have tried placing red filters over windows or using red lighting to reduce feather pecking and cannibalism. This was even the subject of a patent, however, if such a simple solution was effective, it would have been adopted widely by the industry.

It has been suggested that the absence of UV from artificial light sources may have a role in the causation of feather pecking in turkeys. The extent to which the absence of UV from artificial lights compromises poultry and other animal welfare is not yet known. Other poultry species prefer areas illuminated with additional UV, but poultry reared without UV show little indication of being stressed.

Selective Breeding and Genetics

Feather pecking has a heritable component with heritabilities for this trait ranging from 0.07 to 0.56. Lines of hens exhibiting high or low feather pecking activity have been developed by artificial selection with high feather pecking birds showing more feather pecking than low feather pecking birds from the second generation onwards.

Selection for indirect indicators of feather pecking, specifically intact feather cover and livability in multi-bird groups leads, has led to reductions in feather pecking and cannibalism. Considerable additive genetic variation exists for these traits with estimates of heritability ranging from 0.22 to 0.54. A trait has been identified which combines feather pecking and cannibalism leading to severe injury or death in beak-intact birds; this has a high heritability at 0.65.

There has been less work at the molecular level of the genetics of feather pecking. Major genes for feather pecking have been found along with the polygenes. There are markers for severe feather pecking on chromosomes 1, 2, and 10 and also possibly on chromosome 3.

Devices (Bits and Spectacles)

Devices have been developed to reduce or eliminate the damaging effects of feather pecking. These devices require time and skill to fit and therefore have problems of practicality given that commercial flocks usually contain several thousands of birds. Because of this, they are not used widely in modern poultry production, except for gamekeeping.

- Bits or bumpabits are small, plastic circlips, the body of which passes between the maxilla and mandible of the beak and are held in place by the ends of the circlip being placed in the nostrils or nares. Some are held in place by a pin which pierces through the nasal septum. These devices make it difficult for the bird to completely close its beak and grasp the feathers of another individual.

- Spectacles or 'blinders' are pieces of plastic or metal shaped like opaque spectacles and attached to the bird's beak to block its vision. The devices are held in place either with a circlip which enters the nares or a pin which pierces through the nasal septum. These devices are based on the principle that by interfering with the vision of the bird, it is less able to visually locate the feathers of another bird and is therefore less able to grasp and pull the feathers.

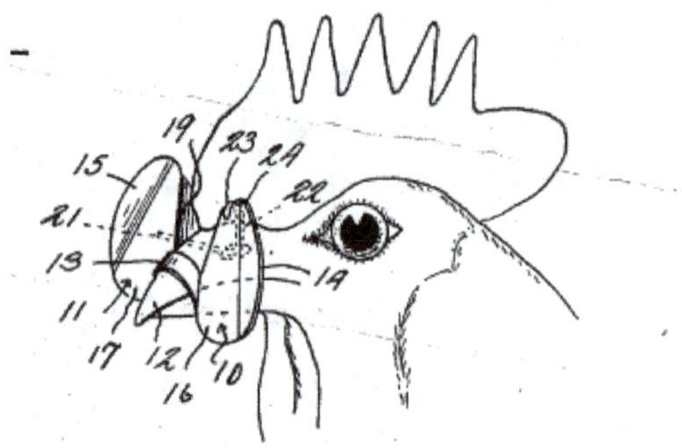

Blinders for poultry - From the U.S. Patent *"Device to prevent picking in poultry"* filed in 1935

Legislation

Legislation regarding these devices in the UK is formulated by Defra.

For laying hens, the relevant literature is the Defra Code of Recommendations for the Welfare of Livestock: Laying Hens. This states *"The Welfare of Livestock (Prohibited Operations) Regulations 1982 (S.I. 1982 No.1884) prohibits...the fitting of any appliance which has the object or effect of limiting vision to a bird by a method involving the penetration or other mutilation of the nasal septum."*

For gamebirds, the relevant legislation is the Defra Code of Practice for the Welfare of Gamebirds Reared for Sporting Purposes. This states *"5.1 The use of management devices or practices that do not allow birds to fully express their range of normal behaviours should not be considered as routine and keepers should work towards the ideal of management systems that do not require these devices. Such devices and practices include mutilations...and the use of bits, spectacles and hoods to prevent feather pecking, egg eating or aggression. Their use should be justified on a flock by flock basis and regularly reviewed in the flock health and welfare plan. Any device that is designed to pierce the nasal septum is illegal."*

Vent Pecking

Vent pecking is an abnormal behaviour of birds performed primarily by commercial egg-laying hens. It is characterised by pecking damage to the cloaca, the surrounding skin and underlying tissue. Vent pecking frequently occurs immediately after an egg has been laid when the cloaca often remains partly everted exposing the mucosa, red from the physical trauma of oviposition or bleeding if the tissue is torn by her laying an oversized egg. Vent pecking clearly causes pain and distress to the bird being pecked. Tearing of the skin increases susceptibility to disease and may become cannibalistic leading to evisceration of the pecked bird and ultimately, death.

Prevalence and Severity

Surveys have shown that 27% of farmers reported seeing damage to the vents of their hens and 36.9% of farmers reported vent pecking had occurred in their previous flock. Whilst farmers attributed 1.3% of mortalities as due to vent pecking the most common findings at autopsy were different types of cannibalism (65.51%), with vent cannibalism (38.57%) the most common. The type of housing system markedly affects the prevalence of vent pecking with 22.5% of hens affected in free-range systems, 10.0% in barn systems, 6.2% in conventional cages and 1.6% in furnished cages, with a similar rank for the severity of vent pecking injuries.

Causation

The causes and development of vent pecking are multifarious.

Risk factors that have been identified as increasing vent pecking include dim lights placed in nest boxes to encourage hens to use the boxes, the diet being changed more than three times during the egg laying period, the use of bell drinkers, and the hens beginning to lay earlier than 20 weeks of age. Vent pecking is associated with indicators of stress, e.g. fluctuating asymmetry, heterophil to lymphocyte ratio, and tonic immobility duration. Vent pecking can be related to disease or immune challenge as it sometimes becomes prevalent in cases of Gumboro disease (Infectious Bursal Disease) and is increased by challenges with the protein antigen, human serum albumin (HuSA). Housing design can influence vent pecking. Mortality caused by cannibalism was reduced when hens had sufficient room to perch all facing the feed trough, thus giving their perch-mates little opportunity to peck at the vent region and increased pecking activity and cannibalistic behaviour can occur due to inadequate height of the perches. Larger group sizes lead to increases in vent pecking suggesting that social learning plays a role.

Infanticide (Zoology)

In animals, infanticide involves the killing of young offspring by a mature animal of its own species, and is studied in zoology, specifically in the field of ethology. Ovicide is the analogous destruction of eggs. Although human infanticide has been widely studied, the practice has been observed in many other species throughout the animal kingdom. These include microscopic rotifers, insects, fish, amphibians, birds and mammals. Infanticide can be practiced by both males and females.

Lion cubs may be killed by males replacing other males in the pride.

Infanticide caused by sexual conflict has the general theme of the killer (often male) becoming the new sexual partner of the victim's parent, which would otherwise be unavailable. This represents a gain in fitness by the killer, and a loss in fitness by the parents of the offspring killed. This is a type of evolutionary struggle between the two sexes, in which the victim sex may have counter-adaptations that reduce the success of this practice. It may also occur for other reasons, such as the struggle for food between females. In this case individuals may even kill closely related offspring.

Filial infanticide occurs when a parent kills its own offspring. This sometimes involves consumption of the young themselves, which is termed filial cannibalism. The behavior is widespread in fishes, and is seen in terrestrial animals as well. Human infanticide has been recorded in almost every culture. A unique aspect of human infanticide is sex-selective infanticide.

Background

Infanticide only came to be seen as a significant occurrence in nature quite recently. At the time it was first seriously treated by Yukimaru Sugiyama, infanticide was attributed to stress causing factors like overcrowding and captivity, and was considered pathological and maladaptive. Classical ethology held that conspecifics (members of the same species) rarely killed each other. By the 1980s it had gained much greater acceptance. Possible reasons it was not treated as a prevalent natural phenomenon include its abhorrence to people, the popular group and species selectionist notions of the time (the idea that individuals behave for the good of the group or species; compare with gene-centered view of evolution), and the fact that it is very difficult to observe in the field.

Infanticide Involving Sexual Conflict

This form of infanticide represents a struggle between the sexes, where one sex exploits the other, much to the latter's disadvantage. It is usually the male who benefits from this behavior, though in cases where males play similar roles to females in parental care the victim and perpetrator may be reversed.

By Males

Hanuman langurs (or gray langurs) are an Old World monkey found in India. They are a social animal, living in groups that consist of a single dominant male and multiple females. The dominant male has a reproductive monopoly within the group, which causes sub-ordinate males to have a much lower fitness value in comparison. To gain the opportunity to reproduce, sub-ordinate males try to take over the dominant role within a group, usually resulting in an aggressive struggle with the existing dominant male. If successful in overthrowing the previous male, unrelated infants of the females are then killed. This infanticidal period is limited to the window just after the group is taken over. Cannibalism, however, has not been observed in this species.

Infanticide not only reduces intraspecific competition between the incumbent's offspring and those of other males but also increases the parental investment afforded to their own young, and allows females to become fertile faster. This is because females of this species, as well as many other mammals, do not ovulate during lactation. It then becomes easier to understand how infanticide evolved. If a male kills a female's young, she stops lactating and is able to become pregnant again. Because of this, the newly dominant male is able to reproduce at a faster rate than without the act of infanticide. As males are in a constant struggle to protect their group, those that express infanticidal behavior will contribute a larger portion to future gene pools.

Similar behavior is also seen in male lions, among other species, who also kill young cubs, thereby enabling them to impregnate the females. Unlike langurs, male lions live in small groups, which cooperate to take control of a pride from an existing group. They will attempt to kill any cubs that are roughly nine months old or younger, though as in other species, the female will attempt to defend her cubs viciously. Males have, on average, only a two-year window in which to pass on their genes, and lionesses only give birth once every two years, so the selective pressure on them to conform to this behavior is strong. In fact it is estimated that a quarter of cubs dying in the first year of life are victims of infanticide. Male mice show great variation in behavior over time. After fertilizing a female, they become aggressive towards mouse pups for three weeks, killing any they come across. After this period however, their behavior changes dramatically, and they become paternal, caring for their own offspring. This lasts for almost two months, but afterwards they become infanticidal once more. It is no coincidence here that the female gestation period is three weeks as well, or that it takes roughly two months for pups to become fully weaned and leave their nest. The proximate mechanism that allows for the correct timing of these periods involves circadian rhythms, each day and night cycle affecting the mouse's internal neural physiology, and disturbances in the duration of these cycles results in different periods of time between behaviors. The adaptive value of this behavior switching is twofold; infanticide removes competitors for when the mouse does have offspring, and allows the female victims to be impregnated earlier than if they continued to care for their young, as mentioned above.

Gerbils, on the other hand, no longer commit infanticide once they have paired with a female, but actively kill and eat other offspring when young. The females of this species behave much like male mice, hunting down other litters except when rearing their own.

Prospective Infanticide

Prospective infanticide is a subset of sexual competition infanticide in which young born *after* the arrival of the new male are killed. This is less common than infanticide of existing young, but can still increase fitness in cases where the offspring could not possibly have been fathered by the new mate, i.e. one gestation or fertility period. This is known to occur in lions and langurs, and has also been observed in other species such as house wrens. In birds, however, the situation is more complex, as female eggs are fertilized one at a time, with a 24-hour delay between each. Males may destroy clutches laid 12 days or more after their arrival, though their investment of around 60 days of parental care is large, so a high level of parental certainty is needed.

By Females

Females are also known to display infanticidal behavior. This may appear unexpected, as the conditions described above do not apply. Males are not always an unlimited resource though - in some species, males provide parental care to their offspring, and females may compete indirectly with others by killing their offspring, freeing up the limiting resource that the males represent. This has been documented in research by Stephen Emlen and Natalie Demong on wattled jacanas (*Jacana jacana*), a tropical wading bird. In the wattled jacana, it is exclusively the male sex that broods, while females defend their territory. In this experiment Demong and Emlen found that removing females from a territory resulted in nearby females attacking the chicks of the male in most cases, evicting them from their nest. The males then fertilized the offending females and cared for their young. Emlen describes how he "shot a female one night, and [...] by first light a new female was already on the turf. I saw terrible things—pecking and picking up and throwing down chicks until they were dead. Within hours she was soliciting the male, and he was mounting her the same day. The next night I shot the other female, then came out the next morning and saw the whole thing again."

Jacana jacana females carry out infanticide.

Infanticide is also seen in giant water bugs. *Lethocerus deyrollei* is a large and nocturnal predatory insect found in still waters near vegetation. In this species the males take care of masses of eggs by keeping them hydrated with water from their bodies. Without a male caring for the eggs like this, they become desiccated and will not hatch. In this species, males are a scarce resource that females must sometimes compete for. Those that cannot find a free male often stab the eggs of a brooding one. As in the above case, males then fertilize this female and care for her eggs. Noritaka Ichikawa has found that males only moisten their eggs during the first 90 seconds or so, after which all of the moisture on their bodies has evaporated. However, they guard the egg masses for as long as several hours at a time, when they could be hunting prey. They do not seem to prevent further evaporation by staying guard, as males that only guarded the nest for short periods were seen to have similar hatching rates in a controlled experiment where there were no females present. It seems rather that males are more successful in avoiding infanticidal females when they are out of the water with their eggs, which might well explain the ultimate cause of this behavior.

Female rats will eat the kits of strange females for a source of nutrition, and to take over the nest for her own litter.

Resource Competition

Black-tailed prairie dogs are colonially-living, harem-polygynous squirrels found mainly in the United States. Their living arrangement involves one male living with four or so females in a territory defended by all individuals, and underground nesting. Black-tails only have one litter per year, and are in estrous for only a single day around the beginning of spring.

The Black-tailed prairie dog (*Cynomys ludovicianus*)

A seven-year natural experiment by John Hoogland and others from Princeton University revealed that infanticide is widespread in this species, including infanticide from invading males and immigrant females, as well as occasional cannibalism of an individual's own offspring. The surprising finding of the study was that by far the most common type of infanticide involved the killing of close kin's offspring. This seems illogical, as kin selection favors behaviors that promote the well-being of closely related individuals. It was postulated that this form of infanticide is more successful than trying to kill young in nearby groups, as the whole group must be bypassed in this case, while within a group only the mother need be evaded. Marauding behavior is evidently adaptive, as infanticidal females had more and healthier young than others, and were heavier themselves as well. This behavior appears to reduce competition with other females for food, and future competition among offspring.

Similar behavior has been reported in the meerkat (*Suricata suricatta*), including cases of females killing their mother's, sister's, and daughter's offspring. Infanticidal raids from neighboring groups also occurred.

Other

Bottlenose dolphins have been reported to kill their young through impact injuries. Dominant male langurs tend to kill the existing young upon taking control of a harem. There has been sightings of infanticide in the leopard population. The males of the Stegodyphus lineatus species of spider have been known to exhibit infanticide as a way to encourage females to mate again.

Costs and Defenses

Costs of the Behavior

While it may be beneficial for some species to behave this way, infanticide is not without risks to the perpetrator. Having already expended energy and perhaps sustained serious wounds in a fight with another male, attacks from females who vigorously defend their offspring may be telling for harem-polygynous males, with a risk of infection. It is also energetically costly to pursue a mother's young, which may try to escape.

Costs of the behavior described in prairie dogs include the risk to an individual of losing their own young while killing another's, not to mention the fact that they are killing their own relatives. In a species where infanticide is common, perpetrators may well be victims themselves in the future, such that they come out no better off; but as long as an infanticidal individual gains in reproductive output by its behavior, it will tend to become common. Further costs of the behavior in general may be induced by counter-strategies evolved in the other sex, as described below.

As a Cost of Social Behavior

Taking a broader view of the black-tailed prairie dog situation, infanticide can be seen as a cost of social living. If each female were to have her own private nest away from others, she would be much less likely to have her infants killed when absent. This, and other costs such as increased spread of parasites, must be made up for by other benefits, such as group territory defense and increased awareness of predators.

An avian example published in *Nature* is acorn woodpeckers. Females nest together, possibly because those nesting alone have their eggs constantly destroyed by rivals. Even so, eggs are consistently removed at first by nest partners themselves, until the entire group lays on the same day. They then cooperate and incubate the eggs as a group, but by this time a significant proportion of their eggs have been lost because of this ovicidal behavior.

Counter-strategies

Because this form of infanticide reduces the fitness of killed individuals' parents, animals have evolved a range of counter-strategies against this behavior. These may be divided into two very different classes - those that tend to prevent infanticide, and those that minimize losses.

Loss Minimization

Some females abort or resorb their own young while they are still in development after a new male takes over; this is known as the Bruce effect. This may prevent their young from being killed after birth, saving the mother wasted time and energy. However, this strategy also benefits the new male. In mice this can occur by the proximate mechanism of the female smelling the odor of the new male's urine.

Preventative adaptations

Infanticide in burying beetles may have led to male parental care. In this species males often cooperate with the female in preparing a piece of carrion, which is buried with the eggs and eaten by the larvae when they hatch. Males may also guard the site alongside the female. It is apparent from experiments that this behavior does not provide their young with any better nourishment, nor is it of any use in defending against predators. However, other burying bugs may try to take their nesting space. When this occurs, a male-female pair is over twice as successful in nest defense, preventing the ovicide of their offspring.

Female langurs may leave the group with their young alongside the outgoing male, and others may develop a *false estrous* and allow the male to copulate, deceiving him into thinking she is actually sexually receptive. Females may also have sexual liaisons with other males. This promiscuous behavior is adaptive, because males will not know whether it is their own offspring they are killing or not, and may be more reluctant or invest less effort in infanticide attempts. Lionesses cooperatively guard against scouting males, and a pair were seen to violently attack a male after it killed one of their young. Resistance to infanticide is also costly, though: for instance, a female may sustain serious injuries in defending her young. At times it is simply more advantageous to submit than to fight.

Infanticide, the destruction of offspring characteristic to many species, has posed so great a threat that there have been observable changes of behavior in respective female mothers; more specifically, these changes exist as preventative measures. A common behavioral mechanism by females to reduce the risk of infanticide of future offspring is through the process of paternity confusion or dilution. In theory, this implies that a female that mates with multiple males will widely spread the assumption of paternity across many males, and therefore make them less likely to kill or attack offspring that could potentially carry their genes. This theory operates under the assumption that the specific males keep a memory of past mates, under a desire to perpetuate their own genes In the Japanese macaque (*macaca fuscata*), female mating with multiple males, or dilution of paternity, was found to inhibit male-to-infant aggression and infanticide eight times less towards infants of females with which they had previously mated. Multi-male mating, or MMM, is recorded as a measure to prevent infanticide in species where young is altricial, or heavily dependent, and where there is a high turnover rate for dominant males, which leads to infanticide of the previous dominant male's young. Examples include, but are not limited to; white-footed mice, hamsters, lions, langurs, baboons, and macaques. Along with mating with multiple males, the mating of females throughout the entirety of a reproductive cycle also serves a purpose for inhibiting the chance of infanticide. This theory assumes that males use information on past matings to make decisions on committing infanticide, and that females subsequently manipulate that knowledge. Females which are able to appear sexually active or receptive at all stages of their cycle, even during pregnancy with another male's offspring, can confuse the males into believing that the subsequent children are theirs. This "pseudo-estrus" theory applies to females within species that do not exhibit obvious clues to each stage of their cycle, such as langurs, rhesus macaques, and gelada baboons. An alternative to paternity confusion as a method of infanticide prevention is paternity concentration. This is the behavior of females to concentrate paternity to one specific dominant male as a means of protection from infanticide at the hands of less-dominant males. This particularly applies to species in which a male has a very long tenure as the dominant male, and faces little instability in this hierarchy. Females choose these dominant males as the best available form of protection, and therefore mate exclusively with this male. This is especially common within small rodents.

Infanticide by Parents and Caregivers

Damselfish may eat their own offspring.

Filial infanticide occurs when a parent kills its own offspring. Both male and female parents have been observed to do this, as well as sterile worker castes in some eusocial animals.

Maternal

Maternal infanticide occurs when newborn offspring are killed by their mother. This is sometimes seen in pigs, a behavior known as savaging, which affects up to 5% of gilts. Similar behavior has been observed in various animals such as rabbits and burying beetles.

Paternal

Paternal infanticide—where fathers eat their own offspring—may also occur. When young bass hatch from the spawn, the father guards the area, circling around them and keeping them together, as well as providing protection from would-be predators. After a few days, most of the fish will swim away. At this point the male's behavior changes: instead of defending the stragglers, he treats them as any other small prey, and eats them.

Worker Caste Killing Young

Honey bees may become infected with a bacterial disease called foul brood, which attacks the developing bee larva while still living in the cell. Some hives however have evolved a behavioral adaptation that resists this disease: the worker bees selectively kill the infected individuals by removing them from their cells and tossing them out of the hive, preventing it from spreading. The genetics of this behavior are quite complex. Experiments by Rothenbuhler showed that the 'hygienic' behavior of the queen was lost by crossing with a non-hygienic drone. This means that the trait must be recessive, only being expressed when both alleles contain the gene for hygienic behavior. Furthermore, the behavior is dependent on two separate loci. A backcross produced a mixed result. The hives of some offspring were hygienic, while others were not. There was also a third type of hive where workers removed the wax cap of the infected cells, but did nothing more. What was not apparent was the presence of a fourth group who threw diseased larvae out of the hive, but did not have the uncapping gene. This was suspected by Rothenbuhler however, who manually removed the caps, and found some hives proceeded to clear out infected cells.

Humans and Infanticide

Family structure is the most important risk factor in child abuse and infanticide. Children who live with both their natural (biological) parents are at low risk for abuse. The risk increases when children live with step-parents or with a single parent. Children living without either parent (foster children) are 10 times more likely to be abused than children that live with both biological parents.

Children who live with a single parent that has a live-in partner are at the highest risk: they are 20 times more likely to be victims of child abuse than children living with both biological parents

Infanticide is a subject that some humans may find discomforting. Cornell University ethologist Glenn Hausfater states that "infanticide has not received much study because it's a repulsive subject [...] Many people regard it as reprehensible to even think about it." Research into infanticide in animals is in part motivated by the desire to understand human behaviors, such as child abuse.

Hausfater explains that researchers are "trying to see if there's any connection between animal infanticide and child abuse, neglect and killing by humans [...] We just don't know yet what the connections are."

Infanticide has been, and still is, practiced by some human cultures, groups, or individuals. In many past societies, certain forms of infanticide were considered permissible, whereas in most modern societies the practice is considered immoral and criminal. It still takes place in the Western world usually because of the parent's mental illness or violent behavior, as well as encompassing the reasons for elective abortion, in addition to some poor countries as a form of population control — sometimes with tacit societal acceptance. Female infanticide, a form of sex-selective infanticide, is more common than the killing of male offspring, especially in cultures where male children are more desirable.

Stereotypy (Non-human)

In animal behavior, stereotypy, stereotypical or stereotyped behavior has several meanings, leading to ambiguity in the scientific literature. A stereotypy is a term for a group of phenotypic behaviours that are repetitive, morphologically identical and which possess no obvious goal or function. These behaviours have been defined as 'abnormal' as the exhibit themselves solely to animals subjected to barren environments, scheduled or restricted feedings, social deprivation and other cases of frustration, but do not arise in 'normal' animals in their natural environments. These behaviors may be maladaptive, involving self-injury or reduced reproductive success, and in laboratory animals can confound behavioral research.

An elephant exhibiting stereotyped trunk swinging and rocking behaviour

Stereotyped behavior can also refer to normal behaviors that show low variation. For example, mammalian chewing cycles or fish capturing prey using suction feeding. Highly stereotyped movements may be due to mechanical constraint (such as the skull of a viper or fish, in which bones are mechanically linked), tight neural control (as in mammalian chewing), or both. The degree of stereotyping may vary markedly between closely related species engaging in the same behavior.

Onset

The display of stereotypies is usually increased in an individual over time, this is due to the changing motivation of the stereotypy. The establishment of the stereotype may be due to a number of factors within a captive environment, however the continuance of the behaviour can be explained by its impact on the basal ganglia and the establishment of a habit in its expression. The interruption or cease of a habit is much more tedious and difficult than that of the initial behaviour. Secondly, as stereotypes develop, they become more readily elicited, so much so that they are no longer just expressed during the original circumstances and may be expressed in the absence of any apparent stress or conflict. The development of the stereotypy into a habit and the difficulty of interrupting said habit explain why it is expected that the frequency of stereotypies increases with age.

It is assumed that the onset of the development of stereotypies is due to poor coping with the animals environment, and is considered one of the most important indicators of long-term animal welfare problems, a prolonged display of stereotypies suggests that the welfare of the animal is in a peril state. The welfare of the mink is considered an important aspect from a management stand point as it has repercussions on the production of the animal. Inadequate welfare has been linked to poor reproductively, poor growth rate.

Examples

Many stereotypies can be induced by confinement; for example, cats pace in zoo cages. Pregnant sows whose feed is restricted bite at their stalls' bars and chew without anything in their mouths. In laboratory rats and mice, grooming is the most common activity other than sleep, and grooming stereotypies have been used to investigate several animal models of anxiety and depression.

Video of a mouse showing distinct amphetamine-induced stereotypy reminiscent of nail biting

Examples of stereotypical behaviors include pacing, rocking, swimming in circles, excessive sleeping, self-mutilation (including feather picking and excessive grooming), and mouthing cage bars.

Stereotypies are seen in many species, including primates, birds, and carnivores. Up to 54% of elephants in zoos display stereotypical behaviors. Stereotypic behaviour in giraffes is also common; they resort to excessive tongue use on inanimate objects, due to a subconscious response to suckle milk from their mother, which many human-reared giraffes and other captive animals do not experience. Stereotypies are well known in stabled horses, usually developing as a result of being confined, particularly with insufficient exercise. They are colloquially called stable vices. They present a management issue, not only leading to facility damage from chewing, kicking, and repetitive motion, but also lead to health consequences for the animal if not addressed.

Stereotypical behaviors are thought to be caused ultimately by artificial environments that do not allow animals to satisfy their normal behavioral needs. Rather than refer to the behavior as abnormal, it has been suggested that it be described as "behavior indicative of an abnormal environment." Stereotypies are correlated with altered behavioral response selection in the basal ganglia.

Stereotypical behavior in laboratory animals can confound behavioral research. It is also seen as a sign of psychological distress in animals, and therefore is an animal welfare issue.

Solutions

Stereotypical behavior can sometimes be reduced or eliminated by environmental enrichment, including larger and more stimulating enclosures, training, and introductions of stimuli (such as objects, sounds, or scents) to the animal's environment. The enrichment must be varied to remain effective for any length of time. Housing social animals such as primates with other members of their species is also helpful. But once the behavior is established, it is sometimes impossible to eliminate due to alterations in the brain.

Stable Vices

Stable vices are stereotypies of equines, especially horses. They are usually undesirable habits that often develop as a result of being confined in a stable with insufficient exercise, boredom, hunger, excess energy or isolation. They present a management issue, not only leading to facility damage from chewing, kicking, and repetitive motion, but also lead to health consequences for the animal if not addressed. They also raise animal welfare concerns.

Placing horses on pasture and the presence of companion animals may both help to reduce stable vices.

Stereotypical behaviors in animals are generally thought to be caused by artificial environments that do not allow animals to satisfy their normal behavioral needs. Rather than refer to these behaviors as abnormal, it has been suggested that they be described as "behavior indicative of an abnormal environment."

It was once thought that stable vices may be learned by observing other horses already performing the behaviors, but studies on the topic to date have failed to establish this as a cause. Stereotypies are correlated with altered behavioral response selection in the basal ganglia. Although a more enriched environment may help minimize or eliminate some stereotypical behavior, once established, it is sometimes impossible to eliminate them due to alterations in the brain.

Examples

Stereotypies in equines are usually placed into one of two classes: Locomotor or Oral. Common stable vices include:

- Wood chewing (lignophagia): Gnawing on wood out of hunger or boredom.
- Cribbing, also called windsucking: When the equine grabs a board or other surface with its teeth, arches its neck, and sucks in air. This can harm the teeth and may lead to colic. Cribbing can be caused either by nervousness or boredom, previously thought to release endorphins in the horse, however, recent research suggests this is a fallacy. Additional research suggests that cribbing increases salivation and may reduce stomach discomfort. There is a direct correlation between diet and cribbing, increasing hay in the ration or feeding more frequent meals appears to help. Cribbing occurs in 2.4-8.3% depending on breed and management.
- Weaving: Rocking back and forth in a repetitive fashion that is correlated to isolation or stall confinement, usually alleviated by pasture turnout possibly a self-stimulating behavior. Problems with weaving can include weight loss and uneven hoof wear, unnatural stress on the legs and lameness.
- Wall kicking: Kicking the walls of its stall with hind legs. This raises the potential of damage both to the equine and to the barn. Usually this is caused by a lack of exercise and boredom. Wall-kicking is one habit that is often acquired by others in the barn once an individual starts doing it.
- Stall-walking or fence-walking: Like weaving, this is a repetitive movement, only the individual paces compulsively. It is usually correlated with isolation or anxiety while awaitng feed. This habit can also lead to weight loss and lameness.
- Pawing or digging: The equine may paw with its front feet. This can lead to abnormal hoof wear and lameness, and may also damage the flooring of the stall. An equine that paws can dig a noticeable hole in a dirt-floored barn in a very short time.

Some behaviors are not classes as stereotypies, but are viewed as undesirable behaviors for health or safety reasons:

- Biting: A nervous or anxious equine may reach out of its stall to bite at passers-by, human or animal. Box stall designs that keep the horse from reaching its head out prevent harm to other animals, but some horses may attempt to bite a handler when the person enters the stall.

- Bolting feed: Eating food too fast without adequate chewing. This can potentially lead to certain problems in the digestive system including choke and colic.
- Masturbation: A male horse, either a stallion or a gelding, will use his abdominal muscles to rhythmically bounce his penis against his belly. Previously believed to be a vice caused by boredom, confinement, or discomfort, masturbation by stallions and geldings is now viewed as a normal behavior.

Horses may engage in a number of undesirable behaviors when being ridden or driven. These are not "stable" vices, but are often classified as "vices" in terms of being behavior that poses a danger to the animal or its handler. These include:

- Head-shaking: Where a horse shakes its head repeatedly for no obvious reason, a condition with many possible causes from insect annoyance, dental problems, allergies, sun exposure or nerve damage.
- Bucking: May be misbehavior or a result of discomfort.
- Rearing: a normal behavior at play, but dangerous around humans, when it is often triggered by fear or pain.
- Bolting or running away

Other equine behaviors that may (or may not) arise from boredom or frustration, but still present management challenges. These include destruction of buckets, mangers, and feed tubs; defecation in the manger or water bucket; dumping water buckets; sloshing feed in water and then scattering it on the ground, and so on. There is little that can be done to stop them, and other than hygienic considerations, they present few health or safety concerns.

Modern Husbandry and the Effects on Behavior

Horses are extremely social creatures, and the process of domestication has not altered their need for social interactions. Also, in the wild, horses are constantly grazing; they are called trickle feeders because they continuously eat small amounts of forage throughout the day, expect the approximately 2 hours that they spent sleeping. Modern equine husbandry sometimes creates conflicts with the horse's natural behaviors; some owners keep their horses confined to a stall with minimal turnout time, little to no social interactions, and sometimes inadequate amounts of roughage. This can be problematic as this system of equine husbandry completely ignores certain basic needs, such as social interactions, foraging, and locomotion. Studies have shown that horses that are offered low quantities of forage and minimal social contact have a higher reported level of stereotypic behaviors such as cribbing, wind sucking, weaving, and other stereotypic behaviors. Social interactions are important to horses; mutual grooming has been shown to reduce heart rate and cortisol levels, therefore reducing stress. Play behavior between two horses aids in the development of the musculo-skeletal system and cardiovascular fitness; play allows practice of reproductive and survival skills. Living in a group also has an adaptive significance, as younger animals living within the herd will learn from the other members of the group.

The amount of forage a horse is given or has access to is extremely important as the equine digestive tract continuously produces acid, therefore the horse's digestive tract must contain food most of time; if a horse is without forage for more that 3 hours, the acid in the digestive tract will build up which can cause ulcers, diarrhea, and potentially colic. Behavioral problems can also develop because the horse is in pain from the ulcers that are a result of the low quantities of forage. The process of chewing produces saliva, which the horse uses as a natural antacid; if the horse has no hay or pasture to chew on, the antacid will not be produced and the horse will find anything to chew on to try and produce saliva, which can be the start of an oral stereotype.

Solutions

In most cases, reducing confinement and providing the animal a more natural setting reduces the incidence of stable vices. There are stopgap "cures" that can be provided in the stall to keep a horse busy or out of trouble, including increased exercise, feeding of larger quantities of lower-quality food (so the animal spends more time eating and less time being bored), feeding more frequently, or cutting back on grain or other high-energy concentrates. Toys such as a ball or empty one-gallon plastic milk jug can be hung in the stall. Sometimes simply giving the animal a companion in the next stall, or even a smaller animal placed in the same stall, also helps a bored or nervous horse.

In extreme cases, a short term fix may include various forms of restraint. However, none of these practices solve the underlying problem, some may raise animal welfare concerns, and the animal will resume its behavior as soon as the restraint is removed. The long-term solution that has the most success is to give the horse less time in the stall and more free turnout time.

References

- Stebbins, Robert C.; Cohen, Nathan W. (1995). A Natural History of Amphibians. Princeton University Press. p. 196. ISBN 0-691-03281-5.
- Dorit, R. L.; Walker, W. F.; Barnes, R. D. (1991). Zoology. Saunders College Publishing. pp. 853–854. ISBN 0-03-030504-7.
- Hausfater, G.; S.B. Hrdy (1984). Infanticide: Comparative and Evolutionary Perspectives. New York, Aldine. ISBN 0-202-02022-3.
- Elgar, M.A.; Crespi, B.J., eds. (1992). Cannibalism: Ecology and Evolution of Cannibalism among Diverse Taxa. Oxford University Press, New York. ISBN 0-19-854650-5.
- Price, Stephen D., et al. The Whole Horse Catalog. Fireside; Rev Upd edition (1998) ISBN 0-684-83995-4, ISBN 978-0-684-83995-0

Permissions

All chapters in this book are published with permission under the Creative Commons Attribution Share Alike License or equivalent. Every chapter published in this book has been scrutinized by our experts. Their significance has been extensively debated. The topics covered herein carry significant information for a comprehensive understanding. They may even be implemented as practical applications or may be referred to as a beginning point for further studies.

We would like to thank the editorial team for lending their expertise to make the book truly unique. They have played a crucial role in the development of this book. Without their invaluable contributions this book wouldn't have been possible. They have made vital efforts to compile up to date information on the varied aspects of this subject to make this book a valuable addition to the collection of many professionals and students.

This book was conceptualized with the vision of imparting up-to-date and integrated information in this field. To ensure the same, a matchless editorial board was set up. Every individual on the board went through rigorous rounds of assessment to prove their worth. After which they invested a large part of their time researching and compiling the most relevant data for our readers.

The editorial board has been involved in producing this book since its inception. They have spent rigorous hours researching and exploring the diverse topics which have resulted in the successful publishing of this book. They have passed on their knowledge of decades through this book. To expedite this challenging task, the publisher supported the team at every step. A small team of assistant editors was also appointed to further simplify the editing procedure and attain best results for the readers.

Apart from the editorial board, the designing team has also invested a significant amount of their time in understanding the subject and creating the most relevant covers. They scrutinized every image to scout for the most suitable representation of the subject and create an appropriate cover for the book.

The publishing team has been an ardent support to the editorial, designing and production team. Their endless efforts to recruit the best for this project, has resulted in the accomplishment of this book. They are a veteran in the field of academics and their pool of knowledge is as vast as their experience in printing. Their expertise and guidance has proved useful at every step. Their uncompromising quality standards have made this book an exceptional effort. Their encouragement from time to time has been an inspiration for everyone.

The publisher and the editorial board hope that this book will prove to be a valuable piece of knowledge for students, practitioners and scholars across the globe.

Index

www.ingramcontent.com/pod-product-compliance
Lightning Source LLC
LaVergne TN
LVHW080923210926
848485LV00004B/717
9781635490251